NEOPROTEROZOIC GEOBIOLOGY AND PALEOBIOLOGY

TOPICS IN GEOBIOLOGY

For detailed information on our books and series please vist: **www.springer.com**

Series Editors:
Neil H. Landman, American Museum of Natural History, New York, New York, landman@amnh.org
Douglas S. Jones, University of Florida, Gainesville, Florida, dsjones@flmnh.ufl.edu

Current volumes in this series

Volume 27: **Neoproterozoic Geobiology and Paleobiology**
Shuhai Xiao and Alan J. Kaufman
Hardbound, IBN 1-4020-520-4, 2006

Volume 26: **First Floridians and Last Mastodons: The Page-Ladson Site in the Aucilla River**
S. David Webb
Hardbound, ISBN 1-4020-4325-2, 2006

Volume 25: **Carbon in the Geobiosphere – Earth's Outer Shell –**
Fred T. Mackenzie and Abraham Lerman
Hardbound, ISBN 1-4020-4044-X, 2006

Volume 24: **Studies on Mexican Paleontology**
Francisco J. Vega, Torrey G. Nyborg, María del Carmen Perrilliat, Marison Montellano-Ballesteros, Sergio R.S. Clleovsa-Ferriz and Sara A Quiroz-Barroso
Hardbound, ISBN 1-4020-3882-8, October 2005

Volume 23: **Applied Stratigraphy**
Eduardo A. M. Koutsoukos
Hardbound, ISBN 1-4020-2632-3, January 2005

Volume 22: **The Geobiology and Ecology of *Metasequoia***
Ben A. LePage, Christopher J. Williams and Hong Yang
Hardbound, ISBN 1-4020-2631-5, March 2005

Volume 21: **High-Resolution Approaches in Stratigraphic Paleontology**
Peter J. Harries
Hardbound, ISBN 1-4020-1443-0, September 2003

Volume 20: **Predator-Prey Interactions in the Fossil Record**
Patricia H. Kelley, Michał Kowalewski, Thor A. Hansen
Hardbound, ISBN 0-306-47489-1, January 2003

Volume 19: **Fossils, Phylogeny, and Form**
Jonathan M. Adrain, Gregory D. Edgecombe, Bruce S. Lieberman
Hardbound, ISBN 0-306-46721-6, January 2002

Volume 18: **Eocene Biodiversity**
Gregg F. Gunnell
Hardbound, ISBN 0-306-46528-0, September 2001

Volume 17: **The History and Sedimentology of Ancient Reef Systems**
George D. Stanley Jr.
Hardbound, ISBN 0-306-46467-5, November 2001

Volume 16: **Paleobiogeography**
Bruce S. Lieberman
Hardbound, ISBN 0-306-46277-X, May 2000

Neoproterozoic Geobiology and Paleobiology

Edited by

SHUHAI XIAO

Department of Geosciences,
Virginia Polytechnic Institute and State University,
Blacksburg, VA 24061, USA

and

ALAN J. KAUFMAN

Department of Geology,
University of Maryland,
College Park, MD 20743, USA

 Springer

A C.I.P. Catalogue record for this book is available from the Library of Congress.

QE
724.5
.N46
2006

ISBN-10 1-4020-5201-4 (HB)
ISBN-13 978-1-4020-5201-9 (HB)
ISBN-10 1-4020-5202-2 (e-book)
ISBN-13 978-1-4020-5202-6 (e-book)

Published by Springer,
P.O. Box 17, 3300 AA Dordrecht, The Netherlands.

www.springer.com

Printed on acid-free paper

Cover illustrations: Multicellular algal fossils from the Neoproterozoic Doushantuo Formation at Weng'an, Guizhou Province, South China.
All photographs courtesy of Dr. Xunlai Yuan at Nanjing Institute of Geology and Paleontology.

All Rights Reserved
© 2006 Springer
No part of this work may be reproduced, stored in a retrieval system, or transmitted
in any form or by any means, electronic, mechanical, photocopying, microfilming, recording
or otherwise, without written permission from the Publisher, with the exception
of any material supplied specifically for the purpose of being entered
and executed on a computer system, for exclusive use by the purchaser of the work.

Aims & Scope Topics in Geobiology Book Series

Topics in Geobiology series treats geobiology – the broad discipline that covers the history of life on Earth. The series aims for high quality, scholarly volumes of original research as well as broad reviews. Recent volumes have showcased a variety of organisms including cephalopods, corals, and rodents. They discuss the biology of these organisms-their ecology, phylogeny, and mode of life – and in addition, their fossil record – their distribution in time and space.

Other volumes are more theme based such as predator-prey relationships, skeletal mineralization, paleobiogeography, and approaches to high resolution stratigraphy, that cover a broad range of organisms. One theme that is at the heart of the series is the interplay between the history of life and the changing environment. This is treated in skeletal mineralization and how such skeletons record environmental signals and animal-sediment relationships in the marine environment.

The series editors also welcome any comments or suggestions for future volumes;

Series Editors:
Douglas S. Jones dsjones@flmnh.ufl.edu
Neil H. Landman landman@amnh.org

Dedication

This work is dedicated to Prof. Zhang Yun (1937-1998), our mentor and friend.

(Photograph by Alan J. Kaufman, 1991)

Preface

The Neoproterozoic Era (1000–542 million years ago) is a geological period of dramatic climatic change and important evolutionary innovations. Repeated glaciations of unusual magnitude occurred throughout this tumultuous interval, and various eukaryotic clades independently achieved multicellularity, becoming more complex, abundant, and diverse at its termination. Animals made their first debut in the Neoproterozoic too. The intricate interaction among these geological and biological events is a centrepiece of Earth system history, and has been the focus of geobiological investigations in recent decades. The purpose of this volume is to present a sample of views and visions among some of the growing numbers of Neoproterozoic workers.

The contributions represent a cross section of recent insights into the field of Neoproterozoic geobiology. **Chapter One** by Porter gives an up-to-date review of Proterozoic heterotrophic eukaryotes, including fungi and various protists. Heterotrophs are key players in Phanerozoic ecosystems; indeed, most Phanerozoic paleontologists work on fossil heterotrophs. However, the fossil record of Proterozoic heterotrophs is extremely meagre. Why? Porter believes that preservation is part of the answer. **Chapter Two** by Huntley and colleagues explore new methods of quantifying the morphological disparity of Proterozoic and Cambrian acritarchs, the vast majority of which are probably autotrophic phytoplankton. They use non-metric multidimensional scaling and dissimilarity methods to analyze acritarch morphologies. Their results show that acritarch morphological disparity appears to increase significantly in the early Mesoproterozoic, with an ensuing long period of stasis followed by renewed diversification in the Ediacaran Period that closed the Neoproterozoic Era. This pattern is broadly consistent with previous compilation of acritarch taxonomic diversity, but also demonstrates that initial expansion of acritarch morphospace appears to predate taxonomic diversification. Using similar methods, Xiao and Dong in **Chapter Three** analyze the morphological disparity of macroalgal fossils, which likely represent macroscopic autotrophs in Proterozoic oceans. The pattern is similar to that of acritarchs: stepwise morphological expansions in both the early Mesoproterozoic and late Neoproterozoic separated by prolonged stasis. What might have caused the morphological stasis of both microscopic and macroscopic autotrophs? The authors speculate that it might have something to do with nutrient limitation.

The following two chapters review the depauperate fossil record of Neoproterozoic animals, or at least fossils that have been interpreted as animals. **Chapter Four** by Bottjer and Clapham places emphasis particularly on the evolutionary paleoecology of benthic marine biotas in the Ediacaran Period. They interpret the paleoecology of Ediacaran fossils in light of increasing evidence of a mat-based world. These authors are particularly intrigued by the non-random association of certain Ediacara fossils (e.g., fronds vs. bilaterians) and the contrasting ecological roles between bilaterian and non-bilaterian tierers in Ediacaran epibenthic communities. They notice that the Avalon (575–560 Ma) and Nama (549–542 Ma) assemblages appear to be dominated by non-bilaterian fronds that stood as tall tierers above the water-sediment interface, while the White Sea assemblage (560–550 Ma) seems to be characterized by flat-lying Ediacara organisms, including such forms as *Dickinsonia* that may be interpreted as mobile animals. It is still uncertain whether all or most Ediacara fossils can be interpreted as animals, but it is clear that evidence of animal activities is preserved as trace fossils in the last moments of Ediacaran time. Jensen, Droser, and Gehling take a step further in **Chapter Five** to comprehensively review the Ediacaran trace fossil record. The interpretation of Ediacaran trace fossils is not as straightforward as one would think. Many Ediacaran body fossils are morphologically simple spheres, discs, tubes, or rods. In many cases, these simple fossils, particularly when preserved as casts and molds, mimic the morphology of trace fossils such as tubular burrows or cnidarian resting traces. Jensen and colleagues do a heroic job of critically reviewing most published claims of Ediacaran trace fossils. They found that many Ediacaran trace fossil-like structures lack the diagnostic features (e.g., sediment disruption) of animal activities, and may be alternatively interpreted as body fossils. Thus, although there are *bona fide* animal traces in the White Sea and Nama assemblages, they conclude that previous estimates of Ediacaran trace fossil "diversity" have been unduly inflated.

Developmental and molecular biologists play a distinct role in understanding animal evolution. In **Chapter Six**, Erwin takes an evo-devo approach to reconstruct what the "urbilaterian"—the common ancestor of protostome and deuterostome animals—would look like. Did it have a segmented body with anterior-posterior, dorsal-ventral, and left-right differentiation? Did it have eyes to see the ancient world? Did it have a through gut system to leave fecal strings in the fossil record? Did it have legs to make tracks? In principle, one can at least achieve a partial reconstruction of the urbilaterian bodyplan based on a robust phylogeny and the phylogenetic distribution of key genetic toolkits. In reality, however, the presence of genetic toolkits does not guarantee the expression of the corresponding morphologies, and homologous genetic toolkits can be recruited to code functionally related, but morphologically distinct and evolutionarily convergent structures. Fortunately, the absence of certain critical genetic toolkits means the absence of corresponding morphologies. Thus, by figuring out what genetic toolkits might have been present in the urbilaterian, Erwin presents a number of ideas about how complex the urbilaterian could have possibly been, thus shedding light on a maximally complex urbilaterian. This is useful for paleontologists who have been searching for the urbilaterian without a search image, but it does not tell paleontologists what geological period they should focus on in their search. Molecular biologists believe that they can fill this gap by comparing homologous gene sequences of different organisms, based on the assumption that divergence at the molecular level follows a clock-like model. Hedges and colleagues present such a molecular timescale in **Chapter Seven**. Hedges and colleagues summarize the molecule-derived divergence times of major clades, including oxygen-generating cyanobacteria and methane-generating euryarchaeotes that have shaped the Earth's surface. In addition, they also present a eukaryote timetree (phylogeny scaled to evolutionary time) in the Proterozoic and give a critical review of the ever complicated models and methods devised to account for the stochastic nature of molecular clocks. Overall, Hedges and colleagues believe that many eukaryote clades, including animals, fungi, and algae, may have a deep history in the Mesoproterozoic–early Neoproterozoic. And they found possible temporal matches between the evolution of geobiologically important clades (e.g., land plants, fungi, etc.) and geological events (e.g., Neoproterozoic ice ages). The field of molecular clock study is still in its infancy, and one would expect more exciting advancements and improvements as it matures over the coming decades.

Another way to date evolutionary and geological events is to correlate relevant strata with geochronometrically constrained rock units. Because index fossils are rare in the Neoproterozoic Era, chemostratigraphic methods using stable carbon isotopes, strontium isotopes, and more recently sulfur isotopes, have been used to correlate Neoproterozoic rocks. In **Chapter Eight**, Halverson presents a Neoproterozoic carbon isotope chemostratigraphic curve based on four well-documented sections. This curve provides a basis on which he considers several key geobiological questions in the Neoproterozoic, including the number and duration of glaciations, and the relationship between widespread ice ages and evolution. In addition to chemostratigraphic data, some distinct sedimentary features have also been used in Neoproterozoic stratigraphic correlation. For example, an enigmatic carbonate is typically found atop Neoproterozoic glacial deposits, and it is characterized by a suite of unusual sedimentary features thought to be useful stratigraphic markers. In particular, Marinoan-style cap carbonates characterized by such features as tepee-like structures, sheet cracks, barite fans, and negative carbon isotope values, are thought to be associated with a synchronous deglaciation event following the Marinoan glaciation in Australia, the Nantuo glaciation in South China, the Ghaub glaciation in Namibia, or the Icebrook glaciation in northwestern Canada. While radiometric dating suggests that some of these cap carbonates may indeed be synchronous, Corsetti and Lorentz in **Chapter Nine** argue that Marinoan-style cap carbonates may be facies variants that occur repeatedly in Neoproterozoic time. Thus, these authors urge caution to be exercised when using cap carbonates as correlation tools.

This is by no means a comprehensive review of recent advancements made by Neoproterozoic workers. Nor does it represent a consensus view of the Neoproterozoic community—or, for that matter, among the contributors to this volume. Diverse opinions and interpretations are the hallmark of a young and vigorous science, and we feel strongly that healthy discussion among different investigators with different world views is an important key to the maturation of Neoproterozoic geobiology.

This project grew from a Pardee keynote symposium ("Neoproterozoic Geobiology: Fossils, Clocks, Isotopes, and Rocks") held at the 2003 Geological Society of America annual meeting in Seattle, USA. We are grateful to the GSA Pardee Foundation and NASA Astrobiology Institute for providing financial support to symposium participants. In addition, we would like to acknowledge the Department of Geosciences, Virginia Polytechnic Institute and State University, for supporting John Huntley, who assisted in formatting the manuscript. We would also like to acknowledge NSF, NASA, NNSFC, PRF, Chinese Academy of Sciences, and Chinese Ministry of Science and Technology for support of our research.

The publication of this volume would not be possible without the help of many individuals. We thank the contributors for the timely submission of their manuscripts, and the reviewers for prompt and constructive evaluation of the manuscripts. We would also like to thank Judith Terpos at Springer Science and John Huntley at Virginia Polytechnic Institute for their editorial assistance.

Finally, we would like dedicate this volume to the memory of our mentor and friend Prof. Zhang Yun (1937-1998) of Beijing University. Yun had a distinguished career in Neoproterozoic paleobiology cut short by a tragic traffic accident. His pioneering work on the Doushantuo Formation represents some of the earliest pages in our ever expanding book of Neoproterozoic paleobiology. We are both fortunate to have been introduced to the Doushantuo Formation and all its mysteries by Yun in a 1991 field trip—a memorable event that launched our integrated paleobiological and geochemical research.

Shuhai Xiao
Blacksburg, Virginia, USA
Alan J. Kaufman
College Park, Maryland, USA
May 8, 2006

Contributors

Fabia U. Battistuzzi Department of Biology and NASA Astrobiology Institute, Pennsylvania State University, University Park, PA 16802, USA. fxb142@psu.edu

Jaime E. Blair Department of Biology and NASA Astrobiology Institute, Pennsylvania State University, University Park, PA 16802, USA. jeb322@psu.edu

David J. Bottjer Department of Earth Sciences, University of Southern California, Los Angeles, CA 90089-0740, USA. dbottjer@usc.edu

Matthew E. Clapham Department of Earth Sciences, University of Southern California, Los Angeles, CA 90089-0740, USA. clapham@usc.edu

Frank A. Corsetti Department of Earth Science, University of Southern California, Los Angeles, CA 90089-0740, USA. fcorsett@usc.edu

Lin Dong Department of Geosciences, Virginia Polytechnic Institute and State University, Blacksburg, VA 24061, USA. lindong@vt.edu

Mary L. Droser Department of Earth Sciences, University of California, Riverside, CA 92521, USA. mary.droser@ucr.edu

Douglas H. Erwin Department of Paleobiology, MRC-121, National Museum of Natural History, Smithsonian Institution, Washington, DC 20013, USA; and Santa Fe Institute, 1399 Hyde Park Rd, Santa Fe, NM 87501, USA. ERWIND@si.edu

James G. Gehling South Australian Museum, South Terrace, 5000 South Australia, Australia. Gehling.Jim@saugov.sa.gov.au

Galen P. Halverson Laboratoire des Mécanismes et Transferts en Géologie, Université Paul Sabatier, 31400 Toulouse, France. (Present Address: Geology and Geophysics, School of Earth and Environmental Sciences, The University of Adelaide, Adelaide 5005, South Australia, Australia.) halverso@lmtg.obs-mip.fr

S. Blair Hedges Department of Biology and NASA Astrobiology Institute, Pennsylvania State University, University Park, PA 16802, USA. sbh1@psu.edu

John Warren Huntley Department of Geosciences, Virginia Polytechnic Institute and State University, Blacksburg, VA 24061, USA. jhuntley@vt.edu

Sören Jensen Area de Paleontologia, Facultad de Ciencias, Universidad de Extremadura, E-06071 Badajoz, Spain. soren@unex.es

Michał Kowalewski Department of Geosciences, Virginia Polytechnic Institute and State University, Blacksburg, VA 24061, USA. michalk@vt.edu

Nathaniel J. Lorentz Department of Earth Science, University of Southern California, Los Angeles, CA 90089-0740, USA. Lorentz@usc.edu

Susannah M. Porter Department of Earth Science, University of California, Santa Barbara, CA 93106, USA. porter@geol.ucsb.edu

Shuhai Xiao Department of Geosciences, Virginia Polytechnic Institute and State University, Blacksburg, VA 24061, USA. xiao@vt.edu

Contents

Published titles in *Topics in Geobiology* Book Series...................... ii
Aims & Scope Topics in Geobiology Book Series.......................... v
Dedication.. vii
Preface... ix
Contributors.. xv

Chapter 1 • The Proterozoic Fossil Record of Heterotrophic Eukaryotes

Susannah M. Porter

1.	Introduction...	1
2.	Eukaryotic Tree..	2
3.	Fossil Evidence for Proterozoic Heterotrophs..............................	4
	3.1 Opisthokonts..	4
	3.2 Amoebozoans...	5
	3.3 Chromalveolates...	7
	3.4 Rhizarians..	9
	3.5 Excavates...	10
	3.6 Summary...	10
4.	Why are Heterotrophs Rare in Proterozoic Rocks?..........................	12
5.	Conclusions..	14
Acknowledgements...		15
References...		15

Chapter 2 • On the Morphological History of Proterozoic and Cambrian Acritarchs

John Warren Huntley, Shuhai Xiao, and Michał Kowalewski

1.	Introduction...	24
2.	Materials and Methods..	25
	2.1 Materials...	25
	2.2 Body Size Analysis..	28
	2.3 Morphological Disparity Analysis....................................	29
	2.3.1 Dissimilarity..	29
	2.3.2 Non-metric Multidimensional Scaling...........................	30
3.	Results..	31
	3.1 Body Size...	31
	3.2 Morphological Disparity...	33
	3.2.1 Dissimilarity..	33
	3.2.2 Non-metric Multidimensional Scaling...........................	35
4.	Discussion...	39
	4.1 Comparative Histories of Morphological Disparity and Taxonomic Diversity	39
	4.2 Linking Morphological Disparity with Geological and Biological Revolutions	40

      4.2.1 Morphological Constraints, Convergence, and Nutrient Stress in the
           Mesoproterozoic 40
      4.2.2 Neoproterozoic Global Glaciations 41
      4.2.3 Ediacara Organisms. 43
      4.2.4 Cambrian Explosion of Animals 44
5. Conclusions 45
Acknowledgements.. 45
References 46
Appendix: SAS/IML Codes... 53

Chapter 3 • On the Morphological and Ecological History of Proterozoic Macroalgae

Shuhai Xiao and Lin Dong

1. Introduction 57
2. A Synopsis of Proterozoic Macroalgal Fossils 60
3. Morphological History of Proterozoic Macroalgae..................... 67
    3.1 Narrative Description 67
    3.2 Quantitative Analysis: Morphospace, Body Size, and Surface/Volume Ratio   70
      3.2.1 Methods... 70
      3.3.2 Results 74
4. Discussion 75
    4.1 Comparison with Acritarch Morphological History 75
    4.2 Surface/Volume Ratio 77
    4.3 Maximum Canopy Height..................................... 80
    4.4 Ecological Interactions with Animals 80
5. Conclusions 82
Acknowledgements.. 83
References 83

Chapter 4 • Evolutionary Paleoecology of Ediacaran Benthic Marine Animals

David J. Bottjer and Matthew E. Clapham

1. Introduction 91
2. A Mat-Based World 92
3. Nature of the Data.. 95
    3.1 Geology and Paleoenvironments................................ 95
    3.2 Lagerstätten 96
    3.3 Biomarkers 97
    3.4 Molecular Clock Analyses 97
4. Evolutionary Paleoecology 98
    4.1 Doushantuo Fauna (?600–570 Mya) 99
    4.2 Ediacara Avalon Assemblage (575–560 Mya)...................... 101
    4.3 Ediacara White Sea Assemblage (560–550 Mya).................... 102
    4.4 Ediacara Nama Assemblage (549–542 Mya)....................... 105

Contents

5. Discussion	108
Acknowledgements	110
References	110

Chapter 5 • A Critical Look at the Ediacaran Trace Fossil Record

Sören Jensen, Mary L. Droser and James G. Gehling

1. Introduction	116
2. Problems in the Interpretation of Ediacaran Trace Fossils	117
2.1 Tubular Organisms	119
2.2 *Palaeopascichnus*-type Fossils	120
3. List of Ediacaran Trace Fossils	120
4. Discussion	135
4.1 True and False Ediacaran Trace Fossils	136
4.1.1 *Archaeonassa*-type trace fossils	136
4.1.2 *Beltanelliformis*-type fossils	136
4.1.3 *Bilinichnus*	137
4.1.4 *Chondrites*	137
4.1.5 *Cochlichnus*	137
4.1.6 *Didymaulichnus*	137
4.1.7 *Gyrolithes*	138
4.1.8 *Harlaniella*	138
4.1.9 *Helminthoidichnites*-type trace fossils	138
4.1.10 *Lockeia*	139
4.1.11 *Monomorphichnus*	139
4.1.12 *Neonereites*	139
4.1.13 *Palaeopascichnus*-type fossils	139
4.1.14 *Planolites-Palaeophycus*	140
4.1.15 "*Radulichnus*"	140
4.1.16 *Skolithos*	141
4.1.17 *Torrowangea*	141
4.1.18 Dickinsonid trace fossils	142
4.1.19 Meniscate trace fossils	142
4.1.20 Star-shaped trace fossils	142
4.1.21 Treptichnids	143
4.2 Ediacaran Trace Fossil Diversity	143
4.3 Stratigraphic Distribution and Broader Implications of Ediacaran Trace Fossils	145
Acknowledgements	147
References	147

Chapter 6 • The Developmental Origins of Animal Bodyplans

Douglas H. Erwin

1. Introduction	160
2. Pre-Bilaterian Developmental Evolution	163

	2.1 Phylogenetic Framework	163
	2.2 Unicellular Development	165
	2.3 Poriferan Development	166
	2.4 Cnidarian Development	167
	2.5 The Acoel Conundrum	171
3.	Development of the Urbilateria	172
	3.1 Anterior-Posterior Patterning and Hox and ParaHox Clusters	172
	3.2 Head Formation and the Evolution of the Central Nervous System	174
	3.3 Eye Formation	176
	3.4 Dorsal-Ventral Patterning	178
	3.5 Gut and Endoderm Formation	178
	3.6 Segmentation	179
	3.7 Heart Formation	181
	3.8. Appendage Formation	182
	3.9 Other Conserved Elements	183
4.	Constructing Ancestors	184
	4.1 Maximally Complex Ancestor	184
	4.2 An Alternative View	186
5.	Conclusions	188
Acknowledgements	189
References	189

Chapter 7 • Molecular Timescale of Evolution in the Proterozoic

S. Blair Hedges, Fabia U. Battistuzzi and Jaime E. Blair

1.	Introduction	199
2.	Molecular Clock Methods	201
3.	Molecular Timescales	203
	3.1 Prokaryotes	203
	3.2 Eukaryotes	205
	3.3 Land Plants	212
	3.4 Fungi	213
	3.5 Animals	215
4.	Astrobiological Implications	217
	4.1 Complexity	217
	4.2 Global glaciations	219
	4.3 Oxygen and the Cambrian explosion	221
5.	Conclusions	221
Acknowledgements	222
References	222

Chapter 8 • A Neoproterozoic Chronology

Galen P. Halverson

1.	Introduction	232
2.	Constructing the Record	233
	2.1 The Neoproterozoic Sedimentary Record	233

Contents

2.2 The δ^{13}C Record	236
2.3 Bases for Correlation	238
3. Review of the Neoproterozoic	242
3.1 The Tonian (1000–720? Ma)	242
3.2 The Cryogenian (720?–635 Ma)	245
3.2.1 The Sturtian Glaciation	245
3.2.2 The Interglacial	248
3.2.3 The Marinoan Glaciation	250
3.3 The Ediacaran Period (635–542 Ma)	253
3.3.1 The Post-Marinoan Cap Carbonate Sequence	253
3.3.2 The Early Ediacaran	254
3.3.3 The Gaskiers Glaciation	258
3.3.4 The Terminal Proterozoic	260
4. Conclusions	261
Acknowledgements	262
References	262

Chapter 9 • On Neoproterozoic Cap Carbonates as Chronostratigraphic Markers

Frank A. Corsetti and Nathaniel J. Lorentz

1. Introduction	273
1.1 "Two Kinds" of Cap Carbonates	276
2. Key Neoproterozoic Successions	277
2.1 Southeastern Idaho	277
2.2 Oman	282
2.3 South China	283
2.4 Namibia	284
2.5 Tasmania	284
2.6 Conterminous Australia	285
2.7 Newfoundland	285
2.8 Northwestern Canada	286
3. Discussion	286
3.1 Global Correlations, Cap Carbonates, and New Radiometric Constraints	286
3.2 Intra-continental Marinoan-style Cap Carbonates ~90 m.y. Apart	288
3.3 Is it Time to Abandon the Terms Sturtian and Marinoan?	290
4. Conclusion	290
References	291
Index	295

Chapter 1

The Proterozoic Fossil Record of Heterotrophic Eukaryotes

SUSANNAH M. PORTER

Department of Earth Science, University of California, Santa Barbara, CA 93106, USA.

1. Introduction	1
2. Eukaryotic Tree	2
3. Fossil Evidence for Proterozoic Heterotrophs	4
3.1 Opisthokonts	4
3.2 Amoebozoans	5
3.3 Chromalveolates	7
3.4 Rhizarians	9
3.5 Excavates	10
3.6 Summary	10
4. Why are Heterotrophs Rare in Proterozoic Rocks?	12
5. Conclusions	14
Acknowledgements	15
References	15

1. INTRODUCTION

Nutritional modes of eukaryotes can be divided into two types: autotrophy, where the organism makes its own food via photosynthesis; and heterotrophy, where the organism gets its food from the environment, either by taking up dissolved organics (osmotrophy), or by ingesting particulate organic matter (phagotrophy). Heterotrophs dominate modern eukaryotic diversity, in fact, autotrophy, which characterizes the algae and land plants, appears to be a derived condition, having evolved several times within the eukaryotes (e.g., Keeling, 2004; although see Andersson and Roger, 2002). Indeed, heterotrophy is a *prerequisite* for autotrophy in eukaryotes, as the plastid—the site of photosynthesis in eukaryotes—was originally acquired via the ingestion of a photosynthetic organism. Thus it may be surprising that the early fossil record of eukaryotes is dominated not by heterotrophs but by algae. Most of the fossils that can be assigned to a modern clade are algal (red, xanthophyte, green, or brown; German, 1981, 1990; Butterfield *et al.*, 1990, 1994; Woods *et al.*, 1998; Xiao *et al.* 1998a, 1998b, 2004; Butterfield, 2000, 2004; see Xiao and Dong, this volume, for a review). Likewise, most taxonomically problematic fossils from the Proterozoic—acritarchs and carbonaceous compressions—are thought to be algal (e.g., Tappan, 1980; Mendelson and Schopf, 1992; Hofmann, 1994; Martin, 1993; Xiao *et al.*, 2002). Even *Grypania*, one of the earliest eukaryotic body fossils (<1.9 Ga), is interpreted as an alga (Han and Runnegar, 1992; Schneider *et al.*, 2002). The presence of red algae in rocks 1200 Ma necessarily implies that heterotrophs* were present by this time, consistent with molecular clock studies that suggest a diversity of heterotrophic clades in Proterozoic oceans (e.g., Wang *et al.*, 1999; Pawlowski *et al.*, 2003; Douzery *et al.*, 2004; Yoon *et al.*, 2004). Yet fossil evidence for Proterozoic heterotrophs is slim. Where are they? Here I review their early fossil record and discuss reasons why fossils of early heterotrophs may be rare.

2. EUKARYOTIC TREE

After much flux, we seem to be converging on a stable phylogeny for eukaryotic organisms (Fig. 1; Baldauf, 2003; Simpson and Roger, 2002; Keeling, 2004; Nikolaev *et al.* 2004; Simpson and Roger, 2004; although see, e.g., Philip *et al.*, 2005). Most eukaryotes fall into one of six major clades: 1) the opisthokonts, containing the animals and fungi and a few unicellular groups; 2) the amoebozoans, containing the lobose amoebae (both naked and testate) and the slime molds; 3) the plants, containing the red and green algae (and the land plants) and a minor group known as the glaucophytes; 4) the chromalveolates, a clade that itself unites two major groups, the alveolates (containing the dinoflagellates, ciliates, and apicomplexans), and the chromists (including the diatoms, the oomycetes, the xanthophyte algae, and the brown algae); 5) the rhizarians, a group

---

* Many members of the Bacteria (=Eubacteria) and Archaea (=Archaebacteria) are also heterotrophic, but I restrict my discussion here to eukaryotic heterotrophs. Thus, when I use the term 'heterotroph', I am referring only to eukaryotic heterotrophs.

characterized by the possession of filose pseudopods, that includes the foraminifera, the (polyphyletic) radiolarians, and the cercozoans; and 6) the excavates, a controversial grouping (Simpson and Roger, 2004) that includes the euglenids and several parasitic taxa such as *Giardia*. Recent gene fusion data suggest that these six clades are divided into two groups: the 'unikonts' (opisthokonts and amoebozoans), and the 'bikonts' (plants, chromalveolates, rhizarians, and excavates), with the root of the eukaryotic tree falling between these two groups (Stechmann and Cavalier-Smith, 2002, 2003).

Heterotrophic taxa are highlighted in Fig. 1. Although many eukaryotes are capable of mixotrophy—acquiring nutrition via photosynthesis and phagotrophy, I will restrict my discussion below to those taxa most or all of whose members are strictly heterotrophic. Thus, I will focus on the early fossil record of only five eukaryotic clades: the opisthokonts, the amoebozoans, the chromalveolates, the rhizarians, and the excavates. With few exceptions, all plants are photosynthetic.

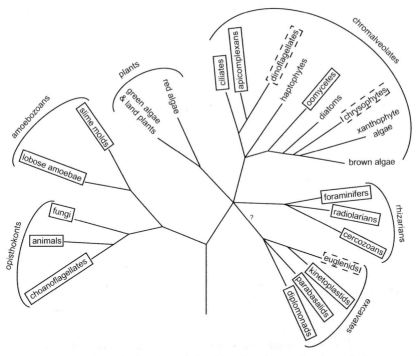

Figure 1. A current view of eukaryote relationships, based on molecular and ultrastructural data (modified from Baldauf 2003; Simpson and Roger, 2002; Keeling, 2004; Nikolaev *et al.*, 2004; Simpson and Roger, 2004). Clades composed primarily of heterotrophs shown in boxes with solid lines; clades with both heterotrophs and autotrophs shown in boxes with dashed lines. A question mark indicates clades that are not strongly supported (Keeling, 2004). Rooting of the tree is based on gene fusion data (Stechmann and Cavalier-Smith 2002, 2003).

3. FOSSIL EVIDENCE FOR PROTEROZOIC HETEROTROPHS

3.1 Opisthokonts

There are two main opisthokont groups: the animals and the fungi. The Proterozoic fossil record of animals is worthy of an extensive review in its own right; I will not discuss it here except to note that the earliest well accepted evidence for animals are ~580 Ma phosphatized embryos from the Doushantuo Formation, China (Xiao et al., 1998b; Xiao and Knoll, 2000; Condon et al., 2005). See papers by Jensen et al. and Bottjer and Clapham, both in this volume, for further information on Proterozoic animals.

The presence of fungi in the Proterozoic Eon is much more controversial. Several authors have noted similarities between certain microfossils and modern fungi, but in none of these reports has a convincing case been made (e.g., Schopf and Barghoon, 1969; Darby, 1974; Timofeev, 1970; Allison and Awramik, 1989; Schopf, 1968). Some Ediacaran taxa have also been interpreted to be fungal. Retallack (1994), for example, argued that because vendobionts exhibit minimal compaction, they cannot represent soft bodied animals like worms or jellyfish, and instead may be fossilized lichens (a symbiotic association between a fungus and an alga). Minimal compaction *has* been observed in some softbodied animals, however (e.g., Hagadorn et al., 2002), and, at least in the Ediacaran biota, could be attributed to unusual "death mask" preservation where early diagenetic minerals form a resistant crust (e.g., Gehling 1999). More recently, Peterson et al. (2003) argued that Ediacaran fossils from Newfoundland, including *Aspidella*, *Charnia*, and *Charniodiscus*, may represent stem-group fungi. Their argument is based primarily on a process of elimination: the fossils are found in sediments deposited below the photic zone and thus cannot be algal; the fossils do not exhibit evidence for escape or defouling behavior despite having been smothered by a thin layer of ash and thus cannot be animals; and the fossils lack evidence for shrinkage—observed in other Ediacaran taxa—inconsistent, again, with an animal interpretation. As the authors admit, however, there is little positive evidence in the form of fungal-specific characters to support a fungal affinity.

Fungi have also been reported from the 551–635 Ma Doushantuo Formation (Yuan et al., 2005). Filaments interpreted to be fungal hyphae occur in lichen-like association with clusters of coccoidal, probably cyanobacterial unicells. A fungal interpretation is based on a combination of characters—dichotomous branching, pyriform terminal structures, absence of sheaths, and narrow diameter (<1μm)—not seen in other filamentous organisms like cyanobacteria, but comparable to features observed in hyphae of glomalean fungi (Yuan *et al.* 2005).

Even earlier evidence for possible Proterozoic fungi comes from organic-walled microfossils preserved in the 723–1077 Ma Wynniatt Formation, Shaler Supergroup, arctic Canada (Fig. 2A; Butterfield, 2005). These beautifully preserved fossils consist of a large central vesicle with branching, septate, filamentous processes apparently capable of secondary fusion (Figs. 2A–B). Secondary cell-cell fusion is found in both the fungi and the red algae (Gregory, 1984; Graham and Wilcox, 2000), and possibly in the brown algae as well (Butterfield, 2005, and references therein). Because the processes are similar to fungal hyphae, however, Butterfield (2005) specifically compared the Wynniatt fossils with fungi, noting that hyphal fusion is a synapomorphy of the basidiomycetes+ascomycetes (Fig. 2C; Gregory, 1984). Butterfield (2005) referred the Wynniatt fossils to the genus *Tappania*, noting similarities with *Tappania* species from the ~1450 Ma Roper Group, Australia (Javaux *et al.*, 2001), and the Meso-Neoproterozoic Ruyang Group, north China (Yin, 1997). Secondary fusion has not been reported in *Tappania*, however, and it is not obvious that the younger and older populations are related.

An additional opisthokont group, the unicellular choanoflagellates, produce siliceous 'baskets' ~10–20 μm in size, and thus, could, in principle, have a fossil record (Leadbetter and Thomsen, 2000). No fossil choanoflagellates have been reported, however, from either Proterozoic or Phanerozoic rocks, although this may reflect a lack of search image as much as a lack of preservation.

3.2 Amoebozoans

Amoebozoans comprise two major groups: the slime molds and the lobose amoebae. Slime molds have a very poor fossil record; there are only two occurrences of fossilized slime molds from Phanerozoic rocks, both in Baltic amber (Eocene in age; Dörfelt *et al.*, 2003, and references therein). *Eosaccharomyces ramosus*, an unusual organic-walled fossil from ~1000 Ma shales of the Lakhanda Formation, Siberia, consists of open, web-like colonies of cells, a structure reminiscent of the aggregating cells of cellular slime molds (Figs. 2D–E; German, 1979; 1990; Bonner, 1967; Stephenson and Stempen, 1994; Knoll, 1996). The amoeboid cells of modern cellular slime molds lack cell walls, however, and thus have a vanishingly small chance of being preserved in shale. Although displaying a similar behavior, *Eosaccharomyces ramosus* itself is not likely to be a slime mold.

Proterozoic fossil evidence for lobose amoebae comes from vase-shaped microfossils (VSMs), a diverse and globally distributed group of middle

Neoproterozoic (~750 Ma) microfossils that also includes species of possible euglyphid amoebae (see below; Porter and Knoll, 2000; Porter *et al.*, 2003). Specifically, three species of VSMs, *Palaeoarcella athanata*, *Melanocyrillium hexodiadema*, and *Hemisphaeriella ornata* (Figs. 2F, 2H–J), possess various combinations of test characters, including an invaginated aperture, regular indentations, and a hemispherical shape, found today only in the Arcellinida, a diverse group of lobose testate amoebae (Figs. 2G, 2K;

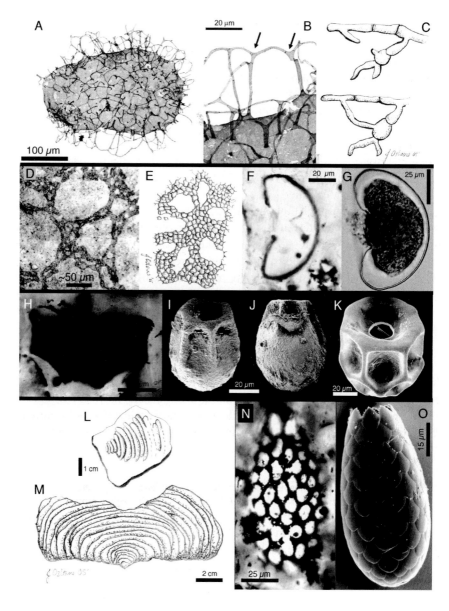

Meisterfeld, 2000a; Porter and Knoll, 2000; Porter et al., 2003). No exact modern analogs can be identified for *M. hexodiadema* and *H. ornata*, but *P. athanata* is indistinguishable from the modern lobose testate amoeban genus, *Arcella*, suggesting this test morphology may have persisted unchanged from Neoproterozoic times until today. Confirmation of a lobose testate amoeban affinity will depend on a better understanding of test evolution in the Arcellinida, a task currently hindered by poor phylogenetic resolution.

3.3 Chromalveolates

Although accumulating evidence suggests that ancestral chromalveolates were photosynthetic (Keeling, 2004), the clade includes several groups that today are either entirely heterotrophic (e.g., apicomplexans, ciliates, and oomycetes), or are a mix of heterotrophic and photosynthetic taxa (e.g., dinoflagellates). It is not clear when these groups lost their ability to photosynthesize (Keeling, 2004), and thus it is possible that early fossil representatives may have been algal. Nevertheless, I will consider their Proterozoic fossil record here.

Apicomplexans, a group composed entirely of intracellular parasites, do not have a fossil record. Fossil ciliates, on the other hand, can be common, particularly in Upper Jurassic and Lower Cretaceous rocks, where their calcareous tests can be useful in biostratigraphy (Tappan, 1993). Ciliate body fossils are not known from Proterozoic rocks, but evidence for the biomarker gammacerane in the ~742–770 Ma Chuar Group, Grand Canyon, suggests they may have been present by this time (Summons et al., 1988; Summons and Walter, 1990). The precursor to gammacerane, tetrahymenol,

---

Figure 2. (on Page 6) Fossils of putative Proterozoic heterotrophic eukaryotes and their modern analogs. (A–B) A probable fungus. Arrows in (B) indicate points of secondary fusion. Wynniatt Formation, Victoria Island, northwestern Canada. Courtesy of N. J. Butterfield. (C) Hyphal fusion in the fungus, *Botrytis elliptica*, modified from Gregory (1984); no scale bar provided, but individual cells are on the order 5 μm in width. (D) *Eosaccharomyces ramosus*, a possible slime mold. Lakhanda Formation, Siberia. Courtesy of A. H. Knoll. (E) Beginning of cell aggregation in a cellular slime mold, modified from Stephenson and Stempen (1994); no scale bar provided, but individual cells are on the order of 10 μm in size (Bonner, 1967). (F, H–J, N) Vase-shaped microfossils from the Chuar Group, Grand Canyon. (F) *Palaeoarcella athanata*, a probable lobose amoeba. (G) *Arcella hemisphaerica*, a modern lobose amoeba. Courtesy of R. Meisterfeld. (H) *Hemisphaeriella ornata*, a probable lobose amoeba. (I–J) *Melanocyrillium hexodiadema*, a probable lobose amoeba. (K) *Arcella conica*, a modern lobose amoeba. Image courtesy of R. Meisterfeld. (L) *Palaeopascichnus*, a possible foraminiferan from Ediacaran rocks. Modified from Seilacher et al. (2003). (M) The modern xenophyophorean foraminiferan, *Stannophyllum*. Modified from Seilacher et al. (2003). (N) The vase-shaped microfossil, *Melicerion poikilon*, a probable filose amoeba. (O) *Euglypha tuberculata*, a modern filose amoeba. Courtesy of R. Meisterfeld.

is not ciliate-specific, however; it is also known to occur in photosynthetic sulfur bacteria (Kleeman *et al.*, 1990), and has even been reported from a fern (Zander *et al.*, 1969; Kamaya *et al.*, 1991). Gammacerane has also been found in the 1.7 Ga Tuanshanzi Formation of China (Peng *et al.*, 1998) but given that there is no fossil evidence for other crown group eukaryotes at this time (Porter, 2004), and, in fact, no undisputed evidence for *any* eukaryotes at this time, it is more conservative to assume that these older biomarkers came from bacteria.

The only claim for Proterozoic oomycetes (Sherwood-Pike, 1991) is based on a single, poorly preserved specimen that was compared by Schopf and Barghoorn (1969) with fungal sporangia. It is possible, however, that other Proterozoic fossils currently interpreted as algae, are actually the remains of oomycetes. Several of the characters found in ~1000 Ma specimens of the fossil *Paleovaucheria*, for example (German, 1981; Woods *et al.*, 1998) are also found in oomycetes: sparsely branching tubes with few septa concentrated near the rounded termini, and circular openings at the tips of the termini (Ingold and Hudson, 1993).

Approximately 50% of extant dinoflagellates are heterotrophic (Dodge and Lee, 2000), and although some of these reflect multiple independent losses of plastids, phylogenetic analyses indicate that dinoflagellates may have been ancestrally heterotrophic (Hackett *et al.*, 2004, and references therein). The earliest undisputed body fossil evidence for dinoflagellates comes from early Triassic rocks (Fensome *et al.*, 1999), but biomarker evidence suggests the group originated at least by early Cambrian time (Moldowan and Talyzina, 1998; Talyzina *et al.*, 2000). Dinoflagellate biomarkers have also been reported from several Proterozoic—and even Archean—units, including the 2.78–2.45 Ga Mount Bruce Supergroup, Pilbara Craton, Australia; the ~1400 Ma McMinn Formation, Roper Group, Australia; the ~1100 Ma Nonesuch Formation, Michigan; the ~800 Ma Bitter Springs Formation, Australia; the ~742–770 Ma Chuar Group, Grand Canyon; and the Ediacaran Pertatataka Formation, Australia (Summons and Walter, 1990; Pratt *et al.*, 1991; Moldowan *et al.*, 2001; Brocks *et al.*, 2003a; see also Moldowan *et al.* 1996). Given its age, the Archean occurrence is attributed to an independent (non-dinoflagellate) origin (Brocks *et al.*, 2003b), and the Proterozoic occurrences have either been interpreted as possible contaminants (Summons and Walter, 1990; Summons *et al.*, 1992) or as dinosteroid precursors that do not by themselves indicate dinoflagellates were present (Moldowan *et al.*, 2001).

Interestingly, the pre-Triassic record of dinosteroid abundance correlates well with that of acritarch diversity, suggesting that many acritarchs may represent dinoflagellate cysts (Moldowan *et al.*, 1996). Indeed, many modern dinoflagellate cysts lack diagnostic characters, and would probably be grouped with the acritarchs if found as fossils (Moldowan et al., 1996, and references therein). Several papers have suggested certain Proterozoic acritarchs might be dinoflagellate cysts (e.g., Tappan, 1980; Butterfield and Rainbird, 1998, although see Butterfield, 2005; Arouri et al., 2000). The most compelling of these is Arouri et al. (2000), which showed that some Ediacaran acanthomorphic acritarchs have chemical and ultrastructural characters consistent with a dinoflagellate affinity (although see Versteegh and Blokker, 2004). Because the taxonomic distribution of these characters is not well documented, however, it is impossible to know whether their occurrence in both fossil and modern groups is due to homology or convergence, and, if due to homology, whether their occurrence reflects a shared derived feature of the dinoflagellates or a plesiomorphic condition.

3.4   Rhizarians

Rhizarians include three major groups, foraminifers, cercozoans, and radiolarians. The last of these is polyphyletic; recent phylogenies suggest that phaeodareans, traditionally grouped with the other radiolarian classes, polycystineans and acantharians, are derived from within cercozoans (Nikolaev et al., 2004). With a few exceptions (e.g., *Paulinella*, chlorarachniophytes), all rhizarians are obligate heterotrophs.

Radiolarians are not known from Precambrian rocks. The earliest fossil evidence for radiolarians is polycystinean skeletons from the Middle Cambrian (Won et al., 1999). Acantharians lack a fossil record—their strontium sulfate skeletons dissolve easily in seawater—and the oldest phaeodareans are Cretaceous (Danelian and Moreira, 2004, and references cited therein).

The earliest undisputed foraminifera are from Early Cambrian rocks (Culver, 1991, 1994; McIlroy et al., 2001), although Seilacher et al. (2003) have made an interesting case that some Ediacaran taxa were giant foraminifera (also see Zhuravlev, 1993). Specifically, Seilacher and his colleagues argue that several Ediacaran trace fossils, including *Palaeopascichnus* (Fig. 2L), *Neonereites*, *Intrites*, and *Yelovichnus*, are xenophyophoreans[*], giant foraminifera up to 25 cm in size that today are known only from abyssal environments (Fig. 2M; Gooday and Tendal, 2000; Pawlowski et al., 2003). They also interpret vendobionts as extinct

---

[*] To be exact, Seilacher et al. (2003) interpret vendobionts as an extinct group of giant rhizopods. As originally construed, however, rhizopods are polyphyletic. The group was recently revised and renamed 'Cercozoa' (Cavalier-Smith, 1998). Presumably Seilacher et al. are interpreting the vendobionts as close relatives of xenophyophoreans.

foraminifers, arguing that the sand-filled, fecal 'skeletons' ('stercomare') found inside the tests of xenophyophoreans may be a modern analog for the sand-filled bodies of some vendobionts (Grazhdankin and Seilacher, 2002).

A recent study suggests that cercozoans may be among the most diverse protozoan groups alive today, comparable in diversity to the ciliates (Bass and Cavalier-Smith, 2004). The majority of cercozoans are zooflagellates, taxa that would be unlikely to fossilize, but the group also includes filose amoebae, some of which possess fossilizable tests. Possible evidence for Proterozoic cercozoans is the 742–770 Ma vase-shaped microfossil, *Melicerion poikilon* (Fig. 2N), thought to be the remains of a filose testate amoeba (Porter and Knoll, 2000; Porter *et al*., 2003). Specifically, *Melicerion* possessed a tear-drop-shaped, aperturate test with circular, regularly arranged, mineralized scales embedded in an organic wall (Porter *et al*., 2003). This character combination is known today only in the euglyphid amoebae, a monophyletic group of filose testate amoebae (Fig. 2O; Meisterfeld, 2000b; Wylezich *et al*., 2002). Some lobose testate amoebae also make tests with mineralized scales, but the scales are different in shape or less regularly arranged (Meisterfeld, 2000b). Interestingly, there is a group of lobose testate amoebae that *do* have circular scales in their tests, but these are not endogenous; i.e. they are acquired by engulfing euglyphid tests and stealing the scales (Gnekow, 1981). Given that there is good evidence for lobose testate amoebae in rocks of this age (see Section 3.2), *Melicerion* could be interpreted as a lobose amoeba, but its strong similarities with euglyphids support a closer tie with cercozoans.

3.5   Excavates

There are no reports of excavate taxa from Proterozoic rocks. Most excavates have extremely low preservation potential, but putative evidence for fossil euglenids in fluvial and nearshore-marine rocks from Ordovician and Silurian strata (Gray and Boucot, 1989) suggests the organic pellicle found in euglenids may be preservable. This is consistent with studies of Lindgren (1981) showing that the lorica of the euglenid *Trachelomonas* is acid-resistant. Possible euglenids are also known, along with kinetoplastids, from amber (Schönborn *et al*., 1999; Poinar and Poinar, 2004), although this preservational window does not extend into the Proterozoic Eon.

3.6   Summary

Table 1 summarizes the fossil evidence for heterotrophic eukaryotes in Proterozoic rocks. There are several reports of heterotrophic taxa from the Proterozoic, but only four of these—animals, fungi, lobose amoebae, and,

Table 1. Fossil evidence for possible heterotrophic protists in Proterozoic (and Archean) rocks. See text for more details.

Taxon	Proterozoic Fossil Evidence	Age (Ma)	Reference
Opisthokonts			
Fungi	Aspidella, Charniodiscus, Charnia, etc.	575–542	Peterson et al., 2003
" "	Fungal hyphae in a lichen-like association	635–551	Yuan et al., 2005
" "	Acritarchs exhibiting secondary cell fusion	>723–1077	Butterfield, 2005
Amoebozoans			
Lobose amoebae	Palaeoarcella athanata, Melanocyrillium hexodiadema, Hemisphaeriella ornata	742–770	Porter et al., 2003
Chromalveolates			
Ciliates	gammacerane (biomarker)	742–770	Summons et al., 1988; Summons and Walter, 1990
" "	" "	~1700	Peng et al., 1998
Dinoflagellates	dinosterane (biomarker)	~540–630	Summons and Walter, 1990
" "	" "	~742–770	Moldowan et al., 2001
" "	" "	~800	Summons and Walter, 1990
" "	" "	~1100	Pratt et al., 1991
" "	" "	~1400	Moldowan et al. 2001
" "	" "	~2780–2450	Brocks et al., 2003a,b
Rhizarians			
Foraminifera	vendobionts, Palaeopascichnus, Neonereites, Intrites, Yelovichnus, etc.	575–542	Zhuravlev, 1993; Seilacher et al., 2003
Cercozoans	Melicerion poilon	742–770	Porter and Knoll, 2000; Porter et al., 2003

probably, filose amoebae—are based on specific characters that are likely to be synapomorphies for the group in question (or for clades within the group). The other reports listed in Table 1 are plausible but either lack specific synapomorphies linking fossils to their modern counterparts or—in the case of biomarker evidence—may be contaminants.

Granted the risks in making generalizations about the sparse Proterozoic fossil record, we can still make a few interesting observations. The first, already noted above, is that although today the diversity of heterotrophs exceeds that of algae, during the Proterozoic the situation seems reversed. The second is that although heterotrophs are ancestral to the algae, the first convincing algal fossils precede the first convincing heterotroph fossils by several hundred million years. Why are heterotrophs rare in Proterozoic rocks?

4. WHY ARE HETEROTROPHS RARE IN PROTEROZOIC ROCKS?

Porter and Knoll (2000) offered two reasons why few, if any, heterotrophs are found in rocks older than ~770–800 Ma (when VSMs first appear). The first is that heterotroph diversity may have been low due to limited primary productivity in Mesoproterozoic oceans. Evidence for limited productivity during this interval comes primarily from theoretical arguments. Anbar and Knoll (2002), for example, have argued that if Mesoproterozoic oceans were anoxic and sulfidic below the mixed layer (Canfield, 1998; Shen et al., 2002, 2003; Arnold et al., 2004; Brocks et al., 2005), then both dissolved iron and molybdenum would have been scarce. As both elements are important components of enzymes responsible for nitrogen fixation and nitrate assimilation, they reason that nitrogen cycling would have been limited in Mesoproterozoic oceans. Further support for a nitrogen-stressed biosphere during this time comes from box models that show that as oxygen levels rose during the early Proterozoic, increasing levels of nitrification and denitrification would have lowered the pool of bioavailable nitrogen (Fennel et al., 2005).

Empirical evidence for limited primary productivity is more problematic. Anbar and Knoll (2002) point out that the average value of $\delta^{13}C$ in Mesoproterozoic carbonates is ~1.5‰ lower than in Paleoproterozoic, Neoproterozoic, and Phanerozoic carbonates, suggesting depressed Mesoproterozoic primary productivity. Nonetheless, late Paleoproterozoic and early Mesoproterozoic $\delta^{13}C$ values hover around 0‰, indicating that organic carbon burial constituted a significant proporation—~20%—of total carbon burial; average values near 3.5‰ from late Mesoproterozoic rocks (Frank et al., 2003) suggest even higher proportions. Of course, because organic carbon burial rates are a function of several factors, including sedimentation rates and redox conditions, they may not be reliable indicators of primary productivity levels at all, high or low. Anbar and Knoll also point out that several Mesoproterozoic basins appear to have had limited depth gradients in the isotopic composition of DIC, consistent with low productivity. Limited gradients could also reflect vigorous ocean mixing, however, or a large DIC reservoir (Bartley and Kah, 2004) that effectively drowned out any signal of $\delta^{13}C$ stratification resulting from high productivity.

Anbar and Knoll (2002) emphasize that the primary organisms affected by nitrogen stress would be eukaryotic algae, which, unlike cyanobacteria, are unable to fix nitrogen or to scavenge bioavailable nitrogen from their surroundings. Thus, even if overall primary productivity was not limited, it is expected that eukaryotic primary productivity was. How would eukaryotic heterotrophs have been impacted? Because they can get bioavailable nitrogen by ingesting organic particles, they are not directly affected by a nitrogen-stressed world. They may have been indirectly affected, however, simply because limited overall primary productivity means limited food supply. But if only eukaryotic algae were negatively impacted, it's not clear that eukaryotic heterotrophs themselves would have been; they could have dined primarily on bacteria, as many do today. Assuming the nutritive content of bacteria and eukaryotic algae is similar, then heterotrophic eukaryotes would have been abundant and diverse in Mesoproterozoic oceans.

More likely, the dearth of heterotrophs prior to ~770 Ma reflects taphonomic bias (Porter and Knoll, 2000). Although both algae and heterotrophs make mineralized structures, with few exceptions (Allison and Hilgert, 1986; Grant, 1990; Horodyski and Mankiewicz, 1990; Watters and Grotzinger, 2001; Wood et al., 2002; Porter et al., 2003), there are no mineralizing eukaryotes—algal or heterotrophic—from the Precambrian. The Precambrian body fossil record thus primarily comprises organic-walled structures, and within these taphonomic limits, algae have an important advantage. Unlike the majority of heterotrophs, which require a flexible membrane for phagocytosis, most algae possess cell walls. Indeed, cell walls have evolved multiple times, suggesting that, as long as the organism does not depend on phagocytosis, having rigid support is advantageous (cf., Leander, 2004). The presence of a cell wall by itself may not impart significant preservational advantages (e.g., Bartley, 1996; de Leeuw and Largeau, 1993), but several algal groups impregnate their walls with highly resistant macromolecules. These include algaenans, which occur in the cell walls or cysts of some green algae, some eustigmatophytes (a group of chromist algae), and the photosynthetic dinoflagellate *Gymnodinium catenatum* (Gelin et al., 1997, 1999; Versteegh and Blokker, 2004); and dinosporin, which occurs in the resting cysts of dinoflagellates (Versteegh and Blokker, 2004). These groups in particular should be well represented among Proterozoic organic-walled fossil protists.

Heterotrophs do make preservable organic-walled structures, however. The fossilized tests of amoebae have been found in a variety of facies, indicating their preservation does not depend on exceptional taphonomic circumstances (Medioli et al. 1990; Porter and Knoll, 2000). Loricae of folliculinid ciliates reported from cherts in Africa indicate these organic-walled structures may also be preserved (Deflandre and Deunff, 1957). Many heterotrophs also make organic-walled cysts, including several naked (non-testate) amoebae (Lee et al., 2000). The degradation-resistance of these structures is not well known, although probable cysts preserved in some fossil testate amoebae suggest it may be relatively good (Martí-Mus and Moczydłowska, 2000; Porter et al., 2003). In addition, fungi, oomycetes, and heterotrophic dinoflagellates have cell walls; either they are osmotrophs, able to transport dissolved organic matter across this rigid boundary, or, in the case of dinoflagellates, they phagocytose by opening their thecal plates and extruding a pseudopod-like structure (Hackett et al., 2004). Their walls are about as resistant as algal cell walls, if not more so (de Leeuw and Largeau, 1993). Finally, highly resistant macromolecules similar to algaenans and dinosporins are known from fungal spores (de Leeuw and Largeau, 1993).

Algae have a taphonomic advantage then, *not* because heterotrophs are inherently *un*preservable, but because more algae make preservable structures than heterotrophs do. Most algae have cell walls, for example, while most heterotrophs do not. If the majority of Precambrian acritarchs are the remains of vegetative cells rather than cysts (Butterfield, 2004), then, statistically speaking, most acritarchs probably are algal. But there is no good reason to think *all* of them are algal (Butterfield, 2005). Cell walls, cysts, and spores of heterotrophs may constitute a sizable—though unrecognized—minority of the Precambrian fossil record.

5. CONCLUSIONS

Although heterotrophic eukaryotes necessarily preceded eukaryotic algae, the latter are much better represented in the Proterozoic fossil record. Convincing evidence exists for only four heterotrophic clades during this time: the fungi, known from ~580 Ma rocks, and possibly from rocks older than 723 Ma; the lobose and filose amoebae, which appear in rocks 742–770 Ma; and the animals, which appear near the close of the Proterozoic Eon. Other Proterozoic body fossils or biomarkers may represent ciliates, dinoflagellates, oomycetes, and foraminifera. The dearth of Proterozoic fossil heterotrophs may reflect low heterotroph diversity caused by limited primary productivity. More likely, however, it reflects a preservational bias among organic-walled fossils: more algae make preservable organic-walled structures than heterotrophs do. Nonetheless, heterotrophs do make preservable structures, and their cysts, spores, and tests probably go unrecognized among the problematic fossils that constitute the bulk of the Precambrian fossil record.

ACKNOWLEDGEMENTS

I am thankful to Shuhai Xiao and Jay Kaufman for inviting me to contribute to this volume, and to Stan Awramik, David Chapman, Linda Kah, Andrew Knoll, David Lamb, Brian Leander, and Alastair Simpson for useful discussions and feedback. Jennifer Osborne provided illustrations, and Shuhai Xiao, Carl Mendelson, and Nick Butterfield provided useful and constructive reviews.

REFERENCES

Allison, C. W., and Awramik, S. M., 1989, Organic-walled microfossils from the earliest Cambrian or latest Proterozoic Tindir Group rocks, northwest Canada, *Precambrian Res.* **43**: 253–294.

Allison, C. W., and Hilgert, J. W., 1986, Scale microfossils from the Early Cambrian of Northwest Canada, *J. Paleont.* **60**(5): 973–1015.

Anbar, A. D., and Knoll, A. H., 2002, Proterozoic ocean chemistry and evolution: a bioinorganic bridge? *Science* **297**: 1137–1142.

Andersson, J. O., and Roger, A. J., 2002, A cyanobacterial gene in nonphotosynthetic protists—an early chloroplast acquisition in eukaryotes? *Curr. Biol.* **12**: 115–119.

Arnold, G. L., Anbar, A. D., Barling, J., and Lyons, T. W., 2004, Molybdenum isotope evidence for widespread anoxia in mid-Proterozoic oceans, *Science*, **304**: 87–90.

Arouri, K. R., Greenwood, P. F., and Walter, M. R., 2000, Biological affinities of Neoproterozoic acritarchs from Australia: microscopic and chemical characterisation, *Org. Geochem.* **31**: 75–89.

Baldauf, S. L., 2003, The deep roots of eukaryotes, *Science* **300**: 1703–1706.

Bartley, J. K., 1996, Actualistic taphonomy of Cyanobacteria: implications for the Precambrian fossil record, *Palaios* **11**: 571–586.

Bartley, J. K., and Kah, L. C., 2004, Marine carbon reservoir, C-org–C-carb coupling, and the evolution of the Proterozoic carbon cycle, *Geology* **32**: 129–132.

Bass, D., and Cavalier-Smith, T., 2004, Phylum-specific environmental DNA analysis reveals remarkably high global biodiversity of Cercozoa (Protozoa), *Int. J. Syst. Evol. Microbiol.* **54**: 2393–2404.

Bonner, J. T., 1967, *Cellular Slime Molds*, Princeton University Press, Princeton, New Jersey.

Bottjer, D. J., and Clapham, M. E., 2006, Evolutionary paleoecology of Ediacaran benthic marine animals. in: *Neoproterozoic Geobiology and Paleobiology* (S. Xiao and A. J. Kaufman, eds.), Springer, Dordrecht, the Netherlands, pp. 91–114.

Brocks, J. J., Buick, R., Logan, G. A., and Summons, R. E., 2003a, Composition and syngeneity of molecular fossils from the 2.78 to 2.45 billion-year-old Mount Bruce

Supergroup, Pilbara Craton, Western Australia, *Geochim. Cosmochim. Acta* **67**: 4289–4319.

Brocks, J. J., Buick, R., Summons, R. E., and Logan, G. A., 2003b, A reconstruction of Archean biological diversity based on molecular fossils from the 2.78 to 2.45 billion-year-old Mount Bruce Supergroup, Hamersley Basin, Western Australia, *Geochim. Cosmochim. Acta* **67**: 4321–4335.

Brocks, J. J., Love, G. D., Summons, R. E., Knoll, A.H., Logan, G. A, Bowden, S. A., 2005, Biomarker evidence for green and purple sulfur bacteria in a stratified Palaeoproterozoic sea, *Nature* **437**: 866–870.

Butterfield, N. J., 2000, *Bangiomorpha pubescens* n. gen., n. sp.: implications for the evolution of sex, multicellularity, and the Mesoproterozoic-Neoproterozoic radiation of eukaryotes, *Paleobiology* **26**: 386–404.

Butterfield, N. J., 2004, A vaucheriacean alga from the middle Neoproterozoic of Spitsbergen: implications for the evolution of Proterozoic eukaryotes and the Cambrian explosion, *Paleobiology* **30**: 231–252.

Butterfield, N. J., 2005, Probable Proterozoic Fungi, *Paleobiology* **31**: 165–182.

Butterfield, N. J., and Rainbird, R. H., 1998, Diverse organic-walled fossils, including "possible dinoflagellates" from the early Neoproterozoic of arctic Canada, *Geology* **26**: 963–966.

Butterfield, N. J., Knoll, A. H., and Swett, K., 1990, A bangiophyte red alga from the Proterozoic of arctic Canada, *Science* **250**: 104–107.

Butterfield, N. J., Knoll, A. H., and Swett, K., 1994, Paleobiology of the Neoproterozoic Svanbergfjellet Formation, Spitsbergen, *Fossils Strata* **34**: 1–84.

Canfield, D. E., 1998, A new model for Proterozoic ocean chemistry, *Nature* **396**: 450–453.

Cavalier-Smith, T., 1998, A revised six-kingdom system of life, *Biol. Rev.* **73**: 203–266.

Condon, D., Zhu, M., Bowring, S., Wang, W., Yang, A., and Jin, Y., 2005, U–Pb ages from the Neoproterozoic Doushantuo Formation, China, *Science* **308**: 95–98.

Culver, S. J., 1991, Early Cambrian Foraminifera from West Africa, *Science* **254**: 689–691.

Culver, S. J., 1994, Early Cambrian Foraminifera from the southwestern Taoudeni Basin, West Africa, *J. Foram. Res.* **24**: 191–202.

Danelian, T., and Moreira, D., 2004, Palaeontological and molecular arguments for the origin of silica-secreting marine organisms, *C. R. Palevol* **3**: 229–236.

Darby, D. G., 1974, Reproductive modes of *Huroniospora microreticulata* from cherts of the Precambrian Gunflint Iron-Formation, *Geol. Soc. Amer. Bull.* **85**: 1595–1596.

de Leeuw, J. W., and Largeau, C., 1993, A review of macromolecular organic compounds that comprise living organisms and their role in kerogen, coal, and petroleum formation, in: *Organic Geochemistry: Principles and Applications* (M. H. Engel and S. A. Macko, eds.), Topics in Geobiology, Plenum Press, New York, pp. 23–72.

Deflandre, G., and Deunff, J., 1957, Sur la presence de cilies fossiles de la familie des Folliculinidae dans un silex du Gabon, *C. R. Hebd. Séances Acad. Sci.* **244**: 3090–3093.

Dodge, J. D., and Lee, J. J., 2000, Phylum Dinoflagellata Bütschli, 1885, in: *An Illustrated Guide to the Protozoa* (J. J. Lee, G. F. Leedale and P. Bradbury, eds.), Society of Protozoologists, Lawrence, Kansas, pp. 656–689.

Dörfelt, H., Schmidt, A. R., Ullman, P., and Wunderlick, J., 2003, The oldest myxogastrid slime mold, *Mycol. Res.* **107**: 123–126.

Douzery, E. J. P., Snell, E. A., Bapteste, E., Delsuc, F., and Philippe, H., 2004, The timing of eukaryotic evolution: does a relaxed molecular clock reconcile proteins and fossils? *Proc. Natl. Acad. Sci. USA* **101**: 15386–15391.

Fennel, K., Follows, M., and Falkowski, P.G., 2005, The co-evolution of the nitrogen, carbon, and oxygen cycles in the Proterozoic ocean, *Am. J. Sci.* **305**: 526–545.

Fensome, R. A., Saldarriaga, J. F., and Taylor, F. J. R., 1999, Dinoflagellate phylogeny revisited: reconciling morphological and molecular based phylogenies, *Grana* **38**: 66–80.

Frank, T. D., Kah, L. C., and Lyons, T. W., 2003, Changes in organic matter production and accumulation as a mechanism for isotopic evolution in the Mesoproterozoic ocean, *Geol. Mag.* **140**: 397–420.

Gehling, J. G., 1999, Microbial mats in terminal Proterozoic siliciclastics: Ediacaran death masks, *Palaios* **14**: 40–57.

Gelin, F., Boogers, I., Noordeloos, A. A. M., Damsté, J. S. S., Riegman, R., and Leeuw, J. W. d., 1997, Resistant biomacromolecules in marine microalgae of the classes Eustigmatophyceae and Chlorophyceae: geochemical implications, *Org. Geochem.* **26**: 659–675.

Gelin, F., Volkman, J. K., Largeau, C., Derenne, S., Damsté, J. S. S., and Leeuw, J. W. D., 1999, Distribution of aliphatic, nonhydrolyzable biopolymers in marine microalgae, *Org. Geochem.* **30**: 147–159.

German, T., 1979, Nakhodki gribov v Rifee (Discoveries of fungi in the Riphean), in: *Paleontologiia Dokembriia i Rannego Kembriia* (B. Sokolov, ed.), Nauka, Leningrad, pp. 129–136.

German, T., 1981, Nitchatye mikroorganizmy Lakhandinskoi svity reki Mai [Filamentous microorganisms in the Lakhanda Formation on the Maya River], *Paleontol. Zh.* **1981**(2): 100–107.

German, T. N., 1990, *Organic World Billion Year Ago*, Nauka, Leningrad.

Gnekow, M. A., 1981, Beobachtungen zur Biologie und Ultrastruktur der moobewohnenden Thecamöbe *Nebela tincta* (Rhizopoda). *Arch. Protistenkd.* **124**: 36–69.

Gooday, A. J., and Tendal, O. S., 2000, Class Xenophyophorea Schulze, 1904, in: *An Illustrated Guide to the Protozoa* (J. J. Lee, G. F. Leedale, and P. Bradbury, eds.), Society of Protozoologists, Lawrence, Kansas, pp. 1086–1097.

Graham, L. E., and Wilcox, L. W., 2000, *Algae*, Prentice Hall, Upper Saddle River, NJ.

Grant, S. W. F., 1990, Shell structure and distribution of *Cloudina*, a potential index fossil for the terminal Proterozoic, *Am. J. Sci.* **290A**: 261–294.

Gray, J., and Boucot, A. J., 1989, Is *Moyeria* a euglenoid?, *Lethaia* **22**: 447–456.

Grazhdankin, D., and Seilacher, A., 2003, Underground Vendobionta from Namibia, *Palaeontology* **45**: 57–78.

Gregory, P. H., 1984, The fungal mycelium: an historical perspective, *Trans. Br. Mycol. Soc.* **82**: 1–11.

Hackett, J. D., Anderson, D. M., Erdner, D. L., and Bhattacharya, D., 2004, Dinoflagellates: a remarkable evolutionary experiment, *Am. J. Bot.* **91**: 1523–1534.

Hagadorn, J. W., Dott, R. H., and Damrow, D., 2002, Stranded on an Upper Cambrian shoreline: Medusae from central Wisconsin, *Geology* **30**: 103–106.

Han, T.-M., and Runnegar, B. 1992, Megascopic eukaryotic algae from the 2.1-billion-year-old Negaunee Iron-Formation, Michigan, *Science* **257**: 232–235.

Hofmann, H. J., 1994, Proterozoic carbonaceous compressions ("metaphytes" and "worms"), in: *Early Life on Earth* (S. Bengtson, ed), Columbia University Press, New York, pp. 342–357.

Horodyski, R. J., and Mankiewicz, C., 1990, Possible Late Proterozoic skeletal algae from the Pahrump Group, Kingston Range, southeastern California, *Am. J. Sci.* **290A**: 149–169.

Ingold, C. T., and Hudson, H. J., 1993, *The Biology of Fungi*, Chapman and Hall, New York.

Javaux, E. J., Knoll, A. H., and Walter, M. R., 2001, Morphological and ecological complexity in early eukaryotic ecosystems, *Nature* **412**: 66–69.

Jensen, S., Droser, M. L., and Gehling, J. G., 2006, A critical look at the Ediacaran trace fossil record. in: *Neoproterozoic Geobiology and Paleobiology* (S. Xiao and A. J. Kaufman, eds.), Springer, Dordrecht, the Netherlands, pp. 115–157.

Kamaya, R., Mori, T., Shoji, H., Ageta, H., Chang, H. C., and Hsu, H. Y., 1991, Fern constituents: triterpenes from *Oleandra wallichii*, *Yakugaku Zasshi (J. Pharmaceutical Soc. Japan)*, **11**: 120–125.

Keeling, P. J., 2004, Diversity and evolutionary history of plastids and their hosts, *Am. J. Bot.* **91**: 1481–1493.

Kleemann, G., Poralla, K., Englert, G., Kjosen, H., Liaaen-jensen, N., Neunlist, S., and Rohmer, M., 1990, Tetrahymanol from the phototrophic bacterium *Rhodopseudomonas palustris*: first report of a gammacerane triterpene from a prokaryote, *J. Gen. Microbiol.* **136**: 2551–2553.

Knoll, A. H., 1996, Archean and Proterozoic paleontology, in: *Palynology: Principles and Applications* (J. Jansonius and D. C. McGregor, eds.), American Association of Stratigraphic Palynologists Foundation, pp. 51–80.

Leadbetter, B. S. C., and Thomsen, H. A., 2000, Order Choanoflagellida, Kent, 1880, *An Illustrated Guide to the Protozoa, Second Edition* (J. J. Lee, G. F. Leedale, and P. Bradbury, eds.), Allen Press, Lawrence, Kansas, pp. 14–38.

Leander, B. S., 2004, Did trypanosomatid parasites have photosynthetic ancestors? *Trends Microbiol.* **12**: 251–258.

Lee, J. J., Leedale, G. F., and Bradbury, P. (eds.), 2000, *An Illustrated Guide to the Protozoa*, Society of Protozoologists, Lawrence, KS.

Lindgren, S., 1981, Remarks on the taxonomy, botanical affinities, and distribution of leiospheres, *Stockh. Contr. Geol.* **38**: 1–20.

Martí Mus, M., and Moczydłowska, M., 2000, Internal morphology and taphonomic history of the Neoproterozoic vase-shaped microfossils from the Visingsö Group, Sweden, *Norsk Geol. Tidsskr.* **80**: 213–228.

Martin, F., 1993, Acritarchs: a review, *Biol. Rev.* **68**: 475–538.

McIlroy, D., Green, O. R., and Brasier, M. D., 2001, Palaeobiology and evolution of the earliest agglutinated Foraminifera: *Platysolenites*, *Spirosolenites* and related forms, *Lethaia* **34**: 13–29.

Medioli, F. S., Scott, D. B., Collins, E. S., and McCarthy, F. M. G., 1990, Fossil thecamoebians: present status and prospects for the future, in: *Paleoecology, Biostratigraphy, Paleoceanography and Taxonomy of Agglutinated Foraminifera* (C. Hemleben *et al.*, eds.), Kluwer Academic, Dordrecht, Netherlands, pp. 813–839.

Meisterfeld, R., 2000a, Order Arcellinida Kent, 1880, in: *An Illustrated Guide to the Protozoa* (J. J. Lee, G. F. Leedale, and P. Bradbury, eds.), Society of Protozoologists, Lawrence, Kansas, pp. 827–860.

Meisterfeld, R., 2000b, Testate amoebae with filopodia, in: *An illustrated guide to the Protozoa* (J. J. Lee, G. F. Leedale, and P. Bradbury, eds,), Society of Protozoologists, Lawrence, Kansas, pp. 1054–1084.

Mendelson, C. V., and Schopf, J. W., 1992, Proterozoic and Early Cambrian acritarchs, in: *The Proterozoic Biosphere* (J. W. Schopf and C. Klein, eds.), Cambridge University Press, Cambridge, pp. 219–232.

Moldowan, J. M., and Talyzina, N. M., 1998, Biogeochemical evidence for dinoflagellate ancestors in the Early Cambrian, *Science* **281**: 1168–1170.

Moldowan, J. M., Dahl, J., Jacobsen, S. R., Huizinga, B. J., Fago, F. J., Shetty, R., Watt, D. S., and Peters, K. E., 1996, Chemostratigraphy reconstruction of biofacies: molecular evidence linking cyst-forming dinoflagellates with pre-Triassic ancestors, *Geology* **24**: 159–162.

Moldowan, J. M., Jacobsen, S. R., Dahl, J., Al-Hajji, A., Huizinga, B. J., and Fago, F. J., 2001, Molecular fossils demonstrate Precambrian origins of dinoflagellates, in: *The Ecology of the Cambrian Radiation* (A. Yu. Zhuravlev and R. Riding, eds.), Columbia University Press, New York, pp. 475–493.

Nikolaev, S. I., Berney, C., Fahrni, J. F., Bolivar, I., Polet, S., Mylnikov, A. P., Aleshin, V. V., Petrov, N. B., and Pawlowski, J., 2004, The twilight of Heliozoa and rise of Rhizaria, an emerging supergroup of amoeboid eukaryotes, *Proc. Natl. Acad. Sci. USA* **101**: 8066–8071.

Pawlowski, J., Holzmann, M., Fahrni, J., and Richardson, S. L., 2003, Small subunit ribosomal DNA suggests that the xenophyophorean *Syringammina corbicula* is a foraminiferan, *J. Eukaryot. Microbiol.* **50**: 483–487.

Peng, P., Sheng, G., Fu, J., and Yan, Y., 1998, Biological markers in 1.7 billion year old rock from the Tuanshanzi Formation, Jixian strata section, North China, *Org. Geochem.* **29**: 1321–1329.

Peterson, K. J., Waggoner, B., and Hagadorn, J. W., 2003, A fungal analog for Newfoundland Ediacaran fossils, *Integr. Comp. Biol.* **43**: 127–136.

Philip, G. A., Creevey, C. J., and McInerney, J. O., 2005, The Opisthokonta and the Ecdysozoa may not be clades: stronger support for the grouping of plant and animal than for animal and fungi and stronger support for the Coelomata than Ecdysozoa, *Molec. Biol. Evol.* **22**: 1175–1184.

Poinar, G., and Poinar, R., 2004, *Paleoleishmania proterus* n. gen., n. sp., (Trypanosomatidae: Kinetoplastida) from Cretaceous Burmese amber, *Protist* **155**: 305–310.

Porter, S. M., 2004, The fossil record of early eukaryotic diversification, *Paleontol. Soc. Papers* **10**: 35–50.

Porter, S. M., and Knoll, A. H., 2000, Testate amoebae in the Neoproterozoic Era: evidence from vase-shaped microfossils in the Chuar Group, Grand Canyon, *Paleobiology* **26**: 360–385.

Porter, S. M., Meisterfeld, R., and Knoll, A. H., 2003, Vase-shaped microfossils from the Neoproterozoic Chuar Group, Grand Canyon: a classification guided by modern testate amoebae, *J. Paleontol.* **77**: 409–429.

Pratt, L. M., Summons, R. E., and Hieshima, G. B., 1991, Sterane and triterpane biomarkers in the Precambrian Nonesuch Formation, North American Midcontinent Rift, *Geochim. Cosmochim. Acta* **55**: 911–916.

Retallack, G. J., 1994, Were the Ediacaran fossils lichens? *Paleobiology* **20**: 523–544.

Schneider, D. A., Bickford, M. E., Cannon, W. F., Sculz, K. J., and Hamilton, M. A., 2002, Age of volcanic rocks and syndepositional iron formations, Marquette Range Supergroup: implications for the tectonic setting of Paleoproterozoic iron formations of the Lake Superior region, *Can. J. Earth Sci.* **39**: 999–1012.

Schönborn, W., Dörfelt, H., Foissner, W., Krienitz, L., and Schäfer, U., 1999, A fossilized microcenosis in Triassic amber, *J. Eukaryot. Microbiol.* **46**: 571–584.

Schopf, J. W., 1968, Microflora of the Bitter Springs Formation, Late Precambrian, central Australia, *J. Paleontol.* **42**: 651–688.

Schopf, J. W., and Barghoorn, E. S., 1969, Microorganisms from the late Precambrian of South Australia, *J. Paleontol.* **43**: 111–118.

Seilacher, A., Grazhdankin, D., and Legouta, A., 2003, Ediacaran biota: the dawn of animal life in the shadow of giant protists, *Paleontol. Res.* **7**: 43–54.

Shen, Y, Canfield, D.E., and Knoll, A.H., 2002, Middle Proterozoic ocean chemistry: evidence from the McArthur Basin, northern Australia, *Am. J. Sci.* **302**: 81–109.

Shen, Y., Knoll, A.H., and Walter, M. R., 2003, Evidence for low sulphate and anoxia in a mid-Proterozoic marine basin, *Nature* **423**: 632–635.

Sherwood-Pike, M., 1991, Fossils as keys to evolution in fungi, *BioSystems* **25**: 121–129.

Simpson, A. G. B., and Roger, A. J., 2002, Eukaryotic evolution: getting to the root of the problem, *Curr. Biol.* **12**: R691–R693.

Simpson, A. G. B., and Roger, A. J., 2004, The real 'kingdoms' of eukaryotes, *Curr. Biol.* **14**: R693–R696.

Stechmann, A., and Cavalier-Smith, T., 2002, Rooting the eukaryote tree by using a derived gene fusion, *Science* **297**: 89–91.

Stechmann, A., and Cavalier-Smith, T., 2003, The root of the eukaryote tree pinpointed, *Curr. Biol.* **13**: R665–R666.

Stephenson, S. L., and Stempen, H., 1994, *Myxomycetes: A Handbook of Slime Molds*, Timber Press, Inc., Portland, Oregon.

Summons, R. E., S. C. Brassell, G. Eglinton, E. Evans, R. J. Horodyski, N. Robinson, and D. M. Ward, 1988, Distinctive hydrocarbon biomarkers from fossiliferous sediment of the Late Proterozoic Walcott Member, Chuar Group, Grand Canyon, Arizona, *Geochim. Cosmochim. Acta* **52**: 2625–2637.

Summons, R. E., Thomas, J., Maxwell, J. R., and Boreham, C. J., 1992, Secular and environmental constraints on the occurrence of dinosterane in sediments, *Geochim. Cosmochim. Acta* **56**: 2437–2444.

Summons, R. E., and Walter, M. R., 1990, Molecular fossils and microfossils of prokaryotes and protists from Proterozoic sediments, *Am. J. Sci.* **290A**: 212–244.

Talyzina, N. M., Moldowan, J. M., Johannisson, A., and Fago, F. J., 2000, Affinities of Early Cambrian acritarchs studied by using microscopy, fluorescence flow cytometry and biomarkers, *Rev. Palaeobot. Palynol.* **108**: 37–53.

Tappan, H., 1980, *The Paleobiology of Plant Protists*, San Francisco.

Tappan, H., 1993, Tintinnids, in: *Fossil Prokaryotes and Protists* (J. H. Lipps, ed.), Blackwell Scientific Publications, Boston, pp. 285–303.

Timofeev, B. V., 1970, Une découverte de phycomycetes dans le Précambrien, *Rev. Palaeobot. Palynol.*, **10**: 79–81.

Versteegh, G. J. M., and Blokker, P., 2004, Resistant macromolecules of extant and fossil microalgae, *Phycol. Res.* **52**: 325–339.

Wang, D., Kumar, S., and Hedges, S., 1999, Divergence time estimates for the early history of animal phyla and the origin of plants, animals and fungi, *Proc. R. Soc. Lond [Biol.]* **266**: 163–171.

Watters, W. A., and Grotzinger, J. P., 2001, Digital reconstruction of calcified early metazoans, terminal Proterozoic Nama Group, Namibia, *Paleobiology* **27**: 159–171.

Won, M. Z., and Below, R., 1999, Cambrian Radiolaria from the Georgina Basin, Queensland, Australia, *Micropaleontology* **45**: 325–363.

Wood, R. A., Grotzinger, J. P., and Dickson, J. A. D., 2002, Proterozoic modular biomineralized metazoan from the Nama Group, Namibia, *Science* **296**: 2383–2386.

Woods, K. N., Knoll, A. H., and German, T., 1998, Xanthophyte algae from the Mesoproterozoic/Neoproterozoic transition: confirmation and evolutionary implications, *Geol. Soc. Amer. Abstr. Progr.* **30**: A232.

Wylezich, C., Meisterfeld, R., Meisterfeld, S., and Schlegel, M., 2002, Phylogenetic analyses of small subunit ribosomal RNA coding regions reveal a monophyletic lineage of euglyphid testate amoebae (Order Euglyphida). *J. Eukaryot. Microbiol.* **49**: 108–118.

Xiao, S., and Dong, L., 2006, On the morphological and ecological history of Proterozoic macroalgae. in: *Neoproterozoic Geobiology and Paleobiology* (S. Xiao and A. J. Kaufman, eds.), Springer, Dordrecht, the Netherlands, pp. 57–90.

Xiao, S., and Knoll, A. H., 2000, Phosphatized animal embryos from the Neoproterozoic Doushantuo Formation at Weng'an, Guizhou, South China, *J. Paleontol.* **74**: 767–788.

Xiao, S., A.H. Knoll, and X. Yuan, 1998a, Morphological reconstruction of *Miaohephyton bifurcatum*, a possible brown alga from the Neoproterozoic Doushantuo Formation, South China, *J. Paleontol.* **72**: 1072–1086.

Xiao, S., Y. Zhang, and A.H. Knoll, 1998b, Three-dimensional preservation of algae and animal embryos in a Neoproterozoic phosphorite, *Nature* **391**: 553–558.

Xiao, S., Yuan, X., Steiner, M., and Knoll, A. H., 2002, Macroscopic carbonaceous compressions in a terminal Proterozoic shale: a systematic reassessment of the Miaohe biota, South China, *J. Paleontol.* **76**: 347–376.

Xiao, S., Knoll, A. H., Yuan, X. L., and Pueschel, C. M., 2004, Phosphatized multicellular algae in the Neoproterozoic Doushantua Formation, China, and the early evolution of the florideophyte algae, *Am. J. Bot.* **91**: 214–227.

Yin, L., 1997, Acanthomorphic acritarchs from Meso-Neoproterozoic shales of the Ruyang Group, Shanxi, China, *Rev. Palaeobot. Palynol.*, **98**: 15–25.

Yoon, H., Hackett, J., Ciniglia, C., Pinto, G., and Bhattacharya, D., 2004, A molecular timeline for the origin of photosynthetic eukaryotes, *Molec. Biol. Evol.* **21**: 809–818.

Yuan, X., Xiao, S., and Taylor, T. N., 2005, Lichen-like symbiosis 600 million years ago, *Science* **308**: 1017–1020.

Zander, J. M., Caspi, E., Pandey, G. N., and Mitra, C., 1969, The presence of tetrahymanol in *Oleandra wallichii*, *Phytochemistry* **8**: 2265–2267.

Zhuravlev, A. Y., 1993, Were Ediacaran Vendobionta multicellulars? *Neues Jahrb. Geol. Paläontol.* **190**: 299–314.

Chapter 2

On the Morphological History of Proterozoic and Cambrian Acritarchs

JOHN WARREN HUNTLEY, SHUHAI XIAO, and MICHAŁ KOWALEWSKI

Department of Geosciences, Virginia Polytechnic Institute and State University, Blacksburg, VA 24061, USA.

1. Introduction	24
2. Materials and Methods	25
2.1 Materials	25
2.2 Body Size Analysis	28
2.3 Morphological Disparity Analysis	29
2.3.1 Dissimilarity	29
2.3.2 Non-metric Multidimensional Scaling	30
3. Results	31
3.1 Body Size	31
3.2 Morphological Disparity	33
3.2.1 Dissimilarity	33
3.2.2 Non-metric Multidimensional Scaling	35
4. Discussion	39
4.1 Comparative Histories of Morphological Disparity and Taxonomic Diversity	39
4.2 Linking Morphological Disparity with Geological and Biological Revolutions	40
4.2.1 Morphological Constraints, Convergence, and Nutrient Stress in the Mesoproterozoic	40
4.2.2 Neoproterozoic Global Glaciations	41
4.2.3 Ediacara Organisms	43
4.2.4 Cambrian Explosion of Animals	44
5. Conclusions	45
Acknowledgements	45
References	46
Appendix: SAS/IML Codes	53

S. Xiao and A.J. Kaufman (eds.), Neoproterozoic Geobiology and Paleobiology, 23–56.
© 2006 Springer.

1. INTRODUCTION

Acritarchs, a group of decay-resistant organic-walled vesicular microfossils, dominate the fossil record of Proterozoic (2500–542 Ma) and Cambrian (542–488 Ma) protists. Most acritarchs from the Proterozoic and Paleozoic are interpreted as unicelled photosynthetic protists, though some may represent multicellular algae (Mendelson, 1987; Butterfield, 2004), and a few have been tentatively interpreted as fungi (Butterfield, 2005). Acritarchs are among the oldest eukaryotes in the fossil record (Zhang, 1986; Yan, 1991) and offer the earliest adequate data to assess the history of protistan biodiversity (Knoll, 1994; Vidal and Moczydlowska-Vidal, 1997).

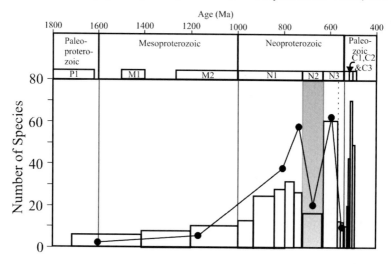

Figure 1. Estimates of acritarch taxonomic diversity during the Phanerozoic and early Paleozoic. Bars are adapted from Knoll (1994). Black circles adapted from Vidal and Moczydlowska-Vidal (1997). Vertical black lines represent Era boundaries. The dashed vertical line to the left of the Neoproterozoic/Paleozoic boundary represents the first appearance of Ediacara organisms. The gray box represents the time of Neoproterozoic global glaciations or Cryogenian. P1, M1, M2, N1, N2, N3, C1, C2 and C3 represent the geochronological bins of our study.

Previous estimates of acritarch diversity suggest that the number of acritarch species was low from the first occurrence in the Paleoproterozoic to as late as the early Neoproterozoic (Fig.1). Acritarch taxonomic diversity began to increase through the Neoproterozoic, but suffered a decline during mid-Neoproterozoic glaciation events. An unprecedented, though short-lived, diversification occurred after these glaciation events, and was then followed by another extinction, concurrent with the rise of macroscopic Ediacara organisms, some of which clearly were metazoans (Fedonkin and

Waggoner, 1997). Acritarch taxonomic diversity subsequently increased in step with animal radiation in the early Cambrian (Knoll, 1994; Vidal and Moczydlowska-Vidal, 1997).

Taxonomic inconsistencies have caused some to question the validity of taxic measures of protistan biodiversity (Butterfield, 2004). The problem is common in paleontology, and can be acute in the study of acritarchs. Evolutionary convergence among simple protists can lead to taxonomic deflation, or an underestimation of diversity, whereas the heteromorphic alternation of generations can lead to taxonomic inflation, or an overestimation of diversity (Butterfield, 2004). However, the problem of taxonomic inconsistency can be partly alleviated by a complementary and concurrent analysis of morphological disparity, as demonstrated by morphometric studies of Phanerozoic plants and animals (Foote and Gould, 1992; Boyce, 2005). The usefulness of morphometric tools in the analysis of Phanerozoic organisms encouraged us to use such strategies to independently address the question of the evolutionary history of acritarchs. In this chapter we present the results of our literature-based investigation of the first 1.3 billion years of morphological evolution in the Group Acritarcha.

2. MATERIALS AND METHODS

2.1 Materials

We used a literature-based morphometric approach to examine the evolutionary history of acritarchs from their first appearance in the Paleoproterozoic through the Cambrian. An extensive literature review, utilizing 50 publications (Table 1), produced a database of species descriptions from 47 stratigraphic intervals representing 778 species occurrences (the occurrence of a species in a lithostratigraphic unit), 247 locations and 1,766 processed rocks samples. The species occurrences were assigned to nine geochronological bins of unequal duration based on our best estimate of the depositional age of the stratigraphic intervals (Table 2).

Size and morphological data were collected from species descriptions and illustrations published in the literature (Table 1). Vesicle diameter was recorded, when reported, from species descriptions and measured from microphotographs of figured specimens. Thirty-one morphological characters were identified to quantify acritarch morphology (Table 3). Every species occurrence in the database was coded for the presence or absence of all 31 morphological characters based on species descriptions of type specimens found in the literature survey. The resulting database of morphological characters was comprised of binary variables, where a present character was scored as one (1) and an absent character was scored as zero (0) (Table 3).

Table 1. Stratigraphic intervals and data sources used in this study.

Bin Age in Ma	Stratigraphic Interval	Reference
C3 (488–500)	Booley Bay	(Moczydlowska and Crimes, 1995)
	Tempe	(Zang and Walter, 1992)
C2 (500–520)	Kaplanosy/Radzyn	(Moczydlowska, 1991)
	Ella Island	(Vidal, 1979)
	Buen	(Vidal and Peel, 1993)
	Læså	(Moczydlowska and Vidal, 1992)
C1 (520–540)	Mazowsze	(Moczydlowska, 1991)
	Dracoisen/Tokammane	(Knoll and Swett, 1987)
	Taozichong	(Yin, 1992)
	Yurtus/Xishanblaq	(Yao *et al.*, 2005)
N3 (540–630)	Lublin	(Moczydlowska, 1991)
	Doushantuo	(Xunlai and Hofmann, 1998; Zhang *et al.*, 1998; Yin, 1999; Zhou *et al.*, 2001; Xiao, 2004b)
	Pertatataka	(Zang and Walter, 1992)
	Khamaka	(Moczydlowska *et al.*, 1993)
	Dongjia	(Yin and Guan, 1999)
	Scotia	(Knoll, 1992)
	Yudoma Complex	(Pyatiletov and Rudavskaya, 1985)
N2 (630–720)	Tanafjord	(Vidal, 1981)
	Tillite	(Vidal, 1979)
N1 (720–1000)	Barents Sea	(Vidal and Siedlecka, 1983)
	Chuar	(Vidal and Ford, 1985)
	Uinta	(Vidal and Ford, 1985)
	Svanbergfjellet	(Butterfield *et al.*, 1994)
	Visingö	(Vidal, 1976b)
	Eleonore Bay	(Vidal, 1976a, 1979)
	Vadso	(Vidal, 1981)
	Tindir	(Allison and Awramik, 1989)
	Wanlong	(Gao *et al.*, 1995)
	Qinggouzi	(Gao *et al.*, 1995)
	Qiaotou	(Gao *et al.*, 1995)
	Draken Conglomerate	(Knoll *et al.*, 1991)

Table 1 (Continued).

	Hunnberg	(Knoll, 1984)
	Liulaobei	(Yin and Sun, 1994)
	Lone Land	(Samuelsson and Butterfield, 2001)
	Mirojedikha	(Hermann, 1990)
	Bitter Springs	(Zang and Walter, 1992)
	Veteranen	(Knoll and Swett, 1985)
	Wynniatt	(Butterfield and Rainbird, 1998)
M2 (1000–1270)	Ruyang	(Yin, 1997)
	Lakhanda	(Hermann, 1990)
	Thule	(Hofmann and Jackson, 1996)
	Baichaoping	(Yan and Zhu, 1992)
	Bylot	(Hofmann and Jackson, 1994)
M1 (1400–1500)	Bangemall	(Buick and Knoll, 1999)
	Roper	(Javaux *et al.*, 2001)
P1 (1625–1800)	Chuanlinggou	(Yan, 1982; Luo *et al.*, 1985; Zhang, 1986; Yan, 1995; Sun and Zhu, 2000)
	Changzhougou	(Luo *et al.*, 1985; Yan, 1991, 1995; Zhang, 1997; Sun and Zhu, 2000)

Table 2. Description of geochronological bins used in this study.

Bin	Bin Description
P1	Paleoproterozoic: 1625–1800 Ma
M1	Mesoproterozoic: 1400–1500 Ma
M2	Mesoproterozoic: 1000–1270 Ma
N1	Pre-Glacial Neoproterozoic: 720–1000 Ma
N2	Cryogenian: 630–720 Ma
N3	Ediacaran: 540–630 Ma
C1	Early Cambrian pre-trilobite: 520–540 Ma
C2	Early Cambrian with trilobites: 500–520 Ma
C3	Middle and Late Cambrian: 488–500 Ma

Table 3. Description of morphological characters and coding used in disparity analyses.

Character	Scoring
Spherical vesicle?*	0=no 1=yes
Ellipsoidal vesicle?*	0=no 1=yes
Polyhedral vesicle?*	0=no 1=yes
Bulb-shaped vesicle?*	0=no 1=yes
Medusoid vesicle?*	0=no 1=yes

Table 3 (Continued).

Barrel-shaped vesicle?*	0=no 1=yes
Enveloping membrane surrounding vesicle?	0=no 1=yes
Costae meshwork surrounding vesicle?	0=no 1=yes
Triangular processes?*	0=no 1=yes
Cylindrical processes?*	0=no 1=yes
Tapered processes?*	0=no 1=yes
Hair-like processes?*	0=no 1=yes
Hemispherical processes?*	0=no 1=yes
Blunt process tips?*	0=no 1=yes
Pointed process tips?*	0=no 1=yes
Capitate process tips?*	0=no 1=yes
Rounded process tips?*	0=no 1=yes
Funnel-shaped process tips?*	0=no 1=yes
Do process tips fuse?	0=no 1=yes
Do processes branch?	0=no 1=yes
Are processes hollow?	0=no 1=yes
Does interior of process communicate with interior of vesicle?	0=no 1=yes
Does vesicle have external plates?	0=no 1=yes
Does vesicle have multicelled appearance?	0=no 1=yes
Do vesicles occur in colonial-like clusters?	0=no 1=yes
Does vesicle have internal body?	0=no 1=yes
Does vesicle have excystment?	0=no 1=yes
Does vesicle have pores?	0=no 1=yes
Does vesicle have crests?	0=no 1=yes
Does vesicle have flange?	0=no 1=yes
Does vesicle surface have concentric ornamentation?	0=no 1=yes

*Some readers might be concerned that these individual characters would be better classified as alternative states of only three characters: vesicle morphology, process morphology, and process tip morphology. However, the binary characters as classified above are not mutually exclusive. The binary characters are necessary to accommodate species that have multiple morphologies; for example, we encountered 7 with multiple vesicle morphologies, 34 with multiple process morphologies, and 46 with multiple process tip morphologies.

2.2 Body Size Analysis

Maximum vesicle diameters were recorded from species descriptions, when available, for all species occurrences. Figure vesicle diameters were measured from microphotographs of acritarchs. Maximum vesicle diameters were used as a proxy for acritarch body size history and figure vesicle diameters were used as a cross-check. Maximum and figure vesicle diameters were log-transformed. Mean maximum vesicle diameters and mean figure vesicle diameters were calculated for each geochronological bin. To estimate 95% confidence intervals (CI) for mean maximum vesicle diameter, maximum diameter values in each bin were resampled with replacement 1000 times (balanced bootstrap module) and mean size values were recomputed (Kowalewski *et al.*, 1998). The percentile approach, or naïve bootstrap (Efron, 1981), was used to calculate 95% confidence intervals from the bootstrapped sampling distributions.

2.3 Morphological Disparity Analysis

The nature of our morphological data matrix lent itself to multiple analytical approaches to investigate the history of disparity in acritarchs. In fact, it is desirable to use multiple methods, when possible, to better understand the morphological history of a clade (Foote, 1997). Therefore, we utilized two methods in this study: 1) an estimation of within geochronological bin dissimilarity and 2) an exploratory non-metric multivariate ordination that simultaneously considered all species occurrences from all geochronological bins. The estimation of within geochronological bin dissimilarity is based upon pairwise comparisons of species within each bin. The resulting measure of dissimilarity is not affected by species occurrences in other bins. The exploratory non-metric multivariate ordination considered species occurrences from all geochronological bins concurrently. Therefore, the resulting values calculated for each species occurrence by the ordination were related to species occurrences from other geochronological bins.

2.3.1 Dissimilarity

Pairwise comparison of character differences between species occurrences was used to calculate mean dissimilarity coefficients for each bin (Sneath and Sokal, 1973; Foote, 1995). Species occurrences were separated *a priori* into their geochronological bins. Pairwise comparison was made between each species occurrence and every other species occurrence in the same geochronological bin, and for each comparison a dissimilarity coefficient was calculated from the number of character differences divided by the total number of characters (31). The mean dissimilarity coefficient was then calculated for each geochronological bin. Pairwise comparisons of character differences were performed using codes written in SAS/IML interactive matrix language (See Appendix).

Balanced bootstrapping techniques were used to assess the analytical error of the mean dissimilarity coefficients. For each bin, dissimilarity coefficients were resampled with replacement 1000 times (Efron, 1981) and mean dissimilarity coefficients were recalculated. Standard errors were calculated from the resulting bootstrapped sampling distribution. Standard errors are considered the appropriate method for assessing analytical error, because phylogenetically related organisms are our units of study, and are therefore not independent observations (Foote, 1994). Bootstrapping and calculation of standard errors were performed using codes written in SAS/IML interactive matrix language (See Appendix).

2.3.2 Non-metric Multidimensional Scaling

Non-metric multidimensional scaling (MDS) is an exploratory multivariate ordination technique used to simplify multidimensional data matrices (Kruskal and Wish, 1978; Schiffman et al., 1981; Marcus, 1990; Roy, 1994). MDS is a particularly attractive method in the case of our data set in that it does not require continuous variables (our variables are binary presence/absence values) and allows for missing values, unlike commonly used parametric techniques such as Principal Components Analysis. MDS was used to create a two dimensional ordination from the original 31 characters of all 778 species occurrences (SAS reported convergence criterion satisfied: 17 iterations were performed, final badness-of-fit: 0.21). Three and four dimensional ordinations were calculated and resulted in the same patterns of variance, but are not reported here. The MDS procedure calculated two scores (dimension one and dimension two) for each species occurrence, which was then assigned to a geochronological bin. For each bin, variances of dimension one and dimension two scores were calculated. The sum of the two variance scores for each time bin is referred to as MDS variance. Correlation coefficients (R) were calculated among MDS loadings and original morphological variables. The MDS procedure was performed using SAS 9.1 (See Appendix).

MDS is different from other ordination techniques in that the primary dimension does not always align with maximum variance of the data. This can make it problematic to relate the dimensions provided by the ordination with the original variables. To alleviate this concern we subjected the MDS scores (which are continuous) to a principal components analysis (PCA). The PCA produced two scores (PCA1 and PCA2) for each species occurrence, which was then assigned to a geochronological bin. For each bin, variances of PCA1 and PCA2 scores were calculated. The sum of the two variance scores for each time bin is referred to as PCA variance. Correlation coefficients (R) were calculated among PCA loadings and original morphological variables. The principal components analysis was performed using PAST 1.33 (Hammer et al., 2001). Interestingly enough, in this case, the primary dimension of the MDS ordination did coincide with maximum variance of the data. Therefore the results of the PCA are similar to those of the MDS (they are a mirror image of one another), and the interpretation of the MDS morphospace in relation to the original variables is reliable.

A randomization with 1000 iterations was performed on the MDS variances to affirm that the overall variance trend was not attributable to varying sample size. The paired MDS scores (dimension one and two scores) of all 778 species occurrences were randomly shuffled into the geochronological bins to replicate the original sampling structure, so that 104 paired MDS scores were placed randomly in the P1 bin, 11 in the M1 bin, *etc* (Table 4). MDS variance was then calculated for the values re-assigned to each bin. This randomization procedure was repeated 1000 times, resulting in a distribution of 1000 variance estimates per bin. We calculated the mean, 2.5 percentile, and 97.5 percentile values from the randomly produced distribution of variance estimates for each bin. The purpose of the randomization was to determine if our observed trend of MDS variance could be explained as a sampling artefact. If the observed values of MDS variance all fell within the 95% CI envelope, then the trend could be explainable as a sampling artefact. If the trend of the MDS variance values fell outside of the envelope, then the observed trend could not be explained as a sampling artefact. The MDS randomization was performed using codes written in SAS/IML interactive matrix language and the SAS/STAT Proc MDS module (See Appendix).

3. RESULTS

3.1 Body Size

The average maximum vesicle diameter of acritarchs displayed non-directional fluctuation through the Proterozoic (Fig. 2). Acritarch body size decreased significantly across the Neoproterozoic-Paleozoic transition, but had increased significantly by the middle/late Cambrian (though not to the size seen in the late Neoproterozoic). The average figure vesicle diameter of acritarchs displayed a similar pattern, though at smaller sizes (Fig. 2). Average maximum diameter and average figure diameter are significantly positively correlated, and figure data generally underestimate maximum vesicle diameter reported in systematic description (Fig. 2 inset). Retrieving body size information from figured specimens appears to be a legitimate approach with Proterozoic and Cambrian acritarchs when one is interested in investigating long-term patterns.

Table 4. Binning structure for morphometric analyses.

Bin	Species Occurrences
P1	104
M1	11
M2	41
N1	248
N2	13
N3	156
C1	54
C2	115
C3	36

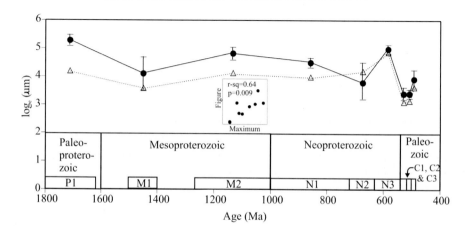

Figure 2. Size history of acritarchs. Solid circles represent log-transformed mean maximum vesicle diameter of acritarchs (from species descriptions) through time with 95% CI calculated from 1000 iteration naïve bootstrapped sampling distribution. Hollow triangles represent log-transformed mean vesicle diameter of acritarchs as measured from figured specimens. Inset shows relationship between maximum reported sizes and sizes of figured specimens. R-sq (R-squared) and p-value from Pearson correlation analysis performed in SAS 9.1.

Table 5. Correlation analyses between measures of disparity and body size.

	Raw Data		First Differences	
	Pearson		Pearson	
	r^2	p	r^2	p
MDS	0.113	0.38	0.180	0.29
Dissimilarity	0.038	0.62	0.247	0.21

The absence of any notable long-term trend in acritarch body size minimizes the likelihood of mistaking spurious disparity trends due to secular changes in body size (e.g., morphological disparity may increase due to diffusive increase in body size or body size range) with shifts in size-invariant shape disparity. Neither of the disparity metrics analyzed show any significant correlation with body size for all possible comparisons, including both raw data and data corrected for autocorrelations by first differencing (Table 5). This discordance between body size and morphospace occupation suggests that the disparity trends discussed below are not an allometric derivative of changes in body size.

3.2 Morphological Disparity

3.2.1 Dissimilarity

Mean dissimilarity was very low in the Paleoproterozoic (<0.02; Fig. 3A). This low value starkly contrasts with the high species per formation values calculated from our database (Fig. 3C). We interpret this stark contrast as severe taxonomic over-splitting in the Paleoproterozoic. Some caution in such an interpretation may be warranted due to the lack of cell wall thickness data in our matrix. However, reports of cell wall thickness in species descriptions are overwhelmingly qualitative (*e.g.*, thick or thin), and are likely not consistently applied between workers. Moreover cell wall thickness is likely highly susceptible to taphonomic processes such as degradation.

A significant increase in mean dissimilarity occurred between the P1 and M1 bin, with an M1 value of 0.08. Mean dissimilarity coefficients reached a plateau during the M1 bin that would remain through the early Neoproterozoic (M2=0.10 and N1=0.09). A slight, yet significant, decrease in dissimilarity occurred during the Cryogenian (N2=0.08 and upper 95% confidence interval <0.09).

A rapid morphological diversification occurred in the early Ediacaran period, resulting in a mean dissimilarity coefficient significantly higher than any seen in previous bins (N3= 0.15). This increase in morphological disparity, though dramatic, was short-lived. The first appearance of the Ediacara organisms (~575 Ma) corresponds in time with a dramatic decrease in acritarch disparity. We did not create a separate geochronological bin (*i.e.*, 575–542 Ma) due to low data density. All known acritarchs from this time are simple sphaeromorphic leiosphaerid-like vesicles. Moreover, it would be impossible to calculate a dissimilarity coefficient (much less mean dissimilarity) in such a bin as our data base contains only one known named

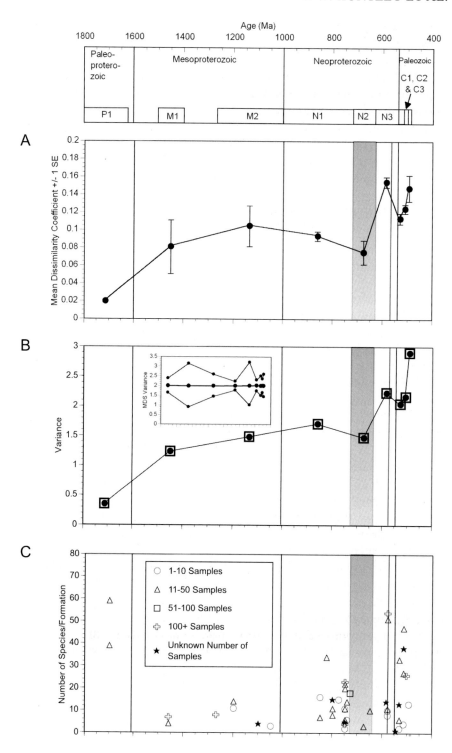

species (*Leiosphaeridia* sp.) in this time interval (Fig. 3C); although other species (*e.g.*, *Bavlinella faveolata*) may also be present in this interval (Germs *et al.*, 1986). Therefore the dramatic decrease in disparity associated with the first appearance of Ediacaran organisms and the rapid diversification seen in the pre-trilobite early Cambrian (C1) is much more dramatic than Fig. 3A suggests.

Mean dissimilarity coefficients increased monotonically through the Cambrian. Pre-trilobite Early Cambrian mean dissimilarity (C1=0.11) reflects the morphological diversification following the late Ediacaran drop in disparity addressed above. Mean dissimilarity coefficients continued to increase significantly through the trilobite-bearing Early Cambrian (C2=0.12) and Middle and Late Cambrian (C3=0.15), achieving the high level of disparity seen in the early Ediacaran (N3).

3.2.2 Non-metric Multidimensional Scaling

Non-metric multidimensional scaling analysis shows significant secular variation in acritarch morphologies (Fig. 3B), and is broadly similar to the dissimilarity pattern. The MDS trend is unlikely a sampling artifact as its overall trajectory falls outside of the randomization's 95% confidence intervals (Fig. 3B inset).

MDS variance was very low in the Paleoproterozoic (P1=0.35; Figs. 3B, 4). This indicator of low disparity is also in stark contrast with high species per formation values (Fig. 3C), and is indicative of taxonomic over-splitting (see above). A significant increase in MDS variance occurred in the early Mesoproterozoic (M1=1.24), signaling the beginning of a disparity plateau that would continue through the early Neoproterozoic (M2=1.49 and N1=1.70). This plateau is apparent in Fig. 3 A–B, but not in Fig. 4. This is because the convex hulls in Fig. 4 to a large extent reflect sampling intensity as well as morphological disparity in each bin.

MDS variance decreased during the Cryogenian (N2=1.48) (Figs. 3B, 4). This morphological contraction, together with taxonomic decrease (Knoll, 1994; Vidal and Moczydlowska-Vidal, 1997; Xiao, 2004a), indicates

---

Figure 3. (on Page 34) History of acritarch disparity. (A) Mean dissimilarity coefficient ± 1 standard error (Note: the standard error brackets for P1 are smaller than the data point.). (B) Variance from multivariate analyses. Black circles are MDS variance. Black squares are PCA variance. Inset graph displays results of MDS randomization. Center line represents mean variance from 1000 iteration randomization. Lower and upper lines represent 95% confidence intervals. (C) Number of species per formation from this study's database, coded according to sampling intensity (number of processed rock samples) of each formation. Vertical black lines represent era boundaries. The gray box represents the Cryogenian. The vertical light gray line at ~575 Ma represents the first appearance of the Ediacara organisms.

possible acritarch extinction during the Cryogenian. Further analysis of MDS plots and loading reveals the restriction of acritarchs from the right-hand side of the morphospace (Fig. 4), suggesting that the Cryogenian acritarch extinction strongly affected acanthomorphic forms. Acritarchs with hollow, cylindrical, blunt-tipped processes are notably absent in the Cryogenian (Knoll, 1994). The post-Cryogenian recovery of acritarchs resulted in the highest MDS variance seen until that time (N3=2.23).

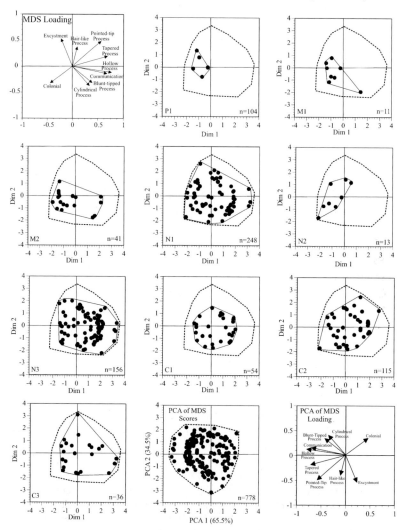

Figure 4. MDS and PCA scatter plots and loading. MDS scatter plots for the nine geochronological bins and the MDS loading chart relating variables to Dim 1 (x-axis) and Dim 2 (y-axis). Solid outlines are convex hulls for bin data. Dashed outlines are convex hulls for pooled data representing maximum realized morphospace. Note how MDS and PCA scatter plots and loadings are mirror images of one another.

Morphological History of Proterozoic and Cambrian Acritarchs 37

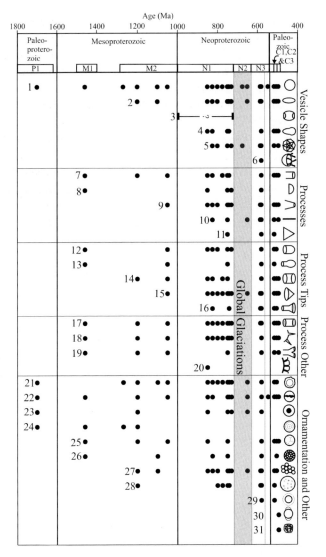

Figure 5. Stratigraphic occurrences of morphological characters utilized in this study: 1) spherical vesicle; 2) ellipsoidal vesicle; 3) barrel-shaped vesicle; 4) bulb-shaped vesicle; 5) polyhedral vesicle; 6) medusoid vesicle; 7) cylindrical process; 8) hemispherical process; 9) tapered process; 10) hair-like process; 11) triangular process; 12) rounded-tip process; 13) capitate-tip process; 14) blunt-tip process; 15) pointed-tip process; 16) funnel-tip process; 17) hollow process; 18) interior of process communicates with interior of vesicle; 19) branching process; 20) processes fuse at tip; 21) enveloping membrane; 22) excystment-like structure; 23) internal bodies in vesicle; 24) concentric ornamentation on vesicle surface; 25) plates on vesicle; 26) multi-celled appearance (vesicles contained in a larger envelope); 27) colonial appearance (aggregation of vesicles); 28) pores in vesicle wall; 29) flange ornamentation; 30) crest ornamentation; 31) costae meshwork surrounding vesicle.

MDS variance decreased between the early Ediacaran and the pre-trilobite Early Cambrian (C1=2.04), concurrent with the diversification of Ediacara organisms. MDS variance increased monotonically through the remainder of the Cambrian in step with the taxonomic diversification of acritarchs and animals (C2=2.16, C3=2.90).

The dissimilarity coefficient and MDS results, described above, are broadly supported by the geochronological distribution of morphological characters (Fig. 5). Paleoproterozoic acritarchs typically had spherical vesicles with the occasional medial split (e.g., *Schizofusa sinica*), enveloping membrane (e.g., *Pterospermopsimorpha pileiformis*), internal bodies (e.g., *Nucellosphaeridium magnum*), or concentric surface ornamentation (e.g., *Thecatovalvia annulata* and *Valvimorpha annulata*). The Mesoproterozoic saw the first appearance of elliptical vesicles (e.g., *Fabiformis baffinensis*), ten process-related characters (e.g., *Shuiyousphaeridium macroreticulatum* and *Tappania plana*), vesicle plates (e.g., *S. macroreticulatum* and *Dictyosphaera delicata*), pores in vesicle walls (e.g., *Tasmanites volkovae*), and multi-celled and colonial appearance (e.g., *Majasphaeridium sp.* and *Satka squamifera*). Of the thirty-one characters identified in this study, fifteen first appeared in the Mesoproterozoic (nine in the M1 bin and six in the M2 bin). Many more acritarch body plans evolved in the Neoproterozoic. Polyhedral vesicles (e.g., *Octoedryxium truncatum*), bulb-shaped vesicles (e.g., *Sinianella uniplicata*), medusoid vesicles (e.g., *Multifronsphaeridium pelorium*), barrel-shaped vesicles (e.g., *Artacellularia kellerii*), triangular and hair-like processes (e.g., *Cymatiosphaera wanlongensis* and *Dasysphaeridium trichotum*), funnel-tipped processes (e.g., *Briareus borealis*), processes that fuse at the tips (e.g., ectophragm acanthomorph from Butterfield and Rainbird 1998), and flange ornamentation about the vesicle equator (e.g., *Simia simica*) all appear for the first time in the Neoproterozoic. Two new morphological characters appeared in the Cambrian: a costae meshwork that surrounds the vesicle (e.g., *Retisphaeridium brayense*) and crest-ornamentation—equatorial ornamentation that is similar to flange but does not circumvent the vesicle resulting in wing-like structures (e.g., *Pterospermella solida*). It should be noted that our data for Cambrian acritarchs were not as exhaustive as our Proterozoic data, and that further investigation would likely reveal more first appearances of characters in the Cambrian than what we report. Another caveat is that the same morphological characters in different taxa or in different geochronological bins may not be homologous. Such simple characters may have evolved multiple times.

The two proxies of morphological disparity used in this study, mean dissimilarity coefficient and MDS variance, resulted in coherent histories of acritarch morphological disparity (Fig. 3). The agreement of the two methods, the independent verification of the significance of the trends found by each method (i.e. randomization for MDS and calculation of standard error for mean dissimilarity), the elimination of allometry as a confounding factor, and the invariance of disparity estimates relative to unequal binning characters (e.g., temporal duration of bin, number of formations per bin, number of sampling localities per bin, number of processed rock samples per bin, and number of species occurrences per bin) (Table 6) all attest to the robustness of our interpreted history of acritarch morphological evolution.

Table 6. Correlation analyses between measures of disparity and unequal binning characters.

	MDS Variance		Mean Dissimilarity Coefficient	
	Spearman		Spearman	
	r^2	p	r^2	p
Duration of Bin	–0.639	0.06	–0.571	0.11
Number of Formations	0.367	0.33	0.428	0.25
Number of Locations	0.092	0.81	0.185	0.63
Number of Samples	0.067	0.86	0.100	0.80
Species Occurrences	0.333	0.38	0.317	0.41

4. DISCUSSION

4.1 Comparative Histories of Morphological Disparity and Taxonomic Diversity

The morphological disparity of acritarchs (as approximated by mean dissimilarity coefficients, MDS variance, and stratigraphic ranges of individual morphological characters) initially increased significantly by the early Mesoproterozoic (Fig. 3, 5). In contrast, the first taxonomic radiation did not occur until the early Neoproterozoic (Fig. 1). This increase in disparity preceded the first major taxonomic radiation by approximately 500 million years. This statement remains true even if the diversity curve (Knoll, 1994; Vidal and Moczydlowska-Vidal, 1997) is updated with more recent data (Xiao *et al.*, 1997; Yin, 1997; Javaux *et al.*, 2001), although the addition of the exuberantly over-split Paleoproterozoic taxa (Fig. 3C) to the diversity curve may significantly change the picture. However, as discussed earlier and implied in previous compilations of acritarch diversity (Knoll, 1994; Vidal and Moczydlowska-Vidal, 1997), such over-splitting is not justified.

The pattern of high morphological disparity early in the history of acritarchs is very similar to patterns seen in the evolution of multi-celled organisms in the Phanerozoic. Many groups of organisms in the Phanerozoic display high morphological disparity early in their history: Cambrian metazoa (Thomas *et al.*, 2000), marine arthropods (Briggs *et al.*, 1992), Paleozoic gastropods (Wagner, 1995), seed plant leafs (Boyce, 2005), and Cenozoic ungulate teeth (Jernvall *et al.*, 1996). Thus, high morphological disparity in the early evolutionary history appears to be a prevailing, although not universal, pattern among many groups (Foote, 1997). As far as we know, this study documents the first example of a similar pattern in protists and in the Precambrian. It is becoming apparent that morphological diversification preceding taxonomic diversification may be a prevailing pattern in eukaryote evolution.

Our comparative analysis of disparity and diversity does differ from the results of other studies. Morphological disparity typically approaches its maximum realized value early in the history of other clades [*e.g.*, Paleozoic crinoids (Foote, 1995)], but our analysis reveals periodic expansions of realized morphospace (Fig. 3). Because the Group Acritarcha is undoubtedly polyphyletic and includes organisms from many phyla or divisions (Butterfield, 2004, 2005), the periodic expansion of acritarch morphospace is best interpreted as a result of the evolutionary appearance of new clades. In particular, fluctuation of acritarch morphospace in the Neoproterozoic and Cambrian may represent the coming and going of different eukaryote groups.

4.2 Linking Morphological Disparity with Geological and Biological Revolutions

4.2.1 Morphological Constraints, Convergence, and Nutrient Stress in the Mesoproterozoic

The ~1500 Ma (M1) morphological expansion was followed by a prolonged plateau of morphological disparity until ~800 Ma (N1). Constraints on protist morphology likely played a dominant role in a significant part of protist history from 1500 Ma to 800 Ma. The increasingly populated morphospace during this period suggests that either the morphological history of acritarchs was characterized by convergent morphologic evolution of phylogenetically unrelated groups or that diversification was restricted within morphologically similar clades. Either way, it is remarkable that the morphological constraints were not overcome for such a long time given that the Group Acritarcha is polyphyletic and thus includes multiple clades (Butterfield, 2004, 2005).

Buick and others (1995) described the Mesoproterozoic as "the dullest time in Earth's history (p.153)" and remarked that "never in the course of Earth's history did so little happen to so much for so long (p.169)". These statements were based upon $\delta^{13}C$ values that hovered around 0‰ with little change for nearly 600 million years (1600–1000 Ma) (Buick *et al.*, 1995; Xiao *et al.*, 1997; Brasier and Lindsay, 1998). The global rate of organic carbon burial relative to inorganic carbon burial, as inferred from the Bangemall Group of northwestern Australia and equivalent carbonates elsewhere, remained unchanged through the Mesoproterozoic, resulting in the static $\delta^{13}C$ pattern. This was ascribed to relatively little environmental and tectonic changes during this most lackluster era (Buick *et al.*, 1995). Tectonic and environmental tranquility would lead to low bioproductivity through nutrient stress such as phosphorus limitation (Brasier and Lindsay, 1998) and/or the dearth of metabolically important trace metals in the Mesoproterozoic oceans (Anbar and Knoll, 2002).

Our results suggest that Buick and others were only partially correct in their depiction of the Mesoproterozoic as being irksome and tedious. Our quantitative measures of acritarch morphological disparity do suggest a long plateau lasting ~600 million years. Similarly, qualitative data suggest that the taxonomic turnover rate of acritarchs during the Mesoproterozoic and early Neoproterozoic was much lower than that of the late Neoproterozoic (Peterson and Butterfield, 2005). However, the first appearance of nearly half the morphological characters considered in this study (15 of 31) occurred during the early Mesoproterozoic, well within the time of subdued $\delta^{13}C$ fluctuations, and the plateau continued into the early Neoproterozoic when the carbon cycle fluctuated moderately. Is this plateau indeed related to Mesoproterozoic nutrient stress? The great temporal overlap between acritarch disparity plateau and Mesoproterozoic geochemical stasis is suggestive of a possible causal relationship, but the apparent mismatch in their initiation and termination raises some concerns. At the present, the mismatch cannot be fully addressed because of poor temporal resolution in acritarch and $\delta^{13}C$ data, as well as poor understanding of the response time (lag time) between the different components of the Earth system.

4.2.2 Neoproterozoic Global Glaciations

The late Neoproterozoic saw perhaps the most dramatic of global climatic events in the history of Earth. It has been hypothesized that multiple global glaciations occurred during this time (~720–630 Ma), even to the extent of glaciers at the equator with tropical sea ice 1 km thick (Kirschvink, 1992; Hoffman *et al.*, 1998; Hoffman and Schrag, 2002). It is reckoned that the "snowball Earth" glaciations lasted for approximately 10 million years (Hoffman et al., 1998; Bodiselitch et al., 2005). With the carbon cycle cut short, due to completely iced-over oceans, the CO_2 concentration in the atmosphere (sourced by volcanic out-gassing) would build up, eventually resulting in greenhouse conditions and deglaciation. The deglaciation events were also likely very dramatic, with wind and waves unlike those seen on Earth before or since (Allen and Hoffman, 2005).

The controversial snowball Earth hypothesis has been criticized on biological grounds (Runnegar, 2000; Corsetti et al., 2003; Olcott et al., 2005). The fossil record clearly indicates that several major photosynthetic clades, including green, red, and chromophyte algae (Butterfield et al., 1994; Butterfield, 2000, 2004), evolved prior to the Cryogenian glaciations. If the snowball model is correct then these three algal clades must have survived the global glaciations, either in sea ice cracks, hydrothermal vents, fresh water melt ponds (Hoffman et al., 1998; Hoffman and Schrag, 2002), or perhaps in an ice-free tropic ocean that may have persisted during the snowball Earth events (Hyde et al., 2000; Runnegar, 2000).

Acritarchs did experience significant change in the Cryogenian. Morphological disparity (Fig. 3) as well as global taxonomic diversity (Fig. 1) decreased significantly in the Cryogenian (N2). It is possible that the Cryogenian suffers from fewer acritarch assemblages reported in the literature; however, Cryogenian acritarch assemblages (Knoll et al., 1981; Vidal, 1981; Vidal and Nystuen, 1990; Yin, 1990) do show lower taxonomic diversity and morphological disparity than older and younger assemblages. Large acritarchs and complex acanthomorphic acritarchs are few in the Cryogenian (Fig. 2, 4). This pattern does suggest that, whether the tropical ocean remained ice-free during the snowball Earth events, eukaryotes did suffer significant loss in the Cryogenian.

Runnegar hypothesized about the biological consequences of the various explanations for Cryogenian glaciations (Runnegar, 2000). A strict snowball scenario would result in an evolutionary bottleneck with the extinction of all but a few eukaryotic lineages. A slushball scenario with ice-free tropical seas would result in a blue-water refugium with the selective filtering of eukaryotic lineages favoring planktonic open ocean forms. He also hypothesized a scenario in which global refrigeration would have had mild impact on the biosphere. Paleoenvironmental analysis appears to suggest that acanthomorphic acritarchs tend to occur in near-shore facies as compared to leiosphaerids (Butterfield and Chandler, 1992). If this paleoecological pattern holds true for all Proterozoic acritarchs, then the selective extinction of acanthomorphic acritarchs during the Cryogenian may be taken as evidence in support of Runnegar's blue-water refugium hypothesis (Runnegar, 2000). It remains to be seen whether benthic algae survived Cryogenian, and, if not, whether post-Cryogenian benthic ecosystem recruited from planktonic algae that did survive glaciations.

4.2.3 Ediacara Organisms

The first macroscopic complex organisms in the fossil record are members of the Ediacara biota and first appeared approximately 575 Ma, within 5 million years after the 580 Ma Gaskiers glaciation that lasted no more than one million years (Narbonne, 1998; Narbonne and Gehling, 2003; Narbonne, 2005). The phylogenetic affinity of many of these organisms is controversial, but whether they represent the ancestors of modern organisms (Runnegar and Fedonkin, 1992) or a failed evolutionary experiment (Seilacher, 1992; Buss and Seilacher, 1994) they certainly indicate a basic ecological restructuring of the world previously dominated by prokaryotes and single-celled eukaryotes (Lipps and Valentine, 2004). The varied body plans of Ediacara organisms suggest equally varied trophic strategies, probably including heterotrophy. Evidence for the presence of heterotrophic consumers includes molluscan-grade bilaterians (Fedonkin and Waggoner, 1997), cnidarian-grade metazoans (Runnegar and Fedonkin, 1992), scratch marks interpreted as radular grazing traces (Seilacher, 1999; Seilacher et al., 2003), epifaunal tiering (Clapham and Narbonne, 2002), and boring of mineralized exoskeletons (Bengston and Zhao, 1992; Hua et al., 2003).

In light of the 580 Ma Gaskiers glaciation and probable consumers in the late Ediacaran (575–542 Ma), it is instructional to explore their possible effects on the primary producers (as represented by most acritarchs). Our data show that acritarch morphological disparity and taxonomic diversity in the late Ediacaran decreased to levels not seen since the Paleoproterozoic (though we didn't construct a separate bin for this time, see Section 3.2.1 and Fig. 3C). During this time, all acritarchs were of simple leiosphaerid-like and *Bavlinella*-like morphologies, and all acritarchs characteristic of early Ediacaran (so called Doushantuo-Pertatataka acritarchs) disappeared.

To test whether the Gaskiers glaciation, the diversification of Ediacara organisms, or perhaps something else caused the disappearance of Doushantuo-Pertatataka acritarchs, we need to sort out the exact temporal relationship between several geobiological events. In South Australia, the appearance of Doushantuo-Pertatataka acritarchs occurred long after the Marinoan glaciation and shortly after the Acraman Impact, which has been estimated to be 580 Ma (Grey et al., 2003). However, the late appearance of Doushantuo-Pertatataka acritarchs in South Australia was probably due to regional, environmental, or preservational biases. In South China, Doushantuo-Pertatataka acritarchs first appeared about 632 Ma (Condon et al., 2005), shortly after the Nantuo glaciation that is considered equivalent to the Marinoan glaciation in South Australia (Xiao, 2004a; Zhou *et al.*, 2004). Condon *et al.* (2005) estimated that Doushantuo-Pertatataka acritarchs persisted at least 50 million years and disappeared somewhere between 580 Ma and 550 Ma. If true, both the Acraman Impact and the Gaskiers glaciation predate, perhaps significantly, the disappearance of Doushantuo-Pertatataka acritarchs. Hence, neither the Acraman Impact nor the Gaskiers glaciation may have directly contributed to the disappearance of Doushantuo-Pertatataka acritarchs.

It is more likely that herbivory by, or other ecological interactions with, Ediacara organisms led to the decline of Doushantuo-Pertatataka acritarchs in the late Ediacaran Period. This hypothesis is distinct from a recent hypothesis proposed by Peterson and Butterfield (2005), who suggest that the origin, not the extinction, of Doushantuo-Pertatataka acritarchs was a consequence of ecological interactions with early eumetazoans. Both hypotheses need to be tested against more precise geochronological data and to be examined for possible taphonomic bias against acritarch preservation in the late Ediacaran Period. If either hypothesis survives more rigorous tests in the future, the origin or extinction of Doushantuo-Pertatataka acritarchs would be the first top-down driven macroevolutionary event recorded in the fossil record (Vermeij, 2004).

4.2.4 Cambrian Explosion of Animals

Perhaps the most dramatic event in the history of life began approximately 540 Ma at the beginning of the Cambrian Period. Almost all known metazoan phyla diverged in the Early–Middle Cambrian (Conway Morris, 1998; Levinton, 2001; Zhuravlev and Riding, 2001; Valentine, 2004). The Cambrian explosion resulted in major ecological restructuring of the biosphere (Zhuravlev and Riding, 2001) and alteration of sedimentation patterns (Bottjer *et al.*, 2000; Droser and Li, 2001).

It has been noted by several observers that acritarch diversity increased in step with animal evolution during the Cambrian explosion (Knoll, 1994; Vidal and Moczydlowska-Vidal, 1997; Butterfield, 2001). So did acritarch morphological disparity (Figs. 1, 3). This implies a close link between these two ecological groups during the radiation. The nature of these links, however, is less clear. It has been argued that morphological diversification of phytoplankton, as shown in acritarch morphology, was an ecological response to the evolution of filter-feeding mesozooplankton in the Cambrian (Butterfield, 1997, 2001). It is also possible that the Cambrian metazoan diversification was driven by morphological and ecological radiation of primary producers including most acritarchs (Moczydlowska, 2001, 2002). Further investigation of this matter, including detailed biostratigraphic studies across complete sections of the Proterozoic–Phanerozoic transition, will help determine which of these scenarios most likely occurred.

5. CONCLUSIONS

- We utilized the published literature to produce an empirical morphospace to describe the evolutionary history of Proterozoic and Cambrian acritarchs.
- Mean acritarch vesicle diameter displayed non-directional fluctuation through the Proterozoic and decreased significantly across the Ediacaran–Cambrian boundary.
- The initial increase of morphological disparity preceded the first taxonomic radiation by approximately 500 million years — a pattern similar to that seen in Phanerozoic multi-celled organisms and perhaps ubiquitous in eukaryote evolution.
- The Mesoproterozoic disparity plateau could be linked to prolonged morphological constraints and convergence related to long-term nutrient stress.
- The selective removal of large and acanthomorphic acritarchs in the Cryogenian may suggest significant impact of extensive Cryogenian glaciations on the evolution of acritarchs.
- The appearance of Ediacara organisms (~575 Ma) altered the Proterozoic trophic structure resulting in a major decrease in acritarch disparity and diversity in the late Ediacaran Period. The late Ediacaran disappearance of Doushantuo-Pertatataka acritarchs (580~550 Ma) that thrived in the early Ediacaran Period may represent a rare case of top-down driven extinction in the fossil record. This hypothesis is distinct from Peterson and Butterfield's (2005) hypothesis that the origin of the eumetazoa (~635 Ma) is linked to the origination of Doushantuo-Pertatataka acritarchs (~632 Ma).
- Acritarch disparity and diversity increased in step with animal evolution during the Cambrian explosion, suggesting close ecological ties between acritarchs and animals.

ACKNOWLEDGEMENTS

We gratefully acknowledge Nick Butterfield, Mike Foote, Kath Grey, Steven Holland, Andrew Knoll, Susannah Porter, and an anonymous reviewer for helpful comments. This research was supported by the National

Science Foundation and the Petroleum Research Fund. This paper is an expanded and updated report building directly on a previous study published recently in *Precambrian Research* (Huntley *et al.*, 2006).

REFERENCES

Allen, P. A., and Hoffman, P. F., 2005, Extreme winds and waves in the aftermath of a Neoproterozoic glaciation, *Nature* **433**: 123–127.

Allison, C. W., and Awramik, S. M., 1989, Organic-walled microfossils from the earliest Cambrian or latest Proterozoic Tindir Group Rocks, Northwest Canada, *Precambrian Res.* **43**: 253–294.

Anbar, A. D., and Knoll, A. H., 2002, Proterozoic ocean chemistry and evolution: A bioinorganic bridge? *Science* **297**: 1137–1142.

Bengston, S., and Zhao, Y., 1992, Predatorial borings in late Precambrian mineralized exoskeletons, *Science* **257**: 367–369.

Bodiselitch, B., Koeberl, C., Master, S., and Reimold, W. U., 2005, Estimating duration and intensity of Neoproterozoic Snowball glaciations from Ir anomalies, *Nature* **308**: 239–242.

Bottjer, D. J., Hagadorn, J. W., and Dornbos, S. Q., 2000, The Cambrian substrate revolution, *GSA Today* **10**: 1–7.

Boyce, C. K., 2005, Patterns of segregation and convergence in the evolution of fern and seed plant leaf morphologies, *Paleobiology* **31**: 117–140.

Brasier, M. D., and Lindsay, J. F., 1998, A billion years of environmental stability and the emergence of eukaryotes: New data from northern Australia, *Geology* **26**: 555–558.

Briggs, D. E. G., Fortey, R. A., and Wills, M. A., 1992, Morphological disparity in the Cambrian, *Science* **256**: 1670–1673.

Buick, R., Des Marais, D. J., and Knoll, A. H., 1995, Stable isotopic compositions of carbonates from the Mesoproterozoic Bangemall Group, northwestern Australia, *Chem. Geol.* **123**: 153–171.

Buick, R., and Knoll, A. H., 1999, Acritarchs and microfossils from the Mesoproterozoic Bangemall Group, northwestern Australia, *J. Paleontol.* **73**: 744–764.

Buss, L. W., and Seilacher, A., 1994, The Phylum Vendobionta: a sister group of the Eumetazoa? *Paleobiology* **20**: 1–4.

Butterfield, N. J., 1997, Plankton ecology and the Proterozoic-Phanerozoic transition, *Paleobiology* **23**: 247–262.

Butterfield, N. J., 2000, Bangiomorpha pubescens n. gen., n. sp.: Implications for the evolution of sex, multicellularity, and the Mesoproterozoic/Neoproterozoic radiation of eukaryotes, *Paleobiology* **26**: 386–404.

Butterfield, N. J., 2001, Ecology and Evolution of Cambrian Plankton. in: *The Ecology of the Cambrian Radiation* (A. Y. Zhuravlev and R. Riding, eds.), Columbia University Press, New York, pp. 200–216.

Butterfield, N. J., 2004, A vaucheriacean alga from the middle Neoproterozoic of Spitsbergen: Implications for the evolution of Proterozoic eukaryotes and the Cambrian explosion, *Paleobiology* **30**: 231–252.

Butterfield, N. J., 2005, Probable proterozoic fungi, *Paleobiology* **31**: 165–182.

Butterfield, N. J., and Chandler, F. W., 1992, Palaeoenvironmental distribution of Proterozoic microfossils, with an example from the Agu Bay Formation, Baffin Island, *Palaeontology* **35**: 943–957.

Butterfield, N. J., Knoll, A. H., and Swett, K., 1994, Paleobiology of the Neoproterozoic Svanbergfjellet Formation, Spitsbergen, *Fossils and Strata* **34**: 1–84.

Butterfield, N. J., and Rainbird, R. H., 1998, Diverse organic-walled fossils, including "possible dinoflagellates," from the early Neoproterozoic of Arctic Canada, *Geology* **26**: 963–966.

Clapham, M. E., and Narbonne, G. M., 2002, Ediacaran epifaunal tiering, *Geology* **30**: 627–630.

Condon, D., Zhu, M., Bowring, S., Wang, W., Yang, A., and Jin, Y., 2005, U–Pb Ages from the Neoproterozoic Doushantuo Formation, China, *Science* **308**: 95–98.

Conway Morris, S., 1998, *The Crucible of Creation: The Burgess Shale and the Rise of Animals*, Oxford University Press, Oxford.

Corsetti, F. A., Awramik, S. M., and Pierce, D., 2003, A complex microbiota from Snowball Earth times: Microfossils from the Neoproterozoic Kingston Peak Formation, Death Valley, USA, *Proc. Nat. Acad. Sci. USA* **100**: 4399–4404.

Droser, M. L., and Li, X., 2001, The Cambrian radiation and the diversification of sedimentary fabrics. in: *The Ecology of the Cambrian Radiation* (A. Y. Zhuravlev and R. Riding, eds.), Columbia University Press, New York, pp. 137–169.

Efron, B., 1981, Nonparametric standard errors and confidence intervals, *Can. J. Statistics* **9**: 139–172.

Fedonkin, M. A., and Waggoner, B. M., 1997, The Late Precambrian fossil Kimberella is a mollusc-like bilaterian organism, *Nature* **388**: 868–871.

Foote, M., 1994, Morphological disparity in Ordovician–Devonian crinoids and the early saturation of morphological space, *Paleobiology* **20**: 320–344.

Foote, M., 1995, Morphological diversification of Paleozoic crinoids, *Paleobiology* **21**: 273–299.

Foote, M., 1997, The evolution of morphological disparity., *Annu. Rev. Ecol. Systematics* **28**: 129–152.

Foote, M., and Gould, S. J., 1992, Cambrian and Recent morphological disparity, *Science* **258**: 1816.

Gao, L., Xing, Y., and Liu, G., 1995, Neoproterozoic micropalaeoflora from Hunjiang area, Jilin Province and its sedimentary environment, *Professional Papers Stratigr. Palaeontol.* **26**: 1–23.

Germs, J. G. B., Knoll, A. H., and Vidal, G., 1986, Latest Proterozoic microfossils from the Nama Group, Namibia (southwest Africa), *Precambrian Res.* **32**: 45–62.

Grey, K., Walter, M. R., and Calver, C. R., 2003, Neoproterozoic biotic diversification: Snowball Earth or aftermath of the Acraman impact? *Geology* **31**: 459–462.

Hammer, O., Harper, D. A. T., and Ryan, P. D., 2001, PAST: Paleontological Statistics Software Package for Education and Data Analysis, *Palaeontol. Electronica* **4**.

Hermann, T. N., 1990, *Organic World Billion Year Ago*, Nauka, Leningrad.

Hoffman, P. F., Kaufman, A. J., Halverson, G. P., and Schrag, D. P., 1998, A Neoproterozoic snowball Earth, *Science* **281**: 1342–1346.

Hoffman, P. F., and Schrag, D. P., 2002, The snowball Earth hypothesis: Testing the limits of global change, *Terra Nova* **14**: 129–155.

Hofmann, H. J., and Jackson, G. D., 1994, Shale-facies microfossils from the Proterozoic Bylot Supergroup, Baffin Island, Canada, *Paleontol. Soc. Mem.* **37**: 1–35.

Hofmann, H. J., and Jackson, G. D., 1996, Notes on the geology and micropaleontology of the Proterozoic Thule Group, Ellesmere Island, Canada and North-West Greenland, *Geol. Survey Can. Bull.* **#495**: 1–26.

Hua, H., Pratt, B. R., and Zhang, L.-Y., 2003, Borings in Cloudina shells: complex predator-prey dynamics in the terminal Neoproterozoic, *Palaios* **18**: 454–459.

Huntley, J. W., Xiao, S., and Kowalewski, M., 2006, 1.3 billion years of acritarch history: An empirical morphospace approach, *Precambrian Res.* **144**: 52–68.

Hyde, W. T., Crowley, T. J., Baum, S. K., and Peltier, W. R., 2000, Neoproterozoic "snowball Earth" simulations with a coupled climate/ice-sheet model, *Nature* **405**: 425–429.

Javaux, E. J., Knoll, A. H., and Walter, M. R., 2001, Morphological and ecological complexity in early eukaryotic ecosystems, *Nature* **412**: 66–69.

Jernvall, J., Hunter, J. P., and Fortelius, M., 1996, Molar tooth diversity, disparity, and ecology in Cenozoic ungulate radiations, *Science* **274**: 1489–1492.

Kirschvink, J. L., 1992, Late Proterozoic low-latitude global glaciation: the snowball Earth. in: *The Proterozoic Biosphere: A Multidisciplinary Study* (J. W. Schopf and C. Klein, eds.), Cambridge University Press, Cambridge, pp. 51–52.

Knoll, A. H., 1984, Microbiotas of the late Precambrian Hunnberg Formation, Nordaustlandet, Svalbard, *J. Paleontol.* **58**: 131–162.

Knoll, A. H., 1992, Microfossils in metasedimentary cherts of the Scotia Group, Prins Karls Forland, western Svalbard, *Palaeontology* **35**: 751–774.

Knoll, A. H., 1994, Proterozoic and Early Cambrian protists: Evidence for accelerating evolutionary tempo, *Proc. Nat. Acad. Sci. USA* **91**: 6743–6750.

Knoll, A. H., Blick, N., and Awramik, S. M., 1981, Stratigraphic and ecologic implications of late Precambrian microfossils from Utah, *Am. J. Sci.* **281**: 247–263.

Knoll, A. H., and Swett, K., 1985, Micropalaeontology of the Late Proterozoic Veteranen Group, Spitsbergen, *Palaeontology* **28**: 451–473.

Knoll, A. H., and Swett, K., 1987, Micropaleontology across the Precambrian–Cambrian boundary in Spitsbergen, *J. Paleontol.* **61**: 898–926.

Knoll, A. H., Swett, K., and Mark, J., 1991, Paleobiology of a Neoproterozoic tidal flat/lagoonal complex: The Draken Conglomerate Formation, Spitsbergen, *J. Paleontol.* **65**: 531–570.

Kowalewski, M., Goodfriend, G. A., and Flessa, K. W., 1998, High-resolution estimates of temporal mixing within shell beds: the evils and virtues of time-averaging., *Paleobiology* **24**: 287–304.

Kruskal, J. B., and Wish, M., 1978, *Multidimensional Scaling. Sage University Paper Series on Quantitative Applications in the Social Sciences*, Sage Publications, London.

Levinton, J. S., 2001, *Genetics, Paleontology, and Macroevolution*, Cambridge University Press, Cambridge.

Lipps, J. H., and Valentine, J. W., 2004, Late Neoproterozoic Metazoa: Wierd, wonderful and ghostly. in: *Neoproterozoic–Cambrian Revolutions* (J. H. Lipps and B. M. Waggoner, eds.), The Paleontological Society Papers, New Haven, CT, pp. 51–66.

Luo, Q., Zhang, Y., and Sun, S., 1985, The eukaryotes in the basal Changcheng System of Yanshan Ranges, *Acta Geol. Sinica* **1985**: 12–16.

Marcus, L., 1990, Traditional Morphometrics. in: *Proceedings of the Michigan Morphometrics Workshop: Special Publication Number 2* (F. J. Rohlf and F. L. Bookstein, eds.), The University of Michigan Museum of Zoology, Ann Arbor, MI, pp. 77–122.

Mendelson, C. V., 1987, Acritarchs. in: *Fossil Prokaryotes and Protists* (J. Lipps, ed.), University of Tennessee, Knoxville, pp. 62–86.

Moczydlowska, M., 1991, Acritarch biostratigraphy of the Lower Cambrian and the Precambrian–Cambrian boundary in southeastern Poland, *Fossils and Strata* **29**: 1–127.

Moczydlowska, M., 2001, Early Cambrian phytoplankton radiations and appearance of metazoans. in: *Cambrian System of South China (Palaeoworld No. 13)* (S. Peng, L. E. Babcock, and M. Zhu, eds.), University of Science and Technology of China Press, Hefei, pp. 293–296.

Moczydlowska, M., 2002, Early Cambrian phytoplankton diversification and appearance of trilobites in the Swedish Caledonides with implications for coupled evolutionary events between primary producers and consumers, *Lethaia* **35**: 191–214.

Moczydlowska, M., and Crimes, T. P., 1995, Late Cambrian acritarchs and their age constraints on an Ediacaran-type fauna from the Booley Bay Formation, Co. Wexford, Eire, *Geol. J.* **30**: 111–128.

Moczydlowska, M., and Vidal, G., 1992, Phytoplankton from the Lower Cambrian Læså Formation on Bornholm, Denmark: biostratigraphy and palaeoenvironmental constraints, *Geol. Mag.* **129**: 17–40.

Moczydlowska, M., Vidal, G., and Rudavskaya, V. A., 1993, Neoproterozoic (Vendian) phytoplankton from the Siberian Platform, Yakutia, *Palaeontology* **36**: 495–521.

Narbonne, G. M., 1998, The Ediacara Biota: A Terminal Neoproterozoic Experiment in the Evolution of Life, *GSA Today* **8**: 1–6.

Narbonne, G. M., 2005, The Ediacara Biota: Neoproterozoic origin of animals and their ecosystems, *Annu. Rev. Earth Planet. Sci.* **33**: 421–442.

Narbonne, G. M., and Gehling, J. G., 2003, Life after snowball: The oldest complex Ediacaran fossils, *Geology* **31**: 27–30.

Olcott, A. N., Sessions, A. L., Corsetti, F. A., Kaufman, A. J., and de Oliviera, T. F., 2005, Biomarker evidence for photosynthesis during neoproterozoic glaciation, *Science* **310**: 471–474.

Peterson, K. J., and Butterfield, N. J., 2005, Origin of the Eumetazoa: Testing ecological predictions of molecular clocks against the Proterozoic fossil record, *Proc. Nat. Acad. Sci. USA* **102**: 9547–9552.

Pyatiletov, V. G., and Rudavskaya, V. V., 1985, Acritarchs of the Yudoma Complex. in: *The Vendian System* (B. S. Sokolov and A. B. Iwanowski, eds.), Springer-Verlag, New York.

Roy, K., 1994, Effects of the Mesozoic Marine Revolution on the taxonomic, morphologic, and biogeographic evolution of a group: Aporrhaid gastropods during the Mesozoic, *Paleobiology* **20**: 274–296.

Runnegar, B., 2000, Loophole for snowball Earth, *Nature* **405**: 403–404.

Runnegar, B. N., and Fedonkin, M. A., 1992, Proterozoic Metazoan body fossils. in: *The Proterozoic Biosphere: A Multidisciplinary Study* (J. W. Schopf and C. Klein, eds.), Cambridge University Press, Cambridge, pp. 369–388.

Samuelsson, J., and Butterfield, N. J., 2001, Neoproterozoic fossils from the Franklin Mountains, northwestern Canada: stratigraphic and paleobiological implications, *Precambrian Res.* **107**: 235–251.

Schiffman, S. S., Reynolds, M. L., and Young, F. W., 1981, *Introduction to Multidimensional Scaling: Theory, Methods, and Applications*, Academic Press, New York.

Seilacher, A., 1992, Vendobionta and Psammocorallia: Lost constructions of Precambrian evolution, *J. Geol. Soc. London* **149**: 607–613.

Seilacher, A., 1999, Biomat-related lifestyles in the Precambrian, *Palaios* **14**: 86–93.

Seilacher, A., Grazhdankin, D., and Legouta, A., 2003, Ediacaran biota: The dawn of animal life in the shadow of giant protists, *Paleontological Res.* **7**: 43–54.

Sneath, P. H. A., and Sokal, R. R., 1973, *Numerical Taxonomy: the Principles and Practice of Numerical Classification*, W.H. Freeman, San Francisco, CA.

Sun, S., and Zhu, S., 2000, Paleoproterozoic eukaryotic fossils from northern China, *Acta Geol. Sinica* **74**: 116–122.

Thomas, R., Shearman, R. M., and Stewart, G. W., 2000, Evolutionary exploitation of design options by the first animals with hard skeletons, *Science* **288**: 1239–1242.

Valentine, J. W., 2004, *On the Origin of Phyla*, The University of Chicago Press, Chicago.

Vermeij, G. J., 2004, *Nature: An Economic History*, Princeton University Press, Princeton.

Vidal, G., 1976a, Late Precambrian acritarchs from the Eleonore Bay Group and Tillite Group in East Greenland, *Grønlands Geologiske Undersøgelse Rapport* **78**.

Vidal, G., 1976b, Late Precambrian microfossils from the Visingsö beds in southern Sweden, *Fossils and Strata* **9**: 1–57.

Vidal, G., 1979, Acritarchs from the upper Proterozoic and Lower Cambrian of East Greenland, *Bull. Geol. Survey Greenland* **No. 134**: 1–55.

Vidal, G., 1981, Micropalaeontology and biostratigraphy of the upper Proterozoic and Lower Cambrian sequences in East Finnmark, northern Norway, *Norges Geologiske Undersokelse Bulletin* **362**: 1–53.

Vidal, G., and Ford, T. D., 1985, Microbiotas from the late Proterozoic Chuar Group (northern Arizona) and Uinta Mountain Group (Utah) and their chronostratigraphic implications, *Precambrian Res.* **28**: 349–389.

Vidal, G., and Moczydlowska-Vidal, M., 1997, Biodiversity, speciation, and extinction trends of Proterozoic and Cambrian phytoplankton, *Paleobiology* **23**: 230–246.

Vidal, G., and Nystuen, J. P., 1990, Micropaleontology, depositional environment, and biostratigraphy of the upper Proterozoic Hedmark Group, Southern Norway, *Am. J. Sci.* **290A**: 170–211.

Vidal, G., and Peel, J. S., 1993, Acritarchs from the Lower Cambrian Buen Formation in North Greenland, *Grønlands Geologiske Undersøgelse Bulletin* **164**: 1–35.

Vidal, G., and Siedlecka, A., 1983, Planktonic, acid-resistant microfossils from the Upper Proterozoic strata of the Barents Sea region of Varanger Peninsula, East Finnmark, northern Norway, *Norges Geologiske Undersokelse Bulletin* **382**: 45–79.

Wagner, P. J., 1995, Testing evolutionary constraint hypotheses with early Paleozoic gastropods, *Paleobiology* **21**: 248–272.

Xiao, S., 2004a, Neoproterozoic glaciations and the fossil record. in: *The Extreme Proterozoic: Geology, Geochemistry, and Climate* (G. S. Jenkins, M. McMenamin, L. E. Sohl, and C. P. McKay, eds.), American Geophysical Union (AGU), Washington DC, pp. 199–214.

Xiao, S., 2004b, New multicellular algal fossils and acritarchs in Doushantuo chert nodules (Neoproterozoic, Yangtze Gorges, South China), *J. Paleontol.* **78**: 393–401.

Xiao, S., Knoll, A. H., Kaufman, A. J., Yin, L., and Zhang, Y., 1997, Neoproterozoic fossils in Mesoproterozoic rocks? Chemostratigraphic resolution of a biostratigraphic conundrum from the North China Platform, *Precambrian Res.* **84**: 197–220.

Yuan X., and Hofmann, H. J., 1998, New microfossils from the Neoproterozoic (Sinian) Doushantuo Formation, Wengan, Guizhou Province, southwestern China, *Alcheringa* **22**: 189–222.

Yan, Y., 1982, *Scizofusa* from the Chuanlinggou Formation of Changcheng System in Jixian County, *Bulletin Tianjin Inst. Geol. Mineral Resources* **6**: 1–7.

Yan, Y., 1991, Shale-facies microflora from the Changzhougou Formation (Changcheng System) in Pangjiapu Region, Hebei, China, *Acta Micropalaeontol. Sinica* **8**: 183–195.

Yan, Y., 1995, Shale facies microfloras from lower Changcheng System in Kuancheng, Hebei, and comparison with those of neighboring areas, *Acta Micropalaeontol. Sinica* **12**: 349–373.

Yan, Y., and Zhu, S., 1992, Discovery of acanthomorphic acritarchs from the Baicaoping Formation in Yongji, Shanxi and its geological significance, *Acta Micropalaeontol. Sinica* **9**: 267–282.

Yao, J., Xiao, S., Yin, L., Li, G., and Yuan, X., 2005, Basal Cambrian microfossils from the Yurtus and Xishanblaq formations (Tarim, north-west China): Systematic revision and biostratigraphic correlation of *Micrhystridium*-like acritarchs from China, *Palaeontology* **48**: 687–708.

Yin, C., 1992, A new algal fossil from Early Cambrian in Qingzhen county, Guizhou Province, China, *Acta Bot. Sinica* **34**: 456–460.

Yin, C., 1999, Microfossils from the Upper Sinian (Late Neoproterozoic) Doushantuo Formation in Changyang, western Hubei, China, *Continental Dynamics* **4**: 1–18.

Yin, L., 1990, Microbiota from middle and late Proterozoic iron and manganese ore deposits in China, *Int. Ass. Sedimentol. Special Pub.* **11**: 109–118.

Yin, L., 1997, Acanthomorphic acritarchs from Meso-Neoproterozoic shales of the Ruyang Group, Shanxi, China, *Rev. Palaeobot. Palynology* **98**: 15–25.

Yin, L., and Guan, B., 1999, Organic-walled microfossils of Neoproterozoic Dongjia Formation, Lushan County, Henan Province, North China, *Precambrian Res.* **94**: 121–137.

Yin, L., and Sun, W., 1994, Microbiota from Neoproterozoic Liulaobei Formation in the Huainan region, Northern Anhui, China, *Precambrian Res.* **65**: 95–114.

Zang, W., and Walter, M. R., 1992, Late Proterozoic and Cambrian microfossils and biostratigraphy, Amadeus Basin, central Australia, *Ass. Australasia Palaeontol. Memoir* **12**: 1–132.

Zhang, Y., Yin, L., Xiao, S., and Knoll, A. H., 1998, Permineralized fossils from the terminal Proterozoic Doushantuo Formation, South China, *Paleontol. Soc. Memoir* **50**: 1–52.

Zhang, Z., 1986, Clastic facies microfossils from the Chuanlinggou Formation (1800 Ma) near Jixian, North China, *J. Micropalaeontol.* **5**: 9–16.

Zhang, Z., 1997, A new Palaeoproterozoic clastic-facies microbiota from the Changzhougou Formation, Changcheng Group, Jixian, north China, *Geol. Mag.* **134**: 145–150.

Zhou, C., Brasier, M. D., and Xue, Y., 2001, Three-dimensional phosphatic preservation of giant acritarchs from the terminal proterozoic Doushantuo Formation in Guizhou and Hubei Provinces, south China, *Palaeontology* **44**: 1157–1178.

Zhou, C., Tucker, R., Xiao, S., Peng, Z., Yuan, X., and Chen, Z., 2004, New constraints on the ages of Neoproterozoic glaciations in South China, *Geology* **32**: 437–440.

Zhuravlev, A. Y., and Riding, R., 2001, *The Ecology of the Cambrian Radiation. Perspectives in Paleobiology and Earth History*, Columbia University Press, New York.

APPENDIX: SAS/IML CODES

Non-parametric Multidimensional Scaling (MDS):
*written by Michał Kowalewski;
*edited by John Huntley October 21, 2004 to reduce all variation to two axes;
*edited by John Huntley January 18, 2005 to accommodate new variables;
%let vars=spher ellips polyhed bulb medusa bar envmem cost triang cylinder taper hair hemi blunt pointed
    capit round funnel branch hollow comm fuse plates multi intbod excyst colony pore crest flange concen;
  %let group=age;
  %let group2='age';
  options pagesize=**1500**;
  data complex2;
  infile cards;
  input Genus $ Species $ spher ellips polyhed bulb medusa bar envmem cost triang cylinder taper hair hemi blunt pointed
    capit round funnel branch hollow comm fuse plates multi intbod excyst colony pore crest flange concen age;
    keep &vars &group;
    cards; *enter data matrix on next line;

  ;
  run;
  proc sort;
  by &group;
  proc transpose data=complex2 out=trans;
  data new;
      set trans;
      if _name_=&group2 then delete;
  proc corr data=new outs=final noprint;
  data mult;
      set final;
      if _TYPE_='CORR';
      drop _type_;
  proc mds data=mult dim=2 fit=1 out=score level=ordinal pineigval;
  data prep1;
      set score;
      if _type_='CRITERION' then delete;
  data all;
      merge prep1 complex2;
      keep dim1 dim2 &vars &group;
  proc print;
  proc univariate data=all noprint;
  var dim1 dim2;
  output out=varall var=var1 var2;
  proc print;
  proc sort data=all;
  by &group;

```
proc univariate data=all noprint;
var dim1 dim2;
by &group;
output out=repvar var=var1 var2;
proc print;
proc corr data=all spearman pearson;
    var dim1 dim2;
    with &vars;
proc plot data=all;
    plot dim2*dim1=&group;
run;
quit;
```

Mean dissimilarity coefficient and assessment of analytical error
```
*written by Michał Kowalewski;
%let times=100;
options linesize=64 pagesize=100; * -pagesize will hold up to 500 lines to avoid pagebreak in dataset;
    data mat;
    infile cards;
    input spher ellips polyhed bulb medusa bar envmem cost triang cylinder taper hair hemi blunt pointed
    capit round funnel branch hollow comm fused plates multi intbod excyst colony pore crest flange concen bin$ age;
    cards; * -enter data matrix on next line;

;
    data mat2;
    set mat;
    if bin='P1'; *- selects bin to be analyzed;
    drop age bin;
    proc iml;
    use mat2;* -specifies SAS dataset to be used in IML;
    read all into X;*-converts SAS dataset into IML matrix named X;
    Y=X;
    START DC(Y,d);
    do i=1 to nrow(Y); * -specifies first taxa to be compared;
      do j=1 to nrow(Y);* -specifies which taxa are to be compared to first taxa;
       if j>i then do;* -prevents redundancy in matrix calculation (i.e. i=taxa 1 compared with j=taxon 3, but when i=taxon 3 it won't be compared with j=taxon 1 or taxon 2;
        a=abs(Y[i,]-Y[j,]); *-calculates the number character differences between species;
        b=sum(a)/ncol(Y)||i||j;*-sum of character differences between all species, divided by number of characters;
        c=c//b;
       end;
      end;
    end;
    d=c[,1];
    n=nrow(d);
    FINISH DC;
```

```
START RANVEC(in,v_out);            *--creates a vector of random integers;
    k=nrow(in);
    v_index=in;
    do i=1 to k;
        rand=floor((k-i+1)*ranuni(0) + 1);
        v_ran=v_ran||v_index;
        v_index=remove(v_index,rand);
    end;
    v_out=v_ran;
FINISH RANVEC;
START MIXUP(X,times,template);     *--creates a template of random values;
    n=nrow(X);
    template=t(1:n)*j(1,times,1);
    do i=1 to times;
        run ranvec(template,out);
        template=t(out);
    end;
    do i=1 to n;
        run ranvec(t(template[i,]),out);
        template[i,]=out;
    end;
FINISH MIXUP;
start mix(X,out);
    z1=x[,1];
    z=x;
    times=&times;
    run DC(z,actd);
    ad=actd||shape(0,nrow(actd),1);
    run mixup(z1,times,rand);
    do i=1 to times;
        mat1=z[rand[,i],];
        run DC(mat1,rd);
        rd1=rd||shape(i,nrow(rd),1);
        rd2=rd2//rd1;
    end;
    out=ad//rd2;
finish mix;
run mix(Y,out);
create new from out; * - this step creates a SAS dataset readable by SAS/STAT procedures;
    append from out;
    close new;
    quit;
    data new1;
    set new;
    coeff=col1;* -names columns in SAS dataset derived from IML procedures above;
    iter=col2;
    keep coeff iter;
    data random;
```

```
set new1;
if iter>0;
data actual;
set new1;
if iter=0;
proc univariate noprint data=actual;
var coeff;
output out=truth n=ncomp mean=tmean median=tmedian min=tmin max=tmax;* -
calculates univariate statistics for pairwise dissimilarity coefficients calculated above in IML
procedures;
proc univariate noprint data=random;
var coeff;
by iter;
output out=rand2 n=n mean=mean;* -calculates univariate statistics for pairwise
dissimilarity coefficients calculated above in IML procedures;
proc univariate noprint data=rand2;
var mean;
output out=rand3 n=iter mean=rmean min=min max=max std=stderr pctlpre=P_
pctlpts=0.5 2.5 97.5 99.5;
data graph;
merge truth rand3;
bias=rmean-tmean;
max=max-bias;
min=min-bias;
L99=P_0_5-bias;
L95=P_2_5-bias;
U95=P_97_5-bias;
U99=P_99_5-bias;
USE=tmean+stderr;
LSE=tmean-stderr;
keep iter ncomp tmean rmean bias L99 L95 U95 U99 USE LSE min max stderr;
proc print;
run;
quit;
```

Chapter 3

On the Morphological and Ecological History of Proterozoic Macroalgae

SHUHAI XIAO and LIN DONG

Department of Geosciences, Virginia Polytechnic Institute and State University, Blacksburg, VA 24061, USA.

---

1. Introduction...	57
2. A Synopsis of Proterozoic Macroalgal Fossils.............................	60
3. Morphological History of Proterozoic Macroalgae........................	67
3.1 Narrative Description..	67
3.2 Quantitative Analysis: Morphospace, Body Size, and Surface/Volume Ratio	70
3.2.1 Methods...	70
3.3.2 Results..	74
4. Discussion..	75
4.1 Comparison with Acritarch Morphological History...............	75
4.2 Surface/Volume Ratio...	77
4.3 Maximum Canopy Height...	80
4.4 Ecological Interactions with Animals.............................	80
5. Conclusions..	82
Acknowledgements...	83
References..	83

---

1. INTRODUCTION

Multicellular or coenocytic, eukaryotic algae that are visible to the unaided eye (i.e. > 1 mm) are usually considered macroalgae. The cut-off between micro- and macroalgae is somewhat arbitrary, but may be of ecological significance because most macroalgae are, with notable exceptions, benthic. Macroalgae are ecologically and biogeochemically important in modern ecosystems. They form dense turfs or giant (~50 m in height) underwater forests in the intersection between the photic zone and the continental shelf. The productivity rate (measured in gram biomass per unit area per unit time) of such benthic macroalgal communities is impressive, —more than ten times greater than that of the open ocean (Bunt, 1975). In addition, algal turfs and forests partition the benthic ecosystem into a myriad of ecological habits, many of which are the grazing, breeding, and encrusting substrates for animals.

Despite the potential geobiological importance of macroalgae, however, the evolution and ecology of macroalgae in the Proterozoic is rarely discussed in the literature. This is partly due to the poor fossil record of Proterozoic macroalgae. Before the rise of calcareous algae in the early Paleozoic (Johnson, 1961; Wray, 1977), the preservation of macroalgae usually occurred in exceptional taphonomic conditions. In fact, most well-preserved Proterozoic macroalgal assemblages, such as the Little Dal assemblage in northwest Canada (Hofmann and Aitken, 1979; Hofmann, 1985), the Liulaobei and Jiuliqiao assemblages of North China (Sun *et al.*, 1986), and the Miaohe assemblage in South China (Xiao *et al.*, 2002), can be considered as Konservat-Lagerstätten that allow the preservation of non-mineralizing organisms (Butterfield, 2003). The exceptional nature of Konservat-Lagerstätten dictates that the stratigraphic completeness of macroalgal fossils is relatively poor although the quality of preservation can be extraordinary.

Another challenge in the study of Proterozoic algae lies in the difficulty of phylogenetic and ecological interpretations. Biochemical and cytological data, which are used routinely in the classification and phylogenetic analysis of modern algae, are not available in algal fossils. Thallus morphology is of limited phylogenetic significance because of pervasive convergence among different algal clades. Only a handful of algal fossils have been phylogenetically resolved into modern clades on the basis of cellular structures (Butterfield *et al.*, 1994; Xiao *et al.*, 1998a; Butterfield, 2000, 2004; Xiao *et al.*, 2004). For simple algal fossils, not only their phylogenetic affinities but also their paleoecology is difficult to constrain. As an example, *Chuaria*-like circular carbonaceous compressions—the most ubiquitous form in Proterozoic shales—have been variously interpreted as floating cyanobacteria colonies (Sun, 1987; Steiner, 1997), as planktonic acritarchs (Ford and Breed, 1973; Vidal and Ford, 1985; Butterfield *et al.*, 1994), as propagules of benthic chlorophyte or xanthophyte algae (Kumar, 2001), as benthic organisms (Butterfield, 1997, 2001), or as distant relatives of metazoans or fungi (Teyssèdre, 2003).

Despite these challenges, however, we can still learn a great deal about the morphological and ecological history of Proterozoic macroalgae at the broadest scale. The pervasive morphological convergence among different algal clades (e.g., chlorophytes, rhodophytes, and phaeophytes) indicates strong physiological and mechanical—as well as developmental and phylogenetic—constraints on algal morphology (Niklas, 2004). Thus, although morphological convergence is a noise in phylogenetic analysis, it is a bonus in ecological analysis of benthic macroalgae; for example, the functional-form model widely used in ecological analysis of modern macroalgal communities emphasizes ecologically important morphological features (e.g., surface/volume ratio) regardless of phylogenetic affinities (Littler and Littler, 1980; Littler and Arnold, 1982; Padilla and Allen, 2000). To explore the morphological and ecological history of Proterozoic macroalgae, we take a simple approach to characterize the morphological complexity, surface/volume ratio, and maximum canopy height of Proterozoic macroalgae. Our data show that the history of macroalgal morphological disparity in the Proterozoic is broadly similar to that of acritarchs (Huntley *et al.*, 2006), showing stepwise increase in the Mesoproterozoic and Ediacaran with a plateau in between. The Ediacaran expansion of macroalgal morphospace was also accompanied by significant increase in thallus surface/volume ratio and maximum canopy height of benthic macroalgal communities.

The causes of the Mesoproterozoic to early Neoproterozoic stasis and the Ediacaran rise in macroalgal morphological disparity are less clear. It is possible that the Mesoproterozoic to early Neoproterozoic plateau may be related to nutrient stress (Brasier and Lindsay, 1998; Anbar and Knoll, 2002) due to bottom-up ecological constraints. Alternatively, morphological evolution of Mesoproterozoic and early Neoproterozoic macroalgae may have been held back by absence of animal grazing pressure, which has been proposed to be a major top-down ecological force that drove the diversification of Ediacaran acritarchs (Peterson and Butterfield, 2005) and perhaps macroalgae. We also discuss the possibility that the Ediacaran rise in surface/volume ratio and morphological disparity may have been driven by decreasing pCO_2 levels after the Cryogenian glaciation. Since thallus surface/volume ratios appear to be positively correlated with bioproduction rate, macroalgae were probably more productive in the Ediacaran than before. If true, the increased bioproductivity may have some impacts on the global carbon cycle and oxygen evolution in the Ediacaran Period. These hypotheses and speculations necessarily need to be tested in the future with more geochemical, paleontological, and geochronological data.

We emphasize the exploratory character of this study and the preliminary nature of our conclusions, because the macroalgal affinity of some carbonaceous compression fossils included in this study may be debatable, and also because the geochronological resolution and stratigraphic completeness of our datadase are rather poor. Nonetheless, this exploratory exercise serves a starting point for more extensive studies of Proterozoic macroalgae in the future, and we hope that it will stimulate paleoecological and geobiological investigation of Proterozoic macroalgae.

2. A SYNOPSIS OF PROTEROZOIC MACROALGAL FOSSILS

Most Proterozoic macroalgae are preserved as carbonaceous compressions. Relatively few macroalgae are preserved in the permineralization windows (i.e., silicification and phosphatization), which are widely open for Proterozoic microorganisms (Schopf, 1968; Knoll, 1985); it is worth mentioning in passing that the contrast between the compression and permineralization windows may represent some major taphonomic biases or environmental heterogeneity. Recently, it has been recognized that some Ediacaran macroalgae may have been preserved as casts and molds, in a way similar to the preservation of classical Ediacara fossils (Droser et al., 2004), but the diversity of these macroalgal fossils awaits systematic documentation.

Hofmann (1994) compiled a comprehensive database of Proterozoic carbonaceous compressions and he classified them into thirteen formally defined families. Several new reports of Proterozoic carbonaceous fossils have been published since 1994 (Chen et al., 1994a; Chen et al., 1994b; Steiner, 1994; Ding et al., 1996; Gnilovskaya et al., 2000; Xiao et al., 2002); however, most of these new fossils can be classified into one of the thirteen families. Because these families were defined on morphological basis, it is likely that some of these families may be polyphyletic. However, as long as we can ascertain that these families represent macroalgae, these morphologically defined families may be to some degree analogous to macroalgal functional-form groups (Littler and Littler, 1980), and they should have ecological if not phylogenetic significance. Four of the thirteen families were considered as likely (Saarinidae and Sabelliditidae) or possible (Sinosabelliditidae and Protoarenicolidae) metazoans, and their nomenclature followed the ICZN rules (Hofmann, 1994). These family names are preserved here for convenience, even though we believe that the sinosabelliditids and protoarenicolids are probably macroalgae. Below we briefly consider the algal affinity of these groups.

Chuariaceae: This group includes the circular compressions *Chuaria* (millimetric diameters; Fig. 1A) and *Beltanelliformis* (centimetric diameters). Both often have concentric wrinkles and sometimes simple splits (Butterfield *et al.*, 1994; Steiner, 1997; Xiao *et al.*, 2002), indicating that in life they were spherical vesicles. Three-dimensionally preserved casts and molds confirm their spherical morphology (Hofmann, 1985; Narbonne and Hofmann, 1987; Yuan *et al.*, 2001). Thus, both genera can be reconstructed as spherical fluid-filled vesicles with a flexible organic wall. This morphological reconstruction is inconsistent with an affinity with cyanobacterial colonies such as *Nostoc* balls (Sun, 1987; Steiner, 1997), where filaments are held in a mucilaginous matrix (Graham and Wilcox, 2000). More likely, both *Chuaria* and *Beltanelliformis* are structurally similar to acritarchs with a coherent and resistant organic wall (Ford and Breed, 1973; Vidal and Ford, 1985; Butterfield *et al.*, 1994). In fact, some *Chuaria*-like compressions have been interpreted as benthic organic vesicles (Butterfield *et al.*, 1994; Butterfield, 1997, 2001), or as planktonic propagules of *Tawuia*-like thalli that are considered as benthic chlorophytes or xanthophytes (Kumar, 2001). Likewise, *Beltanelliformis* has been compared to spherical gametophytes of the benthic coenocytic green alga *Derbesia* (Xiao *et al.*, 2002). *Parachuaria simplicis*, another *Chuaria*-like fossil, has a millimetric circular compression with a subtending filament (Yan *et al.*, 1992; Tang *et al.*, 1997), which may well represent a stipe-like structure that tethered the spherical vesicle to a benthic substrate, in a way similar to *Longfengshania* (Hofmann, 1985; Du and Tian, 1986). Thus, *Chuaria* and *Beltanelliformis* are best considered as benthic or having a benthic stage in their life cycle. It is also probable that they may have been photosynthetic eukaryotes, given that their spherical vesicles have morphological analogues among modern coenocytic algae (e.g., *Derbesia* and *Valonia*), but not among animals or fungi. Thus we tentatively regard chuariaceans as macroalgae. It should be noted, however, that the major patterns of macroalgal morphological history would probably stay even if we had removed chuariaceans from our analysis, because chuariaceans are ubiquitous throughout the entire Proterozoic.

Tawuiaceae: *Tawuia*, the eponymous genus of this group, is reconstructed as a tubular structure with closed and round termini (Hofmann and Aitken, 1979; Hofmann, 1985). Like *Chuaria*, it can be preserved as two-dimensional compressions or three-dimensional molds (Hofmann and Aitken, 1979; Hofmann, 1985). Because all reported populations co-occur with *Chuaria*, *Tawuia* is generally considered ontogenetically or phylogenetically related to *Chuaria* (Duan, 1982; Hofmann, 1985). Recently, Kumar (2001) reported a population of carbonaceous compressions from the Suket Shale of the lower Vindhyan Supergroup in the Rampura-Chittorgarh area, central India. The Suket population, probably between 1600 and 1140 Ma (Kumar, 2001; Rasmussen *et al.*, 2002; Ray *et al.*, 2002; Ray *et al.*, 2003; Sarangi *et al.*, 2004), includes *Chuaria-* and *Tawuia-*like fossils. The termini of several Suket *Tawuia-*like specimens bear circular (*Chuaria-*like) or trapezoidal structures, which Kumar interpreted as compressed spherical cysts and holdfasts, respectively. Kumar gave different taxonomic names to the different parts of the same specimen; the trapezoidal holdfast was described as *Tilsoia* or *Suketea* depending on how it is preserved, the cylindrical stem as *Tawuia*, the spherical cyst as *Chuaria*, and the complete organism was named *Radhakrishnania*. While the identification of the Suket tubular fossils as *Tawuia dalensis* is debatable and the taxonomic practice of Kumar is undesirable, the Suket population does provide a general model by which *Chuaria* and *Tawuia* may be ontogenetically related. This model implies 1) *Tawuia* represents only the benthic stage of a biphasic alga and 2) *Chuaria* and *Tawuia* should have similar geographic, environmental, and stratigraphic distribution. However, these implications are difficult to test, because *Chuaria* is almost certainly a polyphyletic taxon and also because planktonic cysts (i.e., *Chuaria*) can be preserved beyond the geographic and environmental distribution of their benthic vegetative parents (i.e., *Tawuia*). Given that *Tawuia* populations from the type locality (Hofmann and Aitken, 1979; Hofmann, 1985) and elsewhere (Zhang *et al.*, 1991) also contain individuals, including some U-shaped individuals, with a terminal disk at one end, it is probable that *Tawuia* and *Chuaria* may indeed be organ taxa of the same organism. Other *Tawuia-*like fossils, for example *Bipatinella* (Fig. 1B) from the early Neoproterozoic Liulaobei Formation and Shijia Formation in northern Anhui of North China (Zheng *et al.*, 1994) also appear to have terminal swellings. If *Tawuia* and *Chuaria* are indeed organ taxa of the same organism, the combination of characters (a planktonic stage and a benthic stage with holdfast) is most consistent with a macroalgal interpretation for *Tawuia*. Thus, in our compilation, we follow the traditional view that *Tawuia* represents a benthic, tubular macroalga.

Ellipsophysaceae: *Ellipsophysa* (Fig. 1D) and related genera from the Liulaobei, Jiuliqiao, Xiamaling, and Changlongshan formations in North China, are elliptical to oval compressions with a maximum/minimum axis ratio between 1.4 and 2 (Zheng, 1980; Du and Tian, 1986). It is uncertain whether these compression fossils should be classified in the Chuariaceae or in a separate family. Nonetheless, their elliptical/oval morphology is intermediate between *Chuaria* and *Tawuia*, and by analogy they may also be interpreted as macroalgae.

Longfengshaniaceae: *Longfengshania* (Fig. 1C) and *Paralongfengshania* can be reconstructed as algal thalli with an ellipsoidal, ovoidal, or panduroidal vesicle and a subtending stipe (Hofmann, 1985; Du and Tian, 1986). Some specimens preserve a simple discoidal holdfast (for example,

Du and Tian, 1986, plate X, Figs. 2, 8A, 9; plate XI, Figs. 9–11), suggesting a benthic habit. *Longfengshania* was once interpreted as a bryophyte (Zhang, 1988), but this interpretation was disputed because it lacks any diagnostic bryophyte features (Liu and Du, 1991). The simple morphology and marine habitat of *Longfengshania* and *Paralongfengshania* is more consistent with a macroalgal interpretation. Indeed, several modern algae such as *Botrydium* (a xanthophyte), *Botryocladia* (a rhodophyte), and *Valonia* (a chlorophyte), all of which have a balloon-like vesicle tethered to a holdfast or a branch (Abbott, 1999; Graham and Wilcox, 2000), are good interpretive analogues for *Longfengshania* and *Paralongfengshania*.

Grypaniaceae: *Grypania* is a spiral ribbon-like compression fossil that occurr in Paleoproterozoic and Mesoproterozoic rocks (Walter *et al.*, 1976; Du *et al.*, 1986; Walter *et al.*, 1990; Han and Runnegar, 1992). It is reconstructed as a spiral cylindrical organism, probably a photosynthetic alga (Walter *et al.*, 1990; Han and Runnegar, 1992).

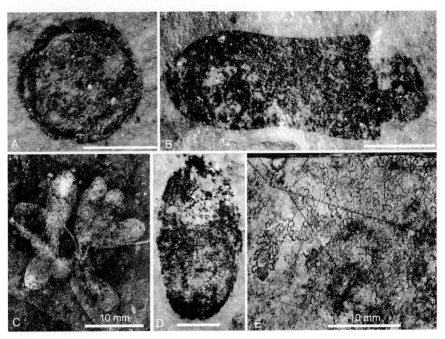

Figure 1. (A) *Chuaria circularis* from the early Neoproterozoic Huaibei Group, North China. (B) *Bipatinella cervicalis* (a *Tawuia*-like fossil) from the early Neoproterozoic Huaibei Group, North China. (C) *Longfengshania stipitata* from the early Neoproterozoic Little Dal Group, northwestern Canada. Photo courtesy of Hans Hofmann. (D) *Ellipsophysa axicula* from the early Neoproterozoic Jiuliqiao Formation, North China. (E) *Seirisphaera zhangii* from the Ediacaran Lantian Formation, South China. Photo courtesy of Chen Meng'e. Scale bar represents 1 mm if not otherwise indicated.

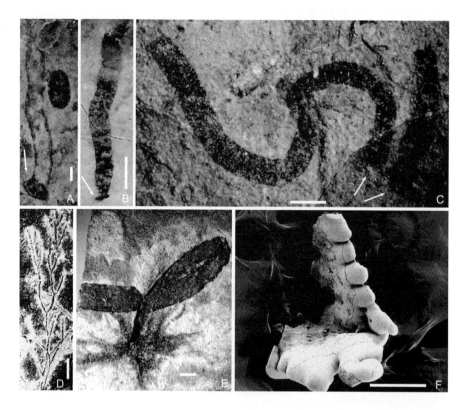

Figure 2. (A–C) Specimens that can be identified as *Protoarenicola baiguashanensis* from the early Neoproterozoic Huaibei Group, North China. Transverse annulations not well preserved in (A) Note discoidal holdfast-like structures (arrows). (C) Courtesy of Xunlai Yuan. (D) *Doushantuophyton lineare* from the Ediacaran Doushantuo Formation, South China. (E) *Baculiphyca taeniata* from the Ediacaran Doushantuo Formation, South China. (F) Phosphatized algal thallus (possibly *Thallophyca ramosa*) from the Ediacaran Doushantuo Formation, South China. Scale bars represent 1 mm.

Eoholyniaceae: Hofmann (1994) created this family to accommodate all branching forms. Some fine filaments, such as *Daltaenia* (Hofmann, 1985) and *Chambalia* (Kumar, 2001), appear to have branches and would be included in this family. However, the junctions of these branching filaments tend to be T-shaped rather than Y-shaped; they could be cyanobacterial branches (e.g. *Fischerella*) and are thus excluded from our analysis. Instead, we focus on carbonaceous fossils with dichotomous, monopodial, or helical branches, because these are more likely eukaryotic algae. A number of carbonaceous compressions from the Ediacaran Doushantuo and Lantian formations, including *Anomalophyton*, *Doushantuophyton* (Fig. 2D), *Enteromorphites*, *Konglingiphyton*, *Longifuniculum*, and *Miaohephyton*

(Chen and Xiao, 1992; Steiner, 1994; Ding *et al.*, 1996; Yuan *et al.*, 1999; Xiao *et al.*, 2002; Yuan *et al.*, 2002), are considered members of this group. Some fan-shaped thalli, such as *Anhuiphyton*, *Flabellophyton*, and *Huangshanophyton* from the Lantian Formation, may also contain rare dichotomously branching filaments (Yan *et al.*, 1992; Chen *et al.*, 1994a; Steiner, 1994; Yuan *et al.*, 1999), but this is difficult to verify because of dense compaction of fine filaments. Nonetheless, the macroscopic thallus size, morphological complexity, and the presence of a holdfast structure (in *Flabellophyton* at least) independently suggest their macroalgal affinity and benthic habit. Thus, these Lantian forms are also considered members of this family.

Sinosabelliditidae and Protoarenicolidae: These two groups are characterized by ribbon-shaped compressions with transverse annulations (Fig. 2A–C). They occur in early Neoproterozoic rocks in North China (Sun *et al.*, 1986), and similar forms have been reported from late Riphean rocks in southern Timan (Gnilovskaya *et al.*, 2000). Some specimens are three-dimensionally preserved with a circular transverse cross section (Zheng, 1980; Wang, 1982; Wang and Zhang, 1984; Xing *et al.*, 1985; Sun *et al.*, 1986; Chen, 1988; Qian *et al.*, 2000), suggesting that they were originally cylindrical tubes. Representative genera are *Sinosabellidites*, *Pararenicola*, *Protoarenicola*, *Parmia*, and many other synonyms (Wang and Zhang, 1984; Xing *et al.*, 1985; Gnilovskaya *et al.*, 2000). *Pararenicola* and *Protoarenicola* appear to bear a proboscis-like structure or a terminal opening in their presumed anterior end. The proboscis-like structure and transverse annulations led some to interpret *Pararenicola* and *Protoarenicola* as possible worm-like animals (Sun *et al.*, 1986; Chen, 1988). *Sinosabellidites* has similar transverse annulations but no terminal opening or proboscis-like structure, and it was considered less likely to be an animal (Sun *et al.*, 1986). It is interesting to note that a number of protoarenicolid specimens (for example, Wang and Zhang, 1984, plate 7, Fig. 2; Xing *et al.*, 1985, plate 39, Fig. 1; Qian *et al.*, 2000) appear to have holdfast-like structures. In fact, several transversely annulated or corrugated tubular fossils from the Doushantuo Formation, including *Cucullus* and *Sinospongia* (Xiao *et al.*, 2002), can be considered members of the Protoarenicolidae (Hofmann, 1994) and they also have holdfast-like structures. Our own observations of protoarenicolids suggest that some of them have a discoidal holdfast structure (Fig. 2A–C). Thus, it is possible that the proboscis-like structures present in a small number of specimens of protoarenicolids (Sun *et al.*, 1986) may be poorly preserved holdfasts or artifacts due to physical tearing of the discoidal holdfast. If confirmed, these observations and interpretations would indicate that protoarenicolids are similar to tawuiaceans described from the Suket Shale in the Vindhyan

Supergroup (Kumar, 2001) in having a holdfast structure. The only major difference is the presence or absence of transverse annulations, which is not a diagnostic animal feature (Sun *et al.*, 1986; Chen, 1988). Thus, the animal interpretation of sinosabelliditids and protoarenicolids is poorly supported. A more likely interpretation is that they were siphonous macroalgae analogous to modern dasycladaleans (Berger and Kaever, 1992).

Moraniaceae, Beltinaceae, Vendotaeniaceae, Saarinidae, and Sabelliditidae: These groups are not included in the current study because their macroalgal affinity is problematic. Moraniaceans, beltinaceans, and vendotaeniaceans may represent bacterial colonies (Walcott, 1919; Vidal, 1989; Hofmann, 1994), although vendotaeniaceans have been interpreted as brown or red algae (Gnilovskaya, 1990; Gnilovskaya, 2003). In addition, beltinaceans and vendotaeniaceans are often fragmented and folded, making it difficult to reconstruct their morphology and paleoecology. Saarinids and sabelliditids have been interpreted as pogonophoran tubes (Sokolov, 1967; Hofmann, 1994); certainly, ultrastructures of *Sabellidites cambriensis* tubes, which consist of interwoven filaments with a diameter of 0.2–0.3 μm (Urbanek and Mierzejewska, 1977; Ivantsov, 1990; Moczydlowska, 2003), have no analogues among modern macroalgae.

Other Macroalgae: *Baculiphyca* (Fig. 2E) from the Doushantuo and Lantian formations in South China was questionably placed in the Protoarenicolidae (Hofmann, 1994). *Baculiphyca* was undoubtedly a benthic macroalga with clavate or blade-like thallus and rhizoidal holdfast but no transverse annulations (Xiao *et al.*, 2002). Thus *Baculiphyca* does not belong to the same family (or functional-form group) as protoarenicolids that are characterized by cylindrical thallus, transverse annulations, and possible discoidal holdfast. Another taxon that was not classified in any of the formally defined families is *Orbisiana* from Vendian rocks in Russia (Sokolov, 1976). *Orbisiana* consists of serial or biserial rings or spheres 0.2-0.9 mm in diameter, and it is probably an algal fossil (Jensen, 2003). Similar fossils (Fig. 1E), preserved as carbonaceous compressions and described as *Catenasphaerophyton* (Yan *et al.*, 1992) or *Seirisphaera* (Chen *et al.*, 1994a), have been known from the Ediacaran Lantian Formation in South China.

Permineralized Macroalgae: In addition to carbonaceous compressions, some permineralized algal fossils can also reach macroscopic size (Fig. 2F). Phosphatized and silicified algae in the Doushantuo Formation (Xiao, 2004; Xiao *et al.*, 2004), for example, can be millimetric in size. However, the overall diversity and abundance of permineralized macroalgae is much lower than carbonaceous ones.

3. MORPHOLOGICAL HISTORY OF PROTEROZOIC MACROALGAE

3.1 Narrative Description

Although many carbonaceous compressions have functional morphologies generally consistent with algal interpretation, their exact phylogenetic affinities are poorly resolved because of pervasive morphological convergence among algae. Possible exceptions include *Miaohephyton bifurcatum* and *Beltanelliformis brunsae* from the Doushantuo Formation; these have been compared, respectively, with fucalean brown algae and the coenocytic green alga *Derbesia* (Xiao *et al.*, 1998a; Xiao *et al.*, 2002). In addition, several microscopic compressions recovered from Proterozoic shales using palynological method are phylogenetically resolved. For example, *Proterocladus major* from the ~750 Ma Svanbergfjellet Formation in Spitsbergen has been interpreted as a clodophoran green alga (Butterfield *et al.*, 1994). *Palaeovaucheria clavata* from the ~1000 Ma Lakhanda Group in southeastern Siberia and *Jacutianema solubila* from the Svanbergfjellet Formation are both interpreted as xanthophyte algae (Hermann, 1990; Butterfield, 2004). Finally, the silicified microfossil *Bangiomorpha pubescens* from the ~1200 Ma Hunting Formation in Arctic Canada has been interpreted as a bangiophyte red alga (Butterfield, 2000), and several phosphatized algae from the Ediacaran Doushantuo Formation have been interpreted as florideophyte red algae (Xiao *et al.*, 2004). These fossils indicate that major algal clades diverged no later than the early Neoproterozoic (Knoll, 1992; Porter, 2004).

However, clade divergence needs not be temporally coupled with morphological, ecological, and taxonomic diversification. Therefore, it is useful to independently characterize important morphological innovations in macroalgal history. We begin by tabulating the temporal distribution of some important macroalgal morphologies (Table 1), followed by a brief summary of macroalgal morphologies in the Proterozoic.

Paleoproterozoic and Mesoproterozoic macroalgae are mostly spherical, ellipsoidal, tomaculate, or cylindrical forms. Carbonaceous compressions similar to *Chuaria*, *Ellipsophysa*, and *Tawuia* are known from the 1800–1700 Ma Changzhougou and Chuanlinggou formations in North China (Hofmann and Chen, 1981; Lu and Li, 1991; Zhu *et al.*, 2000; Wan *et al.*, 2003), although those from the Changzhougou Formation have recently been characterized as pseudofossils (Lamb *et al.*, 2005). *Grypania* and *Grypania*-like fossils have been reported from the ~1900 Ma Negaunee Iron-Formation of Michigan (Han and Runnegar, 1992; Schneider *et al.*, 2002), the Mesoproterozoic Rohtas Formation of central India (Kumar, 1995; Rasmussen *et al.*, 2002; Ray *et al.*, 2002), and the ~1400 Ma Gaoyuzhuang Formation in North China and the Greyson Shale in Montana (Walter *et al.*, 1990); the Indian *Grypania* specimens are distinct in bearing transverse annulations. Abundant carbonaceous compressions occur in the ~1700 Ma Tuanshanzi Formation in the Jixian area (Hofmann and Chen, 1981; Yan, 1995; Zhu and Chen, 1995; Yan and Liu, 1997). Some of the Tuanshanzi fossils have been interpreted as macroalgae with holdfast-stipe-blade differentiation, but their variable morphologies appear to suggest that some of them may be fragmented algal mats. However, *Tawuia*-like fossils from the Mesoproterozoic Suket Shale in central India do appear to have simple discoidal holdfasts (Kumar, 2001).

Table 1. Temporal distribution of important macroalgal features (+: presence; ?: possible presence).

	Paleoproterozoic (2500–1600 Ma)	Mesoproterozoic (1600–1000 Ma)	Early Neoproterozoic (1000–750 Ma)	Ediacaran (635–542 Ma)
Thallus Morphologies				
Spherical	+	+	+	+
Ellipsoidal	+	+	+	+
Tomaculate	+	+	+	+
Cylindrical	+	+	+	+
Conical				+
Fan-shaped				+
Thallus Differentiation				
Holdfast present	?	+	+	+
Discoidal holdfast		+	+	+
Rhizoidal holdfast				+
Stipe	?		+	+
Blade	?			+
Other Features				
Transverse annulation		+	+	+
Dichotomous Branching				+
Monopodial branching				+
Apical meristem				+

Early Neoproterozoic macroalgal assemblages continued to be dominated by simple forms such as *Chuaria*, *Ellipsophysa*, and *Tawuia*. But several morphological innovations did occur in the early Neoproterozoic. These include algal thalli with well-differentiated stipe and holdfast structures (in *Longfengshania* and *Paralongfengshania*), as well as cylindrical thalli with well-defined transverse annulations and holdfast structures (in *Sinosabellidites* and pararenicolids).

Important morphological innovations evolved in the Ediacaran Period. The Doushantuo Formation (635–550 Ma) and equivalent rocks in South China contains diverse macroalgal assemblages (Steiner, 1994; Yuan *et al.*, 1999; Xiao *et al.*, 2002). Doushantuo macroalgae are featured with monopodial and spiral branching (e.g., *Doushantuophyton quyuani* and *Anomalophyton zhangzhongyingi*), true dichotomous branching and apical meristematic growth (e.g., *Doushantuophyton lineare*, *Miaohephyton bifurcatum*, *Konglingiphyton erecta*, and *Enteromorphites siniansis*), rhizoidal holdfasts and flattened blade-like thalli (e.g., *Baculiphyca taeniata*), conical thalli (e.g., *Protoconites minor*), and fan-shaped thalli (e.g., *Longifuniculum dissolutum*, *Anhuiphyton lineatum*, *Flabellophyton strigata*, *Flabellophyton lantianensis*, and *Huangshanophyton fluticulosum*). These Doushantuo macroalgae were first reported (Zhu and Chen, 1984) from uppermost Doushantuo black shale that is less than 10 m below an ash bed dating from 551±1 Ma (Condon *et al.*, 2005). Subsequently, similar fossils have also been found in Doushantuo black shales in southern Anhui (Bi *et al.*, 1988; Yan *et al.*, 1992; Chen *et al.*, 1994a; Yuan *et al.*, 1999) and north-eastern Guizhou (Zhao *et al.*, 2004). More recently, at least one member of the Miaohe biota—*Enteromorphites siniansis*—has also been found in the lower Doushantuo Formation in the Yangtze Gorges area (Tang *et al.*, 2005). The lower Doushantuo Formation is estimated to be between 635 Ma and 580 Ma (Condon *et al.*, 2005). If this estimate is correct, morphological diversification of macroalgae began after the 635 Ma Marinoan glaciation (Hoffmann *et al.*, 2004; Condon *et al.*, 2005) but before the 580 Ma Gaskiers glaciation (Bowring *et al.*, 2003) and perhaps before the diversification of animals (Xiao *et al.*, 1998b; Condon *et al.*, 2005; Narbonne, 2005).

Despite morphological innovations in the Ediacaran, several functional forms of modern macroalgae (Littler and Littler, 1980) have not been observed in any Ediacaran assemblages. These functional forms include very thin sheet-like (leafy), calcareous, and crustose thalli, which are common in modern macroalgal flora (such as *Porphyra*, *Ulva*, and coralline red algae). The lack of leafy thalli in the fossil record may be taphonomic, but the absence of calcareous and crustose thalli in the Precambrian is probably real (Steneck, 1983). Thus, the morphological diversity of Ediacaran macroalgae, although much greater than before, may still be comparatively lower than modern macroalgae.

3.2 Quantitative Analysis: Morphospace, Body Size, and Surface/Volume Ratio

3.2.1 Methods

To quantify the morphological evolution of macroalgae in the Proterozoic, we carried out a morphospace analysis of Proterozoic macroscopic carbonaceous compressions (> 1 mm in maximum dimension, with a few exceptions) that can be reasonably interpreted as macroalgal fossils (see above). Permineralized macroalgae were not included in our quantitative analysis because of the few examples of permineralized macroalgae and because of possible preservational biases between the compression and permineralization windows. After a preliminary analysis, we also excluded in our further analysis carbonaceous compressions from the Paleoproterozoic Tuanshanzi Formation reported by Zhu and colleagues (Yan, 1995; Zhu and Chen, 1995; Yan and Liu, 1997) because at least some of these may be fragmentary microbial mats (see above) and also because their morphologies are unstable.

Table 2. List of characters (or character states) used in quantitative analysis.

Thallus morphology	Thallus differentiation	Other features
1. Spherical	9. Inferred holdfast presence	14. Transverse annulation
2. Ellipsoidal	10. Discoidal holdfast	15. Dichotomous Branching
3. Tomaculate	11. Rhizoidal holdfast	16. Monopodial or spiral branching
4. Ovoid	12. Stipe	17. Coarse branches
5. Cylindrical	13. Blade	18. Delicate branches
6. Conical		19. Colonial appearance (e.g., in some *Beltanelliformis* populations)
7. Fan-shaped		
8. Filamentous		

We collected presence/absence data of 19 morphological characters or character states of 578 carbonaceous compression specimens from 17 published monographs (Table 2 and Table 3). This literature survey was by no means exhaustive, but it included representatives of most macroalgal forms. In our analysis, all characters or character states were treated as binary presence/absence variables. We performed a non-metric multidimensional scaling (MDS) analysis of the pooled data [for a detail description of the MDS method, see (Huntley *et al.*, 2006)]. The MDS analysis allowed us to ordinate all specimens in a two-dimensional space (dimension 1 and dimension 2). The MDS scored specimens were then assigned to four geochronological bins (Paleoproterozoic 1800–1600 Ma; Mesoproterozoic 1600–1000 Ma; early Neoproterozoic 1000–750 Ma; Ediacaran 635–550 Ma) according to their probable age. MDS variances for dimension 1 and dimension 2 were then calculated for each geochronological bin. The sum of dimension 1 and dimension 2 variances is taken as a proxy for morphological disparity in each bin. The sum MDS variances are shown in Figs. 3–4. To test whether the geochronological pattern of MDS variance was due to varying sample intensity in the geochronological bins, we performed a randomization analysis (Huntley *et al.*, 2006). The MDS score pairs associated with each specimen were shuffled randomly into one of the four geochronological bins, but the sample intensity of the geochronological bins was preserved. The MDS variance for each geochronological bin was recalculated. The process was repeated 1000 times, in order to obtain the mean and 95% confidence interval of the MDS

Table 3. List of geochronological bins and source data.

Paleoproterozoic (1800–1700 Ma): 29 specimens, 4 described species	
Changzhougou Fm., 1800–1625 Ma	(Zhu *et al.*, 2000)
Tuanshanzi Fm., 1800–1625 Ma	(Du and Tian, 1986)
Tuanshanzi Fm., 1800–1625 Ma	(Zhu and Chen, 1995), 10 specimens not included in further analysis
Mesoproterozoic (1600–1000 Ma): 46 specimens, 7 described species	
Hongshuizhuang & Gaoyuzhuang Fm., ~1400 Ma	(Du and Tian, 1986; Walter *et al.*, 1990)
Rohtas Fm., 1600–1000 Ma	(Kumar, 1995)
Suket Shale, 1600–1000 Ma	(Kumar, 2001)
Early Neoproterozoic (1000–750 Ma): 422 specimens, 76 described species	
Liulaobei, Jiuliqiao, Jinshanzhai, Shijia, Weiji, & Gouhou Fm., ~850 Ma	(Duan, 1982; Wang and Zhang, 1984; Steiner, 1997)
Wyniatt Fm., 1077–723 Ma	(Hofmann and Rainbird, 1994)
Xiamaling, Changlongshan, & Nanfen Fm., ~850 Ma	(Duan, 1982; Du and Tian, 1986)
Halkal Formation, ~850 Ma	(Maithy and Babu, 1996)
Shihuiding Formation, ~850 Ma	(Zhang *et al.*, 1991; Zhang *et al.*, 1995)
Little Dal Formation	(Hofmann, 1985)
Late Riphean Pav'yuga Formation (?)	(Gnilovskaya *et al.*, 2000)
Ediacaran (635–550 Ma): 91 specimens, 27 described species	
Doushantuo Formation, 635–550 Ma	(Steiner, 1997; Xiao *et al.*, 2002)
Lantian Formation, 635–550 Ma	(Yuan *et al.*, 1999)

variances after randomization (Figs. 3–4). If the observed MDS variances lie beyond the 95% confidence interval, they are unlikely to be explained by differing sampling intensity alone.

To evaluate the impact of the Tuanshanzi compressions (Yan, 1995; Zhu and Chen, 1995; Yan and Liu, 1997), we repeated our analysis with the Tuanshanzi fossils included and the results did not change significantly (compare Fig. 3 and Fig. 4). As the geochronological pattern of MDS variance show no significant difference whether the Tuanshanzi compressions are included or excluded (Figs. 3–4), the Tuanshanzi fossils are excluded in all subsequent analyses (Figs. 5–8) because they are possibly fragmented microbial mats.

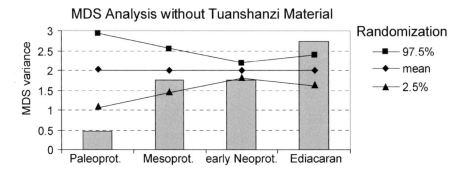

Figure 3. Results of MDS analysis and randomization test with the Tuanshanzi material excluded. Shaded bars represent MDS variances of the four geochronological bins. Filled squares, diamonds, and triangles represent the 97.5% percentile, mean, and 2.5% percentile of the randomization test. Thus, the filled square and triangle bracket the empirically determined 95% confidence interval for each bin. Note that three of the four bins have MDS variances outside the 95% confidence intervals.

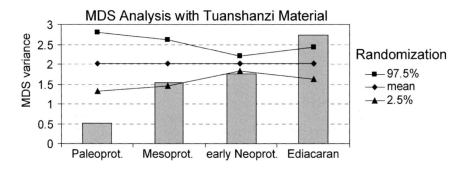

Figure 4. Results of MDS analysis and randomization test with the Tuanshanzi material included. See Fig. 3 for explanation.

Morphological and Ecological History of Proterozoic Macroalgae 73

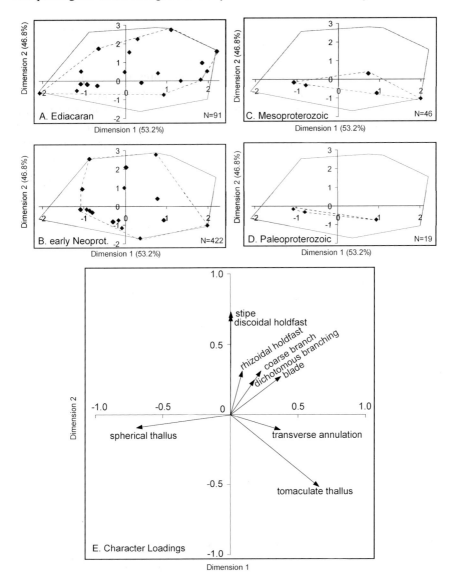

Figure 5. (A–D) Scatter plots showing realized morphospace in each geochronological bin (convex hulls in dashed line) in comparison with occupied morphospace when all Proterozoic data are pooled (convex hulls in solid line). The seemingly small occupied morphospace in Mesoproterozoic as compared with early Neoproterozoic may be related to its smaller sample size. (E) Loading diagram showing characters with significant loadings.

Scatter plots for each geochronological bin are shown in a two-dimensional space (Fig. 5A–D). Correlation coefficients were calculated between the morphological variables and MDS dimension 1 and dimension 2 scores for all species occurrences. R-values from correlation analysis were used to produce a loading chart relating the MDS morphospace to the original morphological characters (Fig. 5E).

As a proxy of body size, we also estimated the maximum dimension (e.g., long axis of an elliptical compression; maximum length of a ribbon-like compression; maximum height of a branching thallus; maximum dimension of a *Longfengshania* thallus including its vesicle and holdfast) of all carbonaceous compression fossils in our database. In addition, we estimated the surface/volume ratio for each specimen in our database, based on three-dimensional reconstructions of the compression fossils (see above). For example, *Chuaria circularis* was modelled as a spherical thallus with a diameter equivalent to its circular compression; *Tawuia dalensis* as a cylindrical thallus with semi-spherical ends; *Longfengshania stipitata* as a spherical to ovoidal vesicle with differentiated stipe and holdfast; and *Doushantuophyton lineare* as terete dichotomous branches with differentiated holdfast. The surface/volume ratio of *Longfengshania* and *Paralongfengshania* was estimated based on the vesicle, because it is likely that only the vesicle was photosynthetic; however, the ratio would not change significantly even if we consider the stipe and holdfast. Similarly, the holdfast of many Doushantuo macroalgae, such as *Baculiphyca taeniata* and *Enteromorphites siniansis*, was not considered in the estimate of surface/volume ratio.

3.2.2 Results

The MDS analysis (Figs. 3–5) shows that macroalgal morphospace increased episodically in the Mesoproterozoic Era and in the Ediacaran Period, confirming the narrative description. This pattern cannot be a sampling artifact because (1) MDS scores show no correlation with bin characters (data density or geochronological duration); and (2) three of the four geochronological bins have morphological disparity outside the 95% confidence interval estimated from randomization analysis (Fig. 3). In addition, a discriminant analysis shows that MDS variances of all pairwise comparisons among the four geochronological bins are significantly different ($p<0.05$), except the early Neoproterozoic vs. Paleoproterozoic comparison ($p=0.10$).

The median of the maximum dimension shows no significant change in the Proterozoic (Fig. 6). However, the range of maximum dimension expanded throughout the Proterozoic. The surface/volume ratio (Fig. 7) appears to have changed little until the Ediacaran, when both the maximum and median surface/volume ratio increased significantly (Wilcoxon test, $p<0.05$).

4. DISCUSSION

4.1 Comparison with Acritarch Morphological History

At the broadest scale, the MDS result appears to be similar to that of Proterozoic acritarchs (Huntley *et al.*, 2006). Morphological disparity of Proterozoic acritarchs increased episodically in the early Mesoproterozoic and in the early Ediacaran, with a long-lasting plateau in between. The acritarch data also show morphospace contraction associated with Cryogenian (750–635 Ma) glaciations and late Ediacaran (575–542 Ma) radiation of Ediacara organisms. These details cannot be tested in the macroalgal data because of the poor geochronological resolution and the absence of macroalgal data in the Cryogenian and latest Ediacaran Period (550–542 Ma).

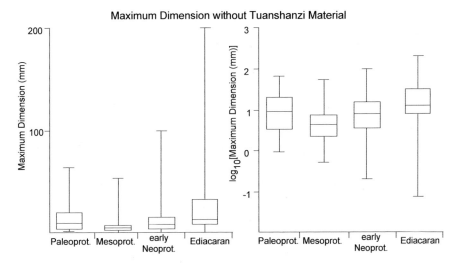

Figure 6. Maximum dimension (in mm) of Proterozoic carbonaceous compressions in linear (left) and $\log_{10}$ scales (right). Box-and-whisker plots show median, lower and upper quartiles, and maximum and minimum values of each geochronological bin.

Given that most acritarchs were probably planktonic photoautotrophs, the first-order match between acritarch and macroalga morphological history is intriguing. The parallel between the morphological histories of Proterozoic acritarchs and macroalgae suggests an external (i.e., environmental or ecological) forcing on the morphological evolution of Proterozoic primary producers—both benthic and planktonic. Huntley *et al.* (2006) hypothesize that the Mesoproterozoic to early Neoproterozoic plateau of acritarch morphospace may be related to nutrient stress and a sluggish carbon cycle in approximately the same geological interval (Brasier and Lindsay, 1998; Anbar and Knoll, 2002). The macroalgal data appear to be consistent with this hypothesis, and would further imply that this environmental forcing affected both the pelagic and benthic realms. To further test this interpretation, more geochronological, chemostratigraphic (Halverson, 2006), paleoenvironmental, and paleontological data are needed to refine the temporal relationship between nutrient stress and algal evolution.

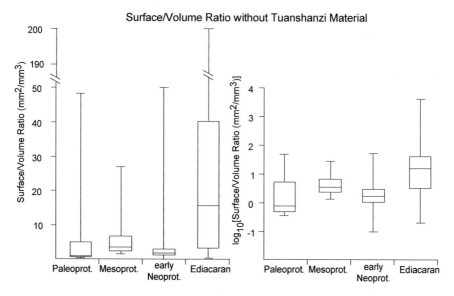

Figure 7. Surface/Volume ratio (in mm^2/mm^3) of Proterozoic carbonaceous compressions in linear (left) and $\log_{10}$ scales (right). Box-and-whisker plots show median, lower and upper quartiles, and maximum and minimum values of each geochronological bin.

Alternatively, this Mesoproterozoic—early Neoproterozoic stasis may be interpreted in ecological terms. It has been recently proposed that the radiation of late Ediacaran large acanthomorphic acritarchs, some of which are interpreted as benthos (Butterfield, 2001), was an ecological response to macrophagous grazing by early eumetazoans which, according to molecular phylogeny and molecular clock data, diverged as benthic animals between 634 and 604 Ma (Peterson and Butterfield, 2005). Using the same argument, was the morphological evolution of macroalgae held back by the absence of animal grazing in the Mesoproterozoic—early Neoproterozoic, and was subsequently accelerated by a major top-down ecological forcing in the Ediacaran when herbivorous metazoans began to evolve? One potential problem with this hypothesis is that the macroalgal morphologies that evolved in the Ediacaran (e.g., dichotomous and monopodial branching, apical meristem, rhizoidal holdfast) do not appear to be effective morphological adaptations to defend against herbivory.

4.2 Surface/Volume Ratio

The surface/volume ratio is an important physiological factor controlling the metabolic rate of modern macroalgae. Mass-specific growth rate, measured as carbon fixed per unit of body mass per unit of time, tends to be greater in macroalgal functional-form groups with higher surface/volume ratio (Littler and Littler, 1980; Littler and Arnold, 1982). This relationship remains true whether the measurements are carried out for phylogenetically related or distant macroalgae (Hanisak et al., 1988; Steneck and Dethier, 1994; Gacia et al., 1996; Stewart and Carpenter, 2003). Clearly, the effect of surface/volume ratio on macroalgal growth rate overrides phylogenetic relatedness and is pervasively convergent. Indeed, comprehensive data compilation shows that log(maximum growth rate) and log(surface/volume ratio) scale linearly over a wide range of surface/volume ratios spanning from unicellular algae, macroalgae, to rooted angiosperms (Fig. 8) (Nielsen and Sand-Jensen, 1990; Nielsen et al., 1996).

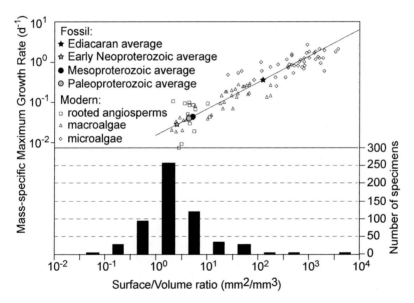

Figure 8. Top: the relationship between surface/volume ratio and maximum growth rate (left vertical scale) of modern photosynthetic eukaryotes [modified from (Nielsen and Sand-Jensen, 1990)]. Mean surface/volume ratios for the four Proterozoic bins are plotted along the regression line, to show the Ediacaran increase in surface/volume ratio. Bottom: surface/volume ratio distribution (right vertical scale) of all Proterozoic macroalgae in our database.

The surface/volume ratios of Proterozoic macroalgae are plotted toward the lower end of modern macroalgae (Fig. 8), but did show a significant increase in the Ediacaran (Fig. 7). This pattern appears to be consistent with the complete absence of some of the extremely fast-growing functional-form groups, such as leafy macroalgae [e.g., *Ulva* or *Porphyra*; (Littler and Arnold, 1982)] in the Proterozoic.

What might have caused the Ediacaran increase in surface/volume ratio? Certainly, the greater surface/volume ratio of Ediacaran macroalgae was introduced by morphological innovations of certain functional-form groups (e.g., delicately branching forms such as *Doushantuophyton*, *Anomalophyton*, and *Glomulus*), which did not appear until the Ediacaran. The question is whether the Ediacaran increase in surface/volume ratio was made possible by a major evolutionary breakthrough that overcame the intrinsic developmental barriers to greater surface/volume ratios, or it was also forced by external selective pressure.

At a fundamental level, the morphogenesis of macroalgae with greater surface/volume ratio (e.g., delicately branching forms and thin leafy forms) requires parenchymatous growth and controlled cell division. The restriction of cell division to a marginal zone of meristematic cells or an apical meristem consisting of one or a few cells appears to be a key innovation in the elaboration of thallus morphology (Graham *et al.*, 2000; Niklas, 2000). Parenchymatous and meristematic growth has been independently achieved in all three macroalgal groups—the chlorophytes, rhodophytes, and phaeophytes, suggesting that it can be achieved with relative ease. The convergent evolution of complex thalli, together with the independent diversification of Ediacaran acritarchs, points to the possible role of external forcing as part of the equation.

Algal growth requires light, nutrient, and CO_2. Modern photosynthesis typically conserves <37% of the energy absorbed as photosynthetically active radiation (Falkowski and Raven, 1996), indicating that macroalgae probably have lived in light saturation even in the Paleoproterozoic when solar luminosity was about 80% of modern level (Kasting *et al.*, 1988). Nutrient availability seems to be an unlikely driver either. Although it has been shown that nutrient uptake by micro- and macroalgae depends on surface/volume ratio (Hein *et al.*, 1995), there is no evidence for greater nutrient availability in Mesoproterozoic oceans than in Ediacaran ones. Quite to the contrary, pelagic oceans of the Mesoproterozoic are thought to have been nutrient-limited because of the low concentration of biologically important elements such as Fe, Mo, and P (Brasier and Lindsay, 1998; Anbar and Knoll, 2002). It is possible that the coastal oceans were decoupled, in terms of nutrient availability, from the pelagic oceans in the Mesoproterozoic—a scenario that would weaken the hypothesis to invoke nutrient stress as a factor holding backing macroalgal morphological disparity in the Mesoproterozoic.

Surface-ocean CO_2, on the other hand, was probably more readily available in the Mesoproterozoic Era than in the Ediacaran Period, given what we know about Proterozoic atmospheric pCO_2 levels (Kaufman and Xiao, 2003). Is it possible that a drop in pCO_2 level in the Cryogenian or Ediacaran Period may have forced macroalgae toward greater surface/volume ratio within their developmental possibilities to compensate for the lower pCO_2 level? There is some evidence of CO_2 limitation in modern macroalgae that do not use carbon concentrating mechanisms to store HCO_3^- as carbon source (Raven, 2003). These algae have to depend on diffusion of CO_2 uptake, and their carboxylation rate is saturated at 25–35 µM [CO_2], while [CO_2] in the surface ocean is only ~10 µM (Hein and Sand-Jensen, 1997). Thus, algal growth in the absence of carbon concentrating mechanisms can be limited by [CO_2] under conditions of light and nutrient saturation. Indeed, controlled experiments show that growth rate of some macroalgae increases moderately with elevated [CO_2] or pCO_2 levels up to 5× present atmospheric level (Gao et al., 1993; Hein and Sand-Jensen, 1997; Kübler et al., 1999). Thus, it appears that both carbon concentrating mechanisms and greater surface/volume ratios could have been physiological and morphological responses to decreasing pCO_2 levels in the Ediacaran (Graham and Wilcox, 2000).

To the extent that macroalgal morphological diversification in the Ediacaran may have been driven by top-down ecological forcing by animal grazers (see 4.1), it is also possible that the Ediacaran increase in surface/volume ratio may have been caused by the same ecological process, because macroalgal surface/volume ratio may be coupled with morphological disparity. However, delicate macroalgal thalli with greater surface/volume ratio and faster growth rate (e.g., *Ulva*) tend to poorly defend against metazoan grazing (Littler and Littler, 1980; Steneck and Dethier, 1994), and thus would not be the predicted outcome of herbivory forcing.

Whatever the cause, greater surface/volume ratios of Ediacaran macroalgae may have had significant consequence on the global carbon cycle. Eukaryotic phytoplankton and macroalgae are important autotrophs in coastal environments where most organic carbon burial occurs in modern oceans. Thus, macroalgal bioproductivity could have considerable impact on the carbon cycle. A uniformitarian interpretation of the Proterozoic surface/volume data suggests that, on average, Ediacaran macroalgae were more than an order of magnitude more productive than those came before (Fig. 8). Did more productive Ediacaran macroalgae (and perhaps microalgae as well?) contribute to a larger dissolved organic carbon reservoir (Rothman *et al.*, 2003), more volatile carbon cycle, and perhaps the eventual rise of oxygen level in the Ediacaran? Here again, our ability to answer these questions is limited by the poor temporal resolution of the Proterozoic geological and paleontological record.

4.3   Maximum Canopy Height

Vertically oriented benthic organisms evolved in the Mesoproterozoic or earlier. If the presence of holdfasts in some of the Tuanshanzi compression fossils is confirmed, macroalgal canopy height was already millimeters to centimeters in the Paleoproterozoic (Yan, 1995; Zhu and Chen, 1995; Yan and Liu, 1997). *Bangiomorpha pubescens* from the Mesoproterozoic Hunting Formation has holdfast structures and was up to 2 mm in height (Butterfield, 2000). *Tawuia*-like fossils from the Mesoproterozoic Suket Shale also appear to bear holdfast structures (Kumar, 2001) and they could reach up to 14 mm in height (note that scales in Fig. 8 and Fig. 11 of Kumar, 2001 were incorrect). *Longfengshania stipitata* from early Neoproterozoic rocks (Hofmann, 1985; Du and Tian, 1986) has well-preserved holdfasts and were centimetric in height. Early Neoproterozoic *Pararenicola huaiyuanensis* and *Protoarenicola baiguashanensis* were interpreted as possible animal fossils (Sun *et al.*, 1986); however, new material (Fig. 2A–C) indicates that these carbonaceous compressions may represent holdfast-bearing, benthic macroalgae with a centimetric canopy height (Qian *et al.*, 2000).

The Ediacaran Period experienced a significant expansion of macroalgal canopy height. Some of the holdfast-bearing forms from the Doushantuo Formation, such as *Baculiphyca taeniata*, were decimetric in height (Xiao *et al.*, 2002). Maximum dimension of Proterozoic carbonaceous compressions, regardless whether they are benthic or planktonic, also shows a sharp increase in the Ediacaran Period (Fig. 6). Given that many specimens in our database are benthic macroalgae (with or without preserved holdfasts), the maximum dimension data can be taken as suggestive evidence that maximum canopy height was greater in the Ediacaran Period than before. The simultaneous increase in both maximum dimension and surface/volume ratio of Ediacaran macroalgae indicates greater morphological complexity, consistent with our morphometric analysis that shows a significant Ediacaran increase in MDS variance (Figs. 3–5).

4.4   Ecological Interactions with Animals

Ecological interactions among living organisms form a complex network (Fig. 9). The nature of ecological interactions includes competition, predation, symbiosis, parasitism, herbivory, and many others. Very little is known about ecological interactions in the Proterozoic ecosystem (Fig. 9). Among the few examples of ecological interactions in the Proterozoic are predation on *Cloudina* animals (Bengtson and Yue, 1992; Hua *et al.*, 2003) and lichen-like algal-fungal symbiosis (Yuan *et al.*, 2005), both are preserved in Ediacaran rocks.

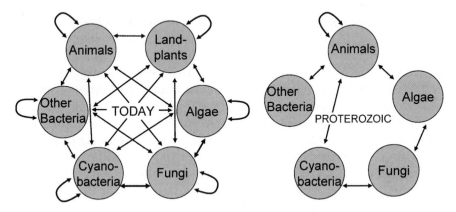

Figure 9. Organismal interactions in modern ecosystems (left) compared with what we know about ecological interactions in the Proterozoic (Bengtson and Yue, 1992; Seilacher, 1999; Yuan *et al.*, 2005). Animal-algal interactions are indirectly inferred based on arguments presented by Peterson and Butterfield (2005), not on direct fossil evidence. Modified from (Taylor *et al.*, 2004).

Is there paleontological evidence for Proterozoic macroalga-animal interactions? Herbivory is an important form of macroalga-animal interaction, but so far we have identified no direct evidence for herbivory in Proterozoic carbonaceous compression fossils. Proterozoic macroalgal fossils, at least those >550 Ma in age, typically do not have wounds, particularly healed wounds. There might be a taphonomic issue here; after all, healed wounds are not usually preserved in the meagre fossil record of Phanerozoic macroalgae either. The complete absence of crustose calcareous algae in the Proterozoic, however, is a true signal. In fact, some phosphatized algae from the Doushantuo Formation are probably phylogenetically related to modern calcareous coralline algae but lacked biocalcification (Xiao *et al.*, 2004). Insofar as biocalcification is an effective protection against herbivory and calcareous coralline algae depend on herbivore denudation to prevent epiphyte colonization (Steneck, 1983), the lack of calcareous algae in the Proterozoic is circumstantial evidence for the absence of herbivory, at least Phanerozoic-style herbivory; incidentally, this inference is also consistent with conclusions derived from phylogenetic arguments that herbivores appeared relatively late among animal groups (Vermeij and Lindberg, 2000). In addition, the dominance of simple discoidal and delicate rhizoidal holdfasts in Proterozoic benthic macroalgae, as well as the concurrent absence of robust holdfasts (e.g., in modern seaweeds such as *Laminaria* and *Caulerpa*), indicates that animal bioturbation in normal marine soft substrates was relatively weak and that the microbially dominated substrates were firmer and less soupy prior to ~550 Ma (Seilacher, 1999; Bottjer *et al.*, 2000; Droser *et al.*, 2002).

The indirect evidence for the insignificance of herbivory and bioturbations can be interpreted in three different ways: animals did not evolve until 550 Ma; they were microscopic (millimetric or smaller) prior to 550 Ma, hence leaving unrecognizable traces; or they were macroscopic but not effective herbivores or burrowers. Regardless, the limited evidence seems to suggest that macroalga-animal interactions were comparatively weak or unrecognizable in the fossil record. This is particularly true for Phanerozoic-style herbivory, but it remains to be seen whether other forms of macroalga-animal interactions, for example parasitism, commensalisms, and herbivory on microalgae, played a significant role in the ecological evolution in the Ediacaran.

5. CONCLUSIONS

This chapter critically reviews the macroalgal affinity of Proterozoic carbonaceous compression fossils, particularly the sinosabelliditids and protoarenicolids, presents results of quantitative analysis of Proterozoic macroalgal morphologies, and discusses their paleoecological implications. The analysis reveals that the morphological history of Proterozoic macroalgae is similar to that of Proterozoic acritarchs. At the broadest scale, macroalgal morphological disparity in the Paleoproterozoic was low but increased in the Mesoproterozoic Era and in the Ediacaran Period, with a prolonged plateau in between. It is hypothesized that the morphological plateau in the Mesoproterozoic and early Neoproterozoic may be related to nutrient stress or/and the lack of ecological forcing by animals.

The Ediacaran increase in macroalgal morphological disparity during 635–550 Ma is coupled with concurrent increase in thallus surface/volume ratio and maximum canopy height of benthic macroalgal communities. A uniformitarian interpretation of this pattern suggests that Ediacaran macroalgal communities, in comparison with earlier ones, were dominated by taller and more complex benthic algae that grew faster. It is hypothesized that the elaboration of macroalgal morphology and increase in surface/volume ratio may have been driven by lower pCO_2 levels and/or by herbivory forcing in the Ediacaran Period. Regardless, benthic macroalgal communities probably have been some of the highly productive areas since the Ediacaran, and they may be an important piece in the puzzle of Ediacaran carbon cycle, carbon isotope excursions, and oxygen evolution. As a final remark, we would like to reiterate the preliminary nature of these conclusions, which should be tested in the future with an expanded database, better resolved geochronology, and further improved phylogenetic interpretations.

ACKNOWLEDGEMENTS

We would like to acknowledge the National Science Foundation and the Petroleum Research Fund for financial support. Meng'e Chen, Hans J. Hofmann, and Xunlai Yuan kindly provided photographs of Proterozoic carbonaceous compression fossils. Mike Kowalewski and John Huntley helped in the quantitative analysis of morphological data. Nick Butterfield, Sören Jensen, and Stan Awramik provided constructive comments on an earlier version of this paper.

REFERENCES

Abbott, I. A., 1999, *Marine Red Algae of the Hawaiian Islands*, Bishop Museum Press, Honolulu, Hawaii.
Anbar, A. D., and Knoll, A. H., 2002, Proterozoic ocean chemistry and evolution: A bioinorganic bridge? *Science* **297**: 1137–1142.
Bengtson, S., and Yue, Z., 1992, Predatorial borings in late Precambrian mineralized exoskeletons, *Science* **257**: 367–369.
Berger, S., and Kaever, M. J., 1992, *Dasycladales: An Illustrated Monograph of a Fascinating Algal Order*, Georg Thieme Verlag, Stuttgart.
Bi, Z., Wang, X., Zhu, H., Wang, Z., and Ding, F., 1988, The Sinian of southern Anhui, *Prof. Papers Strat. Palaeont.* **19**: 27–60.
Bottjer, D. J., Hagadorn, J. W., and Dornbos, S. Q., 2000, The Cambrian substrate revolution, *GSA Today* **10**: 1–7.
Bowring, S., Myrow, P., Landing, E., Ramezani, J., and Grotzinger, J., 2003, Geochronological constraints on terminal Neoproterozoic events and the rise of metazoans, *Geophys. Res. Abs.* **5**: 13219.
Brasier, M. D., and Lindsay, J. F., 1998, A billion years of environmental stability and the emergence of eukaryotes: New data from northern Australia, *Geology* **26**: 555–558.
Bunt, J. S., 1975, Primary productivity of marine ecosystems. in: *Primary Productivity of the Biosphere* (H. Lieth and R. H. Whittaker, eds.), Springer, New York, pp. 169–183.
Butterfield, N. J., 1997, Plankton ecology and the Proterozoic-Phanerozoic transition, *Paleobiology* **23**: 247–262.

Butterfield, N. J., 2000, *Bangiomorpha pubescens* n. gen., n. sp.: Implications for the evolution of sex, multicellularity, and the Mesoproterozoic/Neoproterozoic radiation of eukaryotes, *Paleobiology* **26**: 386–404.

Butterfield, N. J., 2001, Ecology and evolution of Cambrian plankton. in: *The Ecology of the Cambrian Radiation* (A. Y. Zhuravlev and R. Riding, eds.), Columbia University Press, New York, pp. 200–216.

Butterfield, N. J., 2003, Exceptional fossil preservation and the Cambrian Explosion, *Integr. Comp. Biol.* **43**: 166–177.

Butterfield, N. J., 2004, A vaucheriacean alga from the middle Neoproterozoic of Spitsbergen: Implications for the evolution of Proterozoic eukaryotes and the Cambrian explosion, *Paleobiology* **30**: 231–252.

Butterfield, N. J., Knoll, A. H., and Swett, K., 1994, Paleobiology of the Neoproterozoic Svanbergfjellet Formation, Spitsbergen, *Fossils and Strata* **34**: 1–84.

Chen, J., 1988, Precambrian metazoans of the Huai River drainage area (Anhui, E. China): their taphonomic and ecological evidence, *Senkenbergiana Lethaea* **69**: 189–215.

Chen, M., Lu, G., and Xiao, Z., 1994a, Preliminary study on the algal macrofossils—Lantian Flora from the Lantian Formation of Upper Sinian in southern Anhui, *Bull. Inst. Geol., Acad. Sinica* **No. 7**: 252–267.

Chen, M., and Xiao, Z., 1992, Macrofossil biota from upper Doushantuo Formation in eastern Yangtze Gorges, China, *Acta Palaeontol. Sinica* **31**: 513–529.

Chen, M., Xiao, Z., and Yuan, X., 1994b, A new assemblage of megafossils–Miaohe biota from Upper Sinian Doushantuo Formation, Yangtze Gorges, *Acta Palaeontol. Sinica* **33**: 391–403.

Condon, D., Zhu, M., Bowring, S., Wang, W., Yang, A., and Jin, Y., 2005, U–Pb ages from the Neoproterozoic Doushantuo Formation, China, *Science* **308**: 95–98.

Ding, L., Li, Y., Hu, X., Xiao, Y., Su, C., and Huang, J., 1996, *Sinian Miaohe Biota*, Geological Publishing House, Beijing.

Droser, M. L., Gehling, J. G., Rice, D., Mrofka, D. D., and Kennedy, M. J., 2004, Ecology of the Ediacaran explosion, *GSA Annu. Meeting Abs. with Prog.* **36**(5): 521–522.

Droser, M. L., Jensen, S., and Gehling, J. G., 2002, Trace fossils and substrates of the terminal Proterozoic–Cambrian transition: Implications for the record of early bilaterians and sediment mixing, *Proc. Nat. Acad. Sci. USA* **99**: 12572–12576.

Du, R., and Tian, L., 1986, *The Macroalgal Fossils of the Qingbaikou Period in the Yanshan Range*, Hebei Science and Technology Press, Shijiazhuang.

Du, R., Tian, L., and Li, H., 1986, Discovery of megafossils in the Gaoyuzhuang Formation of the Changchengian System, Jixian, *Acta Geol. Sinica* **60**: 115–120.

Duan, C., 1982, Late Precambrian algal megafossils *Chuaria* and *Tawuia* in some areas of eastern China, *Alcheringa* **6**: 57–68.

Falkowski, P. G., and Raven, J. A., 1996, *Aquatic Photosynthesis*, Blackwell Science, Malden, MA.

Ford, T. D., and Breed, W. J., 1973, The problematical Precambrian fossil *Chuaria*, *Palaeontology* **16**: 535–550.

Gacia, E., Littler, M. M., and Littler, D. S., 1996, The relationships between morphology and photosynthetic parameters within the polymorphic genus *Caulerpa*, *J. Exp. Mar. Biol. Ecol.* **204**: 209–224.

Gao, K., Aruga, Y., Asada, K., and Kiyohara, M., 1993, Influence of enhanced CO_2 on growth and photosynthesis of the red algae *Gracilaria* sp. and *G. chilensis*, *J. App. Phycol.* **5**: 563–571.

Gnilovskaya, M. B., 1990, Vendotaenids—Vendian metaphytes. in: *The Vendian System. vol. 1, Paleontology* (B. S. Sokolov and A. B. Iwanowski, eds.), Springer-Verlag, Berlin, pp. 138–147.

Gnilovskaya, M. B., 2003, The Oldest Tissue Differentiation in Precambrian (Vendian) Algae, *Paleontol. J.* **37**: 196–204.

Gnilovskaya, M. B., Veis, A. F., Bekker, A. Y., Olovyanishnikov, V. G., and Raaben, M. E., 2000, Pre–Ediacarian fauna from Timan (Annelidomorphs of the late Riphean), *Stratigr. Geol. Correlation* **8**: 327–352.

Graham, L. E., Cook, M. E., and Busse, J. S., 2000, The origin of plants: Body plan changes contributing to a major evolutionary radiation, *Proc. Nat. Acad. Sci. USA* **97**: 4535–4540.

Graham, L. E., and Wilcox, L. E., 2000, *Algae*, Prentice-Hall, Upper Saddle River, NJ.

Halverson, G., 2006, A Neoproterozoic chronology. in: *Neoproterozoic Geobiology and Paleobiology* (S. Xiao and A. J. Kaufman, eds.), Springer, Dordrecht, the Netherlands, pp. 231–271.

Han, T.-M., and Runnegar, B., 1992, Megascopic eukaryotic algae from the 2.1 billion-year-old Negaunee Iron-Formation, Michigan, *Science* **257**: 232–235.

Hanisak, M. D., Littler, M. M., and Littler, D. S., 1988, Significance of macroalgal polymorphism: Intraspecific tests of the functional-form model, *Mar. Biol.* **99**: 157–165.

Hein, M., Pedersen, M. F., and Sand-Jensen, K., 1995, Size-dependent nitrogen uptake in micro- and macroalgae, *Mar. Ecol. Prog. Ser.* **118**: 247–253.

Hein, M., and Sand-Jensen, K., 1997, CO_2 increases oceanic primary production, *Nature* **388**: 526–527.

Hermann, T. N., 1990, *Organic World Billion Year Ago*, Nauka, Leningrad.

Hoffmann, K.-H., Condon, D. J., Bowring, S. A., and Crowley, J. L., 2004, U–Pb zircon date from the Neoproterozoic Ghaub Formation, Namibia: Constraints on Marinoan glaciation, *Geology* **32**: 817–820.

Hofmann, H., and Chen, J., 1981, Carbonaceous megafossils from the Precambrian (1800 Ma) near Jixian, northern China, *Can. J. Earth Sci.* **18**: 443–447.

Hofmann, H. J., 1985, The mid-Proterozoic Little Dal macrobiota, Mackenzie Mountains, north-west Canada, *Palaeontology* **28**: 331–354.

Hofmann, H. J., 1994, Proterozoic carbonaceous compressions ("metaphytes" and "worms"). in: *Early Life on Earth* (S. Bengtson, ed.), Columbia University Press, New York, pp. 342–357.

Hofmann, H. J., and Aitken, J. D., 1979, Precambrian biota from the Little Dal Group, Mackenzie Mountains, northwestern Canada, *Can. J. Earth Sci.* **16**: 150–166.

Hofmann, H. J., and Rainbird, R. H., 1994, Carbonaceous megafossils from the Neoproterozoic Shaler Supergroup of Arctic Canada, *Palaeontology* **37**: 721–731.

Hua, H., Pratt, B. R., and Zhang, L., 2003, Borings in Cloudina Shells: Complex Predator-Prey Dynamics in the Terminal Neoproterozoic, *Palaios* **18**: 454–459.

Huntley, J. W., Xiao, S., and Kowalewski, M., 2006, On the morphological history of Proterozoic and Cambrian acritarchs. in: *Neoproterozoic Geobiology and Paleobiology* (S. Xiao and A. J. Kaufman, eds.), Springer, Dordrecht, the Netherlands, pp. 23–56.

Ivantsov, A. Y., 1990, New data on the ultrastructure of sabelliditids (Pogonophora?), *Paleontologicheskii Zhurnal* **24**: 125–128.
Jensen, S., 2003, The Proterozoic and earliest Cambrian trace fossil record: patterns, problems and perspectives, *Integr. Comp. Biol.* **43**: 219–228.
Johnson, J. H., 1961, *Limestone-building Algae and Algal Limestones*, Colorado School of Mines, Golden.
Kasting, J. F., Toon, O. B., and Pollack, J. B., 1988, How climate evolved on the terrestrial planets, *Sci. Am.* **258**(2): 90–97.
Kaufman, A. J., and Xiao, S., 2003, High CO_2 levels in the Proterozoic atmosphere estimated from analyses of individual microfossils, *Nature* **425**: 279–282.
Knoll, A. H., 1985, Exceptional preservation of photosynthetic organisms in silicified carbonates and silicified peats, *Philosophical Transactions of the Royal Society of London, Series B. Biological Sciences* **311**: 111–122.
Knoll, A. H., 1992, The early evolution of eukaryotes: A geological perspective, *Science* **256**: 622–627.
Kübler, J. E., Johnston, A. M., and Raven, J. A., 1999, The effects of reduced and elevated CO_2 and O_2 on the seaweed *Lomentaria articulata*, *Plant Cell Environ.* **22**: 1303–1310.
Kumar, S., 1995, Megafossils from the Mesoproterozoic Rohtas Formation (the Vindhyan Supergroup), Katni area, central India, *Precambrian Res.* **72**: 171–184.
Kumar, S., 2001, Mesoproterozoic megafossil *Chuaria–Tawuia* association may represent parts of a multicellular plant, Vindhyan Supergroup, Central India, *Precambrian Res.* **106**: 187–211.
Lamb, D. M., Awramik, S. M., and Zhu, S., 2005, Paleoproterozoic eukaryotes from the Changcheng Group, North China, *GSA Annu. Meeting Abs. with Prog.* **37**(7): 305.
Littler, M. M., and Arnold, K. E., 1982, Primary productivity of marine macroalgal functional-form groups from southwestern North America, *J. Phycol.* **18**: 307–311.
Littler, M. M., and Littler, D. S., 1980, The evolution of thallus form and survival strategies in benthic marine macroalgae: Field and laboratory tests of a functional form model, *Am. Naturalist* **116**: 25–44.
Liu, Z., and Du, R., 1991, Morphology and systematics of *Longfengshania*, *Acta Palaeontol. Sinica* **30**: 106–114.
Lu, S., and Li, H., 1991, A precise U–Pb single zircon age determination for the volcanics of the Dahongyu Formation, Changcheng System in Jixian, *Bull. Chinese Acad. Geol. Sci.* **22**: 137–146.
Maithy, P. K., and Babu, R., 1996, Carbonaceous macrofossils and organic-walled microfossils from the Halkal Formation, Bhima Group, Karnataka, with remarks on age, *The Palaeobotanist* **45**: 1–6.
Moczydlowska, M., 2003, Earliest Cambrian putative bacterial nanofossils, *Mem. Ass. Australasian Palaeontol.* **29**: 1–11.
Narbonne, G. M., 2005, The Ediacara Biota: Neoproterozoic origin of animals and their ecosystems, *Annu. Rev. Earth and Planet. Sci.* **33**: 421–442.
Narbonne, G. M., and Hofmann, H. J., 1987, Ediacaran biota of the Wernecke Mountains, Yukon, Canada, *Palaeontology* **30**: 647–676.
Nielsen, S. L., Enriquez, S., Duarte, C. M., and Sand-Jensen, K., 1996, Scaling maximum growth rates across photosynthetic organisms, *Funct. Ecol.* **10**: 167–175.
Nielsen, S. L., and Sand-Jensen, K., 1990, Allometric scaling of maximal photosynthetic growth rate to surface/volume ratio, *Limnol. Oceanogr.* **35**: 177–181.

Niklas, K. J., 2000, The evolution of plant body plans: A biomechanical prespective, *Ann. Bot.* **85**: 411–438.

Niklas, K. J., 2004, Computer models of early land plant evolution, *Annu. Rev. Earth and Planet. Sci.* **32**: 47–66.

Padilla, D. K., and Allen, B. J., 2000, Paradigm lost: Reconsidering functional form and group hypotheses in marine ecology, *J. Exp. Mar. Biol. Ecol.* **250**: 207–221.

Peterson, K. J., and Butterfield, N. J., 2005, Origin of the Eumetazoa: Testing ecological predictions of molecular clocks against the Proterozoic fossil record, *Proc. Nat. Acad. Sci. USA* **102**: 9547–9552.

Porter, S. M., 2004, The fossil record of early eukaryotic diversification. in: *The Paleontological Society Papers 10: Neoproterozoic–Cambrian Biological Revolutions* (J. H. Lipps and B. Waggoner, eds.), Paleontological Society, New Haven, pp. 35–50.

Qian, M., Yuan, X., Wang, Y., and Yan, Y., 2000, New material of metaphytes from the Neoproterozoic Jinshanzhai Formation in Huaibei, North Anhui, China, *Acta Palaeontol. Sinica* **39**: 516–520.

Rasmussen, B., Bose, P. K., Sarkar, S., Banerjee, S., Fletcher, I. R., and McNaughton, N. J., 2002, 1.6 Ga U–Pb zircon age for the Chorhat Sandstone, lower Vindhyan, India: Possible implications for early evolution of animals, *Geology* **30**: 103–106.

Raven, J. A., 2003, Inorganic carbon concentrating mechanisms in relation to the biology of algae, *Photosynthesis Res.* **77**: 155–171.

Ray, J. S., Martin, M. W., Veizer, J., and Bowring, S. A., 2002, U–Pb zircon dating and Sr isotope systematics of the Vindhyan Supergroup, India, *Geology* **30**: 131–134.

Ray, J. S., Veizer, J., and Davis, W. J., 2003, C, O, Sr and Pb isotope systematics of carbonate sequences of the Vindhyan Supergroup, India: age, diagenesis, correlations and implications for global events, *Precambrian Res.* **121**: 103–140.

Rothman, D. H., Hayes, J. M., and Summons, R., 2003, Dynamics of the Neoproterozoic carbon cycle, *Proc. Nat. Acad. Sci. USA* **100**: 8124–8129.

Sarangi, S., Gopalan, K., and Kumarb, S., 2004, Pb–Pb age of earliest megascopic, eukaryotic alga bearing Rohtas Formation, Vindhyan Supergroup, India: Implications for Precambrian atmospheric oxygen evolution, *Precambrian Res.* **132**: 107–121.

Schneider, D. A., Bickford, M. E., Cannon, W. F., Schulz, K. J., and Hamilton, M. A., 2002, Age of volcanic rocks and syndepositional iron formations, Marquette Range Supergroup: Implications for the tectonic setting of Paleoproterozoic iron formations of the Lake Superior region, *Can. J. Earth Sci.* **39**: 999–1012.

Schopf, J. W., 1968, Microflora of the Bitter Springs Formation, Late Precambrian, central Australia, *J. Paleontol.* **42**: 651–688.

Seilacher, A., 1999, Biomat-related lifestyles in the Precambrian, *Palaios* **14**: 86–93.

Sokolov, B. S., 1967, Drevneyshiye pognofory [The oldest Pogonophora], *Dok. Akad. Nauk SSSR* **177**(1): 201–204 (English translation, page 252–255).

Sokolov, B. S., 1976, The Earth's organic world on the path toward Phanerozoic differentiation, *Vestn. Akademiya Nauk SSSR* **1**: 126–143.

Steiner, M., 1994, Die neoproterozoischen Megaalgen Südchinas, *Berliner geowiss. Abh.* **15**: 1–146.

Steiner, M., 1997, *Chuaria circularis* Walcott 1899—"megasphaeromorph acritarch" or prokaryotic colony? *Acta Univ. Carolinae Geol.* **40**: 645–665.

Steneck, R. S., 1983, Escalating herbivory and resulting adaptive trends in calcareous algal crusts, *Paleobiology* **9**: 44–61.

Steneck, R. S., and Dethier, M. M., 1994, A functional-group approach to the structure of algal-dominated communities, *Oikos* **69**: 476–498.

Stewart, H. L., and Carpenter, R. C., 2003, The effects of morphology and water flow on photosynthesis of marine macroalgae, *Ecology* **84**: 2999–3012.

Sun, W., 1987, Palaeontology and biostratigraphy of Late Precambrian macroscopic colonial algae: *Chuaria* Walcott and *Tawuia* Hofmann, *Palaeontographica Abt. B* **203**: 109–134.

Sun, W., Wang, G., and Zhou, B., 1986, Macroscopic worm-like body fossils from the Upper Precambrian (900–700Ma), Huainan district, Anhui, China and their stratigraphic and evolutionary significance, *Precambrian Res.* **31**: 377–403.

Tang, F., Yin, C., and Gao, L., 1997, A new idea of metaphyte fossils from the late Sinian Doushantuo stage at Xiuning, Anhui Province, *Acta Geol. Sinica* **71**: 289–296.

Tang, F., Yin, C., Liu, Y., Wang, Z., and Gao, L., 2005, Discovery of macroscopic carbonaceous compression fossils from the Doushantuo Formation in eastern Yangtze Gorges, *Chinese Sci. Bull.* **50**: 2632–2637.

Taylor, T. N., Klavins, S. D., Krings, M., Taylor, E. L., Kerp, H., and Hass, H., 2004, Fungi from the Rhynie chert: a view from the dark side, *Trans. Roy. Soc. Edinburgh: Earth Sci.* **94**: 457–473.

Teyssèdre, B., 2003, *Chuaria, Tawuia, Longfengshania*: Trois classes de fossiles précambriens pour un même taxon [*Chuaria, Tawuia, Longfengshania*: Three classes of Precambrian fossils for the same taxon], *Comptes Rendus Palevol* **2**: 503–508.

Urbanek, A., and Mierzejewska, G., 1977, The fine structure of zooidal tubes in Sabelliditida and Pogonophora with reference to their affinity, *Acta Palaeontol. Polonica* **22**: 223–240.

Vermeij, G. J., and Lindberg, D. R., 2000, Delayed herbivory and the assembly of marine benthic ecosystems, *Paleobiology* **26**: 419–430.

Vidal, G., 1989, Are late Proterozoic carbonaceous megafossils metaphytic algae or bacteria?, *Lethaia* **22**: 375–379.

Vidal, G., and Ford, T. D., 1985, Microbiotas from the late Proterozoic Chuar Group (northern Arizona) and Uinta Mountain Group (Utah) and their chronostratigraphic implications, *Precambrian Res.* **28**: 349–389.

Walcott, C. D., 1919, Cambrian Geology and Paleontology IV: Middle Cambrian algae, *Smithsonian Misc. Coll.* **67**: 217–260.

Walter, M. R., Du, R., and Horodyski, R. J., 1990, Coiled carbonaceous megafossils from the middle Proterozoic of Jixian (Tianjin) and Montana, *Am. J. Sci.* **290A**: 133–148.

Walter, M. R., Oehler, J. H., and Oehler, D. Z., 1976, Megascopic algae 1300 million years old from the Belt Supergroup, Montana: A reinterpretation of Walcott's *Helminthoidichnites*, *J. Paleontol.* **50**: 872–881.

Wan, Y. S., Zhang, Q. D., and Song, T. R., 2003, SHRIMP ages of detrital zircons from the Changcheng System in the Ming Tombs area, Beijing: Constraints on the protolith nature and maximum depositional age of the Mesoproterozoic cover of the North China Craton, *Chinese Sci. Bull.* **48**: 2500–2506.

Wang, G., 1982, Late Precambrian Annelida and Pogonophora from the Huainan of Anhui Province, *Bull. Tianjin Inst. Geol. & Mineral Resources, Chinese Acad. Geol. Sci.* **No. 6**: 9–22.

Wang, G., and Zhang, S., 1984, *Research on the Upper Precambrian of Northern Jiangsu and Anhui Provinces*, Anhui Press of Science and Technology, Hefei, Anhui.

Wray, J. L., 1977, *Calcareous Algae*, Elsevier, Amsterdam.

Xiao, S., 2004, New multicellular algal fossils and acritarchs in Doushantuo chert nodules (Neoproterozoic, Yangtze Gorges, South China), *J. Paleontol.* **78**: 393–401.

Xiao, S., Knoll, A. H., and Yuan, X., 1998a, Morphological reconstruction of *Miaohephyton bifurcatum*, a possible brown alga from the Doushantuo Formation (Neoproterozoic), South China, and its implications for stramenopile evolution, *J. Paleontol.* **72**: 1072–1086.

Xiao, S., Knoll, A. H., Yuan, X., and Pueschel, C. M., 2004, Phosphatized multicellular algae in the Neoproterozoic Doushantuo Formation, China, and the early evolution of florideophyte red algae, *Am. J. Bot.* **91**: 214–227.

Xiao, S., Yuan, X., Steiner, M., and Knoll, A. H., 2002, Macroscopic carbonaceous compressions in a terminal Proterozoic shale: A systematic reassessment of the Miaohe biota, South China, *J. Paleontol.* **76**: 345–374.

Xiao, S., Zhang, Y., and Knoll, A. H., 1998b, Three-dimensional preservation of algae and animal embryos in a Neoproterozoic phosphorite, *Nature* **391**: 553–558.

Xing, Y., Duan, C., Liang, Y., and Cao, R., 1985, *Late Precambrian Palaeontology of China, People's Republic of China Ministry of Geology and Mineral Resources, Geological Memoirs, Series 2, Number 2*, Geological Publishing House, Beijing.

Yan, Y., 1995, Discovery and preliminary study of megascopic algae (1700 Ma) from the Tuanshanzi Formation in Jixian, China, *Acta Micropalaeontol. Sinica* **12**: 107–126.

Yan, Y., Jiang, C., Zhang, S., Du, S., and Bi, Z., 1992, Research of the Sinian System in the region of western Zhejiang, northern Jiangxi, and southern Anhui provinces, *Bull. Nanjing Inst. Geol. Mineral Res., Chinese Acad. Geol. Sci.* **Supplementary Issue 12**: 1–105.

Yan, Y., and Liu, Z., 1997, Tuanshanzian macroscopic algae of 1700 Ma b. p. from Changcheng System of Jixian, China, *Acta Palaeontol. Sinica* **36**: 18–41.

Yuan, X., Li, J., and Cao, R., 1999, A diverse metaphyte assemblage from the Neoproterozoic black shales of South China, *Lethaia* **32**: 143–155.

Yuan, X., Xiao, S., Li, J., Yin, L., and Cao, R., 2001, Pyritized chuarids with excystment structures from the late Neoproterozoic Lantian Formation in Anhui, South China, *Precambrian Res.* **107**: 251–261.

Yuan, X., Xiao, S., and Taylor, T. N., 2005, Lichen-like symbiosis 600 million years ago, *Science* **308**: 1017–1020.

Yuan, X., Xiao, S., Yin, L., Knoll, A. H., Zhou, C., and Mu, X., 2002, *Doushantuo Fossils: Life on the Eve of Animal Radiation*, China University of Science and Technology Press, Hefei, China.

Zhang, R., Feng, S., Ma, G., Xu, G., and Yan, D., 1991, Late Precambrian macroscopic fossil algae from Hainan Island, *Acta Palaeontol. Sinica* **30**: 115–125.

Zhang, R., Yao, H., and Yan, D., 1995, Study on systematic position of *Tawuia* in Shilu Group in Hainan Island and depositional environment, *Palaeoworld* **6**: 1–14.

Zhang, Z., 1988, *Longfengshania* Du emend.: An earliest record of Bryophyte-like fossils, *Acta Palaeontol. Sinica* **27**: 416–426.

Zhao, Y. L., Chen, M., Peng, J., Yu, M. Y., He, M. H., Wang, Y., Yang, R. J., Wang, P. L., and Zhang, Z. H., 2004, Discovery of a Miaohe-type Biota from the Neoproterozoic Doushantuo formation in Jiangkou County, Guizhou Province, China, *Chinese Sci. Bull.* **49**: 2224–2226.

Zheng, W., 1980, A new occurence of fossil group *Chuaria* from the Sinian System in north Anhui and its geological meaning, *Bull. Tianjin Inst. Geol. & Mineral Resources, Chinese Acad. Geol. Sci.* **1**: 49–69.

Zheng, W., Mu, Y., Zheng, X., Wang, J., and Xin, L., 1994, Discovery of carbonaceous megafossils from Upper Precambrian Shijia Formation, north Anhui and its biostratigraphic significance, *Acta Palaeontol. Sinica* **33**: 455–471.

Zhu, S., and Chen, H., 1995, Megascopic multicellular organisms from the 1700-million-year-old Tuanshanzi Formation in the Jixian area, North China, *Science* **270**: 620–622.

Zhu, S., Sun, S., Huang, X., He, Y., Zhu, G., Sun, L., and Zhang, K., 2000, Discovery of carbonaceous compressions and their multicellular tissues from the Changzhougou Formation (1800 Ma) in the Yanshan Range, North China, *Chinese Sci. Bull.* **45**: 841–846.

Zhu, W., and Chen, M., 1984, On the discovery of macrofossil algae from the late Sinian in the eastern Yangtze Gorges, south China, *Acta Botanica Sinica* **26**: 558–560.

Chapter 4

Evolutionary Paleoecology of Ediacaran Benthic Marine Animals

DAVID J. BOTTJER and MATTHEW E. CLAPHAM

Department of Earth Sciences, University of Southern California, Los Angeles, CA 90089-0740, USA.

1. Introduction	91
2. A Mat-Based World	92
3. Nature of the Data	95
3.1 Geology and Paleoenvironments	95
3.2 Lagerstätten	96
3.3 Biomarkers	97
3.4 Molecular Clock Analyses	97
4. Evolutionary Paleoecology	98
4.1 Doushantuo Fauna (?600–570 Mya)	99
4.2 Ediacara Avalon Assemblage (575–560 Mya)	101
4.3 Ediacara White Sea Assemblage (560–550 Mya)	102
4.4 Ediacara Nama Assemblage (549–542 Mya)	105
5. Discussion	108
Acknowledgements	110
References	110

1. INTRODUCTION

Paleobiological study of the Phanerozoic and the Precambrian largely developed as separate cultures during the 20th century. This was primarily because the presence of common body fossils with mineralized skeletons

S. Xiao and A.J. Kaufman (eds.), Neoproterozoic Geobiology and Paleobiology, 91–114.
© 2006 Springer.

typical of the Phanerozoic allowed certain types of science, such as biostratigraphy and benthic paleoecology, to be done that was not feasible in the Precambrian. However, as fossils have increasingly been documented from the late Neoproterozoic, approaches more typical of the Phanerozoic have become possible, leading recently to the definition of the Ediacaran Period as a formal name for the terminal Neoproterozoic interval (Knoll et al., 2004). This time period is bounded below by the Marinoan Snowball Earth glaciation, and above by the Cambrian, and is characterized by diverse suites of largely soft-bodied fossils, exemplified by the Doushantuo and Ediacara biotas.

This push back from the Phanerozoic has also allowed for the first time studies on the evolutionary paleoecology of Ediacaran benthic assemblages (e.g., Clapham and Narbonne, 2002; Droser et al., 2006). In particular this is because we are beginning to understand the biological affinities of these fossils better (e.g., Fedonkin and Waggoner, 1997; Seilacher, 1999; Narbonne, 2004), although differences of opinion remain (Xiao et al., 2000; Peterson et al., 2003; Seilacher et al., 2003; Bengtson and Budd, 2004; Chen et al., 2004; Grazhdankin, 2004). This promises to be a growing field of research in the future, as we begin to apply some of the approaches developed in the Phanerozoic (e.g., Brenchley and Harper, 1998; Allmon and Bottjer, 2001) to the record of animal life in the Ediacaran. This contribution, which focuses on ecological aspects of Ediacaran benthic marine animal life, synthesizes the early stages of a research program that will eventually integrate the two worlds of Precambrian and Phanerozoic evolutionary paleoecology.

2. A MAT-BASED WORLD

Much Phanerozoic benthic paleoecology has focused on the effects of bioturbation and the nature of substrates. One of the biggest differences between Phanerozoic and Proterozoic benthic ecosystems is that while vertical bioturbation and all its effects upon seafloor characteristics are nearly ubiquitous in the Phanerozoic, only limited horizontal bioturbation was present in the Ediacaran (Seilacher, 1999; Bottjer et al., 2000). Thus, a distinguishing feature of the Ediacaran is that it was a time where seafloors were commonly dominated by microbial mats (Gehling, 1999; Hagadorn and Bottjer, 1999; Wood et al., 2003; Droser et al., 2005; Gehling et al., 2005; Dornbos et al., 2006; Droser et al., 2006). The most typical expression of this transition is the widespread occurrence of stromatolites in carbonates during the Proterozoic and their decline at the end of the Precambrian, likely due to the emergence of animals (Awramik, 1971). A recent reanalysis of

Awramik's database confirms that the greatest decline in stromatolite form diversity is indeed in the Ediacaran, culminating in the Early Cambrian (Olcott *et al.*, 2002). Although other factors were also involved (Grotzinger and Knoll, 1999), since this was a time of appearance and radiation of animal groups, it implies that the stromatolite form diversity decline was strongly affected by increasing development of passive and/or active interactions between evolving animals and microbial sedimentary structures. Some part of this was likely caused by the increasing disruption due to more intense bioturbation.

Recent research has shown that there is also substantial evidence for the common presence of seafloor microbial mats in siliciclastic settings during the Neoproterozoic (Hagadorn and Bottjer, 1997; Gehling, 1999; Hagadorn and Bottjer, 1999). Despite their widespread occurrence in Proterozoic carbonates, mat structures in siliciclastics are less dramatic than the stromatolites found in carbonates, so they have received less attention. Their subdued appearance is because siliciclastic settings are non-mineral-precipitating seafloor environments, so vertical dimensions of such structures are commonly on the millimeter scale. Microbially-mediated structures in siliciclastic settings have been given intriguing names, reflecting the lack of understanding, until recently, of their origin. Thus, we have Ediacaran wrinkle structures (Fig. 1A–C) and "elephant skin" (Fig. 1D) from a variety of depositional environments. Much of what we know about the distribution of mats in Ediacaran strata is from the presence of trace fossils, such as the scratch-like trace *Radulichnus* that can be found associated with *Kimberella* (Fig. 1D), a possible stem-group mollusc that is found in Ediacaran-aged deposits from both Russia and Australia. These widespread structures are interpreted as the feeding traces produced by *Kimberella* as it scraped the mat-covered seafloor. Other diffuse impressions show movement of *Yorgia* and *Dickinsonia*, and possibly also their feeding activities, on mat-covered sediment (Fedonkin, 2003; Gehling *et al.*, 2005).

Interpretations of the taphonomy of the Ediacara biota by Gehling (1999) also demonstrate the typical occurrence of microbial mats on siliciclastic Ediacaran seafloors. For example, Ediacaran bedding surfaces at Mistaken Point, Newfoundland, are commonly covered by a red coating of limonitic "rust," implying an enrichment by organic matter and hence the presence of a mat. Such features are also found in Lower Cambrian marine facies which contain bedding planes representing seafloors once extensively covered by microbial mats (Dornbos *et al.*, 2004). Some relatively unweathered Mistaken Point outcrops of bedding planes show the presence of pyrite, indicating that the red "rust" is oxidized from pyrite, which likely formed as a decomposition product of the mat organic matter (Gehling *et al.*, 2005).

The presence of crinkly carbonaceous laminae within siltstone intervals at Mistaken Point confirms the widespread distribution of microbial mats (Wood et al., 2003). In addition, the abundance of discoidal fossils is likely an indicator that the disc was a holdfast structure embedded in the seafloor below a mat-bound surface (Seilacher, 1999; Wood et al., 2003). Strata in which the Doushantuo biota of southwest China are found have not been studied extensively for the presence of microbial structures, but preliminary investigations indicate the presence of microbial mats on Doushantuo seafloors (Dornbos et al., 2006). Despite the increasing presence of animals through the Ediacaran, the common presence of microbial mats strongly influenced benthic organism ecology, and hence their morphology as well as taphonomy (Gehling, 1999; Seilacher, 1999).

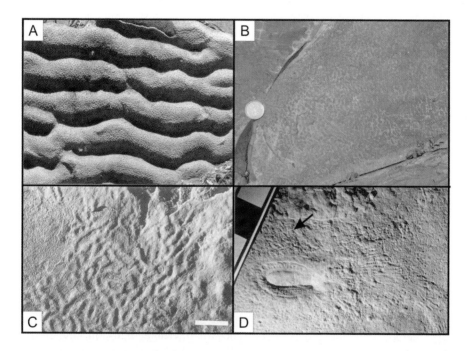

Figure 1. Evidence for mat-dominated Neoproterozoic benthic ecosystems. (A) Wrinkle structures preserved on the surface of wave ripples, Rawnsley Quartzite, Ediacara Hills, South Australia. From Selden and Nudds (2004). (B) Wrinkle structures in deep-water turbiditic siltstone, Drook Formation, Mistaken Point, Newfoundland. Coin is 19 mm diameter. Photo courtesy of M. L. Fraiser. (C) Wrinkle structures from the late Ediacaran Wyman Formation, Silver Peak Range, Nevada. Scale bar 1 cm. From Hagadorn and Bottjer (1999). (D) *Kimberella* fossil and associated *Radulichnus* grazing trace on microbial "elephant skin" (arrow) bedding surface, White Sea, Russia. Scale bar divisions are 1 cm. From Fedonkin (2003).

3. NATURE OF THE DATA

A broad variety of data from both earth and biological sciences is incorporated in the analysis of Ediacaran paleobiology and evolutionary paleoecology. The habitats of Ediacaran animals can be reconstructed through geological analysis of depositional environments. The fossils themselves are almost exclusively found in Lagerstätten where soft tissues are preserved as molds and casts, although the presence of animals in ancient environments can sometimes be inferred through studies of preserved organism-specific organic molecules, or biomarkers. Molecular biology and the molecular clock can be used to estimate when the ancestors of extant higher taxa first appeared in geological time.

3.1 Geology and Paleoenvironments

The Ediacaran Period encompasses the time from the Marinoan Snowball Earth glaciation to the base of the Cambrian (Knoll *et al.*, 2004). Thus, this interval represents a transitional time from typical Proterozoic conditions, typified by low oxygen levels in the deep ocean, abundant microbialites, and an absence of metazoans, to the fully oxygenated Phanerozoic oceans containing widespread metazoans and restricted microbialites. The most intense Snowball Earth glaciations (Hoffman and Schrag, 2002), which occurred primarily between 720 Ma and 635 Ma and immediately preceded the Ediacaran Period, were the most significant paleoenvironmental events of the Neoproterozoic and may have inhibited metazoan evolution during their duration (Runnegar, 2000). The final Neoproterozoic glacial episode, during the Ediacaran at 580 Ma, predates the appearance of large metazoan fossils by less than 5 million years (Bowring *et al.*, 2003; Narbonne and Gehling, 2003).

In addition to the biotic restrictions from the extreme climatic fluctuations, oxygen levels would have exerted a fundamental control on the evolution of metazoans, especially megascopic animals, in the Neoproterozoic. There is evidence that the deep oceans were largely anoxic during the Mesoproterozoic (Arnold *et al.*, 2004) and that the levels of dissolved oxygen increased significantly only during the Neoproterozoic (Shields *et al.*, 1997; Canfield, 1998). The presence of large Ediacara fossils in deep slope settings confirms that oxygen levels had increased by the late Neoproterozoic, at least in basins where substantial contour currents were present (Dalrymple and Narbonne, 1996; Wood *et al.*, 2003).

Neoproterozoic metazoans lived in a wide range of sedimentary environments, from deep slope and basinal settings to shallow shelf and deltaic environments. Most Ediacaran fossils, such as those found in the classic localities of Ediacara (Australia), the White Sea in Russia, and Namibia, occur in shallow-water depositional environments, including storm-influenced shelf settings, prodeltaic environments, and distributary mouth bars (Saylor *et al.*, 1995; Grazhdankin and Ivantsov, 1996; Gehling, 2000; Grazhdankin, 2004). Phosphorite facies of the Doushantuo Formation (Guizhou Province, China), which contains animal eggs and embryos, probable sponges and stem-group cnidarians, and the oldest known bilaterian fossils, were also deposited in shallow, nearshore marine environments (Dornbos *et al.*, 2006). In contrast, diverse and abundant Ediacarans lived on a deep-water slope, well below the photic zone, in the Avalonian localities of Mistaken Point and Charnwood Forest, England (Wood *et al.*, 2003). Other deep-water Ediacaran fossil localities include northwest Canada, where the fossils are preserved on the base of siliciclastic turbidites deposited on the continental slope (Dalrymple and Narbonne, 1996), and the Olenek Uplift in Siberia, where the fossils occur in slope and basinal carbonaceous limestones of the Khatyspyt Formation (Knoll *et al.*, 1995).

3.2 Lagerstätten

Because almost all of the fossils in the Ediacaran do not have a mineralized skeleton, paleobiologists rely upon the presence of Lagerstätten (Bottjer *et al.*, 2002) to understand faunas from this time. Thus, it is only from Lagerstätten that we can view the Doushantuo biota and its microscopic world and the Ediacara biota and its macroscopic world. Taphonomy, although not the focus of this contribution (but see Bottjer *et al.*, 2002), is paramount. There are different types of preservational processes at work in forming each of the known Ediacaran Lagerstätten (Bottjer, 2005; Narbonne, 2005). The Doushantuo biota of animals is preserved as phosphatized microfossils (Xiao *et al.*, 2000; Dornbos *et al.*, 2005). In contrast, fossils of the Ediacara biota are preserved largely as flattened two-dimensional impressions, but recently an increasing number of specimens with three-dimensional preservation have been found (e.g., Grazhdankin and Seilacher, 2002; Narbonne, 2004; Xiao *et al.*, 2005). The taphonomic pathways leading to preservation of Ediacaran fossils are varied, but the common thread is the requirement of early lithification to preserve the delicate structures or impressions. In most cases this is accomplished by the presence of microbial mats, which for the Doushantuo biota likely sealed the surfaces of storm deposits allowing high concentrations of phosphate to build up rapidly leading to very early phosphatization (Bottjer, 2005; Dornbos *et al.*, 2006). For the Ediacara biota, microbial mats produced a firm substrate that contributed to rapid lithification during early diagenesis, acting as a "death mask" recording the fossil moulds (Gehling, 1999). This preservational style ("Flinders-style" of Narbonne, 2005) is typical of localities in the White Sea and at Ediacara itself, where the fossils are preserved as impressions on the base of storm beds (Gehling, 1999). Avalonian localities display "Conception-style" preservation where the role of the microbial mat in early diagenesis has been replaced by a volcanic ash layer; the lithification of ash layers records the fossils as impressions on the upper surface of bedding planes (Narbonne, 2005). The final important preservational style for the Ediacara biota is "Nama-style" preservation, most typical of localities in Namibia, where the fossils are preserved as three-dimensional casts, seemingly without the aid of microbial mats to enhance early lithification (Narbonne, 2005).

3.3 Biomarkers

Molecular fossils from hydrocarbons have begun to be used to indicate the presence of prokaryotes and eukaryotes in ancient environments. This area of study is particularly useful when searching for the presence of organisms with no mineralized body parts, as is usually characteristic of the Precambrian. Early results from biomarker studies reveal the likely presence of eukaryotes long before the Ediacaran (Brocks *et al.*, 1999) and a diverse plankton community during the Neoproterozoic (Olcott *et al.*, 2005), although the only biomarker evidence yet available for metazoans can only demonstrate the presence of at least sponges throughout the Ediacaran (McCaffrey *et al.*, 1994).

3.4 Molecular Clock Analyses

Most molecular clock studies indicate the presence of sponges, cnidarians and bilaterians during the Ediacaran from the time of the Doushantuo biota (possibly as old as 600 mya) to the beginning of the Cambrian (542 mya) (e.g., Peterson *et al.*, 2004; Peterson and Butterfield, 2005). As with the case of biomarkers, these studies are useful to consider because they serve as good indicators of what extant clades may be present but are not preserved. However, the geological range of other Ediacaran organisms, for example rangeomorphs (Narbonne, 2004) or Dengying "quilts" (Xiao *et al.*, 2005), cannot be determined using this approach because these organisms likely represent extinct clades.

4. EVOLUTIONARY PALEOECOLOGY

The nature of Ediacara biota preservation, recording soft-bodied fossils instantaneously buried underneath or within events beds (tempestites or ashfalls), is ideal for paleoecological studies as the fossil assemblages are often preserved as census assemblages of *in situ* organisms. The soft-bodied nature of the Ediacara biota precludes time-averaging and there is no sedimentary or taphonomic evidence for transport in most localities. Namibian localities (and Nama-style preservation in general), where fossils are entombed within storm beds, likely suffer from transportation and possibly sorting by size or resistance to fragmentation; however, it is still possible to reconstruct, in a qualitative sense, the ecological structure of those communities. In contrast, localities from Avalonia, Australia, and the White Sea are ideal for detailed, quantitative reconstruction of the ecological structure of Ediacara biota communities, including the tiering structure and within-community diversity and relative abundance distributions.

Sedimentological studies of the Doushantuo Formation indicate a relatively high degree of transport and sorting of Doushantuo biota microfossils, within shallow, nearshore marine environments (Dornbos *et al.*, 2006). In addition to transportation, the small size of the fossils and thus the necessity for sampling by thin section analysis preclude ecological counts, so that paleoecological analysis of the Doushantuo biota is constrained to a more limited level of understanding than is possible for the Ediacara biota.

We now have learned enough about the Ediacara biota that we can define a variety of time-restricted assemblages (Narbonne, 2005). The simple temporal succession of the different Ediacara biota assemblages is, however, somewhat confounded by the geographic, environmental, and to some extent taphonomic, differences among these localities (Waggoner, 1999; Grazhdankin, 2004). It is likely that the ultimate composition of a given Ediacaran locality is dependent on all of these factors to varying degrees, and apparent temporal patterns in the ecological structure of Ediacaran communities must be interpreted with caution as they may instead reflect primarily environmental control or taphonomic overprint.

We can also reconstruct the living position of most Ediacaran organisms in relation to the substrate with some confidence, specifically whether they stood upright in the water column, rested directly on the substrate, or were motile as in the case of the bilaterians. What is less clear are their trophic relationships, largely because the specific taxonomic affinities of most Ediacaran organisms remain unknown (Narbonne, 2004, 2005). The conservative assumption, employed here, is that bilaterians were mobile (e.g., *Vernanimalcula*, Chen *et al.*, 2004), with some being grazers (e.g., *Kimberella*, Fedonkin and Waggoner, 1997), while others (e.g., *Dickinsonia* and *Yorgia*; Fedonkin, 2003) may have absorbed nutrients from the underlying decomposing microbial mat substrate (Gehling et al., 2005). Many others fed from the water column, either as true suspension feeders, microcarnivores, or by absorbing nutrients directly (but, see Grazhdankin and Seilacher, 2002, who interpret forms such as *Pteridinium* as infaunal).

Tiering is the distribution of benthic organisms above and below the substrate, and is a paleoecological structure commonly determined for Phanerozoic assemblages (e.g., Watkins, 1991; Taylor and Brett, 1996; Yuan et al., 2002). The Phanerozoic summary diagrams on tiering (Bottjer and Ausich, 1986; Ausich and Bottjer, 2001) have included only suspension-feeders, although other tiering studies have included additional trophic groups. Some Ediacaran assemblages are most likely composed solely of non-bilaterian suspension-feeders, but others include bilaterians. The following presentation of proposed Ediacaran tiering relationships will include all animals of each assemblage represented by body and trace fossils, based on quantitative counts of individual communities where possible.

4.1 Doushantuo Fauna (?600–570 Mya)

The Doushantuo phosphorites of southwestern China, constrained between 635 and 551 Ma (Condon et al., 2005) but probably 600–570 Ma in age (Barfod et al., 2002; Condon et al., 2005), are famous for phosphatized eggs and embryos (Fig. 2A) (Xiao et al., 1998; Xiao and Knoll, 2000). Other reported animal fossils include tiny adult sponges (Fig. 2C) (Li et al., 1998), stem cnidarians (Fig. 2D) (Xiao et al., 2000; Chen et al., 2002) and a tiny bilaterian (Fig. 2B) (Chen et al., 2004; Bottjer, 2005). These fossil occurrences are consistent with molecular clock analysis (Peterson et al., 2004; Peterson and Butterfield, 2005) that predicts the presence of cnidarians, sponges, and stem-group bilaterians. This assemblage of animal fossils was likely primarily microscopic—no macroscopic fauna has been found for this time in other rocks, although there may be an age overlap with the earliest Ediacara biota. Adult animals in this assemblage were at most several millimeters tall. Thus, there would not have been any tiering on the macroscopic scale as currently defined. However, the number of forms indicates that they utilized a variety of food sources, and likely lived in an ecosystem of some complexity. These tiny adult sponges and stem cnidarians may have lived partially inserted into the seafloor as mat stickers (Seilacher, 1999). In sufficient densities, these meadows of tiny sponges and stem cnidarians would have caused the microbial mat to assume a "fuzzy" appearance (Chen et al., 2002).

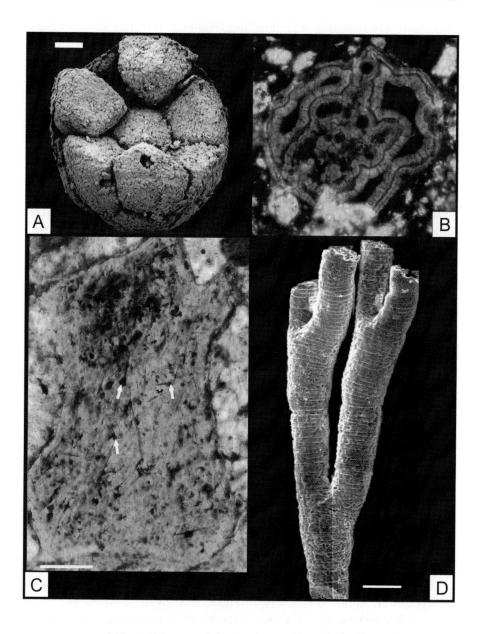

Figure 2. Representative members of the Doushantuo fauna. (A) Phosphatized embryo. Scale 100 μm. From Xiao and Knoll (2000). (B) *Vernanimalcula*, a putative adult bilaterian. Field of view approximately 120 μm. From Chen *et al.* (2004). (C) Microscopic sponge, arrows denote spicules. Scale bar 100 μm. From Li *et al.* (1998). (D) *Sinocyclocyclicus*, a putative adult stem-group cnidarian. Scale bar 140 μm. From Xiao *et al.* (2000).

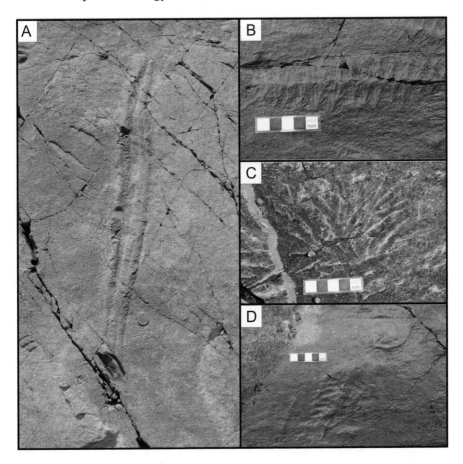

Figure 3. Representative members of the Ediacara Avalon Assemblage. (A) *Charnia wardi* frond from the Drook Formation, Pigeon Cove. Coin diameter 24 mm. (B) "Spindle" rangeomorph form from the Mistaken Point Formation, Mistaken Point. (C) *Bradgatia* rangeomorph from the Mistaken Point Formation, Green Head, Spaniard's Bay. (D) *Charniodiscus spinosus* from the Mistaken Point Formation, Mistaken Point.

4.2   Ediacara Avalon Assemblage (575–560 Mya)

The Avalon assemblage contains the oldest-known Ediacara fossils (indeed the oldest-known megascopic metazoans), is the only diverse deep-water locality, and is the only region where Ediacaran fossils are preserved beneath volcanic ash layers. The assemblage is numerically dominated by enigmatic fractally-organized organisms such as *Charnia*, *Bradgatia*, and "spindles" (Fig. 3), grouped into a biological clade called "Rangeomorpha" (Narbonne, 2004). The average numerical abundance of rangeomorphs within communities at Mistaken Point is 75%, with many communities containing greater than 90% rangeomorphs (Clapham *et al.*, 2003). For example, rangeomorphs comprise 99.7% of the 1488 specimens found on the "D surface" at Mistaken Point (Clapham *et al.*, 2003). The only important non-rangeomorph taxa in the Avalonian assemblage are the frond *Charniodiscus* (Fig. 3D) (Laflamme *et al.*, 2004), the pustular, but possibly non-metazoan, discoidal fossil *Ivesheadia* (Boynton and Ford, 1995), and the rare conical *Thectardis* (Clapham *et al.*, 2004). The preserved fossil communities contain between 3–12 forms (approximately equivalent to genera) (Clapham *et al.*, 2003), although recent taxonomic studies indicate that the classic "E Surface" may contain as many as 15–18 genera (M. Laflamme, pers comm., 2005). Macroscopic bilaterian body fossils and trace fossils are absent from these oldest Ediacara assemblages, either because large bilaterians had not yet evolved or were restricted to shallow settings.

Because of the great morphological diversity displayed by rangeomorphs, ranging from recumbent sheets, to bush-like forms, to tall frondose shapes (Narbonne, 2004), Avalonian assemblages have a well developed tiering structure (Fig. 6A) (Clapham and Narbonne, 2002). As in highly tiered Phanerozoic assemblages, greater than 90% of the individuals occupy the lowest tier at less than 8 cm above the seafloor. The characteristic taxon of this tier at Mistaken Point is the "spindle" rangeomorph form (Fig. 3B), in conjunction with small specimens of the "pectinate" or "comb" rangeomorph form and *Bradgatia* (Fig. 3C), and very small fronds (e.g., *Charniodiscus*, *Charnia*, "duster" rangeomorphs). The intermediate tier (8–22 cm) is numerically dominated by the bush-like rangeomorph *Bradgatia*, the "pectinate" form, and small frondose specimens, whereas the upper tier (22–35 cm) exclusively contains frondose forms such as *Charnia* and *Charniodiscus* (Fig. 6A). Rare taxa, such as *Charnia wardi* (Fig. 3A) and the "Xmas tree" fossil (Fig. 6A), demonstrate that the maximum height attained by these earliest Ediacara fossils (1 m or even greater) was similar to that of the tallest Phanerozoic marine invertebrates.

4.3   Ediacara White Sea Assemblage (560–550 Mya)

The White Sea Assemblage, represented by fossils from the White Sea in Russia and from the Flinders Ranges in South Australia, includes many of the archetypal Ediacara fossils (e.g., *Dickinsonia;* Fig. 4A). Radiometric dating of ash layers from Russia constrains the age of this assemblage to ca. 560–550 Ma (Martin *et al.*, 2000), and sedimentological investigation has shown that the fossils lived in a variety of shallow marine environments, from relatively distal lower shoreface to onshore distributary mouth bars (Grazhdankin and Ivantsov, 1996; Gehling, 2000; Grazhdankin, 2004).

Some fossil localities display Nama-style preservation, but the dominant type of preservation is Flinders-style (Narbonne, 2005), preserved beneath storm deposits. The White Sea Assemblage represents the peak of Ediacara diversity, including some frondose forms known from the earlier Avalon Assemblage (*Charniodiscus, Charnia*), abundant bilaterian fossils [*Dickinsonia, Kimberella* (Fig. 4C), *Yorgia*, and other putative bilaterians such as *Spriggina* (Fig. 4B) and "vendomiids"], and enigmatic forms such as *Parvancorina* and *Tribrachidium* (Fig. 4D). In addition to bilaterian body fossils, trace fossils appear for the first time in the White Sea Assemblage. In contrast to the Avalon Assemblage, rangeomorphs are significantly less abundant in the White Sea Assemblage, represented only by rare specimens of *Charnia* and *Rangea* (Grazhdankin, 2004).

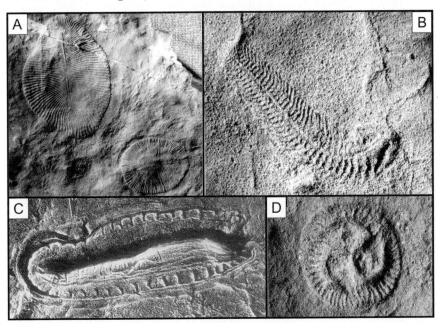

Figure 4. Representative members of the Ediacara White Sea Assemblage. (A) *Dickinsonia* from Ediacara, South Australia. Length of largest specimen is 13 cm. Photo courtesy of B. Runnegar, from Bottjer (2002). (B) *Spriggina* from Ediacara, South Australia. Length of specimen is 4 cm. Photo courtesy of B. Runnegar, from Bottjer (2002). (C) *Kimberella* from the White Sea, Russia. Length of specimen is 8 cm. From Fedonkin and Waggoner (1997). (D) *Tribrachidium* from Ediacara, South Australia. Specimen is about 20 mm across. From Selden and Nudds (2004).

Quantitative counts of bedding-plane assemblages reveal the presence of several different community types in the White Sea Assemblage from Russia (Grazhdankin and Ivantsov, 1996). Two small bedding planes are strongly dominated by discoidal holdfasts (*Aspidella*), comprising 92.5–100% of the specimens. Other taxa present include *Dickinsonia, Parvancorina*, and

Kimberella. Another assemblage contains only *Dickinsonia* (76%) and *Tribrachidium* (24%). The final assemblage studied by Grazhdankin and Ivantsov (1996) is a 6.6 m^2 bedding plane containing a diverse community dominated by *Kimberella* (35.4%) and *Tribrachidium* (14.6%). *Dickinsonia*, *Parvancorina*, and *Aspidella* are also present but each comprise less than 6% of the assemblage. Presence-absence data collected by Grazhdankin (2004) indicates that the community diversity of White Sea fossil assemblages ranges from 1–9 forms, comparable to the diversity of Mistaken Point communities. Quantitative community paleoecology of the Australian White Sea assemblages also documents a wide range of paleocommunity types with taxonomic richness of 2–11 forms (Droser *et al.*, 2006). Some bedding planes were extremely dominated by *Aspidella* frond holdfasts (99.3% of bed MM b1) whereas others contained abundant *Dickinsonia* (84.6% of bed CG db), *Arkarua* (76.2% of bed CH 29.14) or *Palaeophragmodictya* (90.2% of bed BathT3) (Droser *et al.*, 2006).

The tiering structure of the White Sea assemblage appears to be less developed than that of Mistaken Point, although the taphonomic bias against frondose fossils obscures the exact tiering relationship (Droser *et al.*, 2006). The abundant discoidal fossils present in certain White Sea communities likely represent holdfasts of frondose organisms (such as *Charniodiscus*) but, although there is a positive relationship between frond height and disc size, it is not possible to estimate the frond height from the size of the basal disc alone. Nevertheless, specimens of *Charniodiscus arboreus* from Australia have a frond size of approximately 30 cm (Glaessner and Wade, 1966), indicating that the upper tiers of Ediacaran communities were present in some assemblages. A qualitatitve tiering diorama, compiled from quantitative data in Grazhdankin and Ivantsov (1996) and Droser *et al.* (2006) and qualitative data in Grazhdankin (2004), is presented in Fig. 6B. Two broad tiers are present: a lower tier of benthic grazers and attached presumed suspension-feeders, and an upper tier of frondose taxa. The lower tier includes probable suspension feeders such as *Tribrachidium* and other taxa such as *Parvancorina* but is dominated by putative bilaterians such as *Kimberella* and *Dickinsonia*. Fronds present in the upper tier include *Charnia* and *Charniodiscus*. The relative abundance of organisms in these two tiers varies among communities, but the overall presence-absence data presented by Grazhdankin (2004) implies that upper tier frondose organisms are generally rare in relation to the lower tier bilaterians.

Quantitative data presented by Grazhdankin and Ivantsov (1996) and Droser *et al.* (2006) document an apparent exclusionary relationship between frondose taxa and bilaterians, with some communities strongly to exclusively dominated by discoidal holdfasts and others strongly to exclusively dominated by bilaterians and other non-frondose taxa. Some

Australian communities tend to have superdominant frondose forms (99.3% of bed MM b1) or putative grazers such as *Dickinsonia* (84.6% of bed CG db). However, *Charniodiscus* and *Dickinsonia* do co-occur in abundance in bed BuG A (comprising 52.5% and 26.2%, respectively). In Russia, communities that contain both frond holdfasts and bilaterians tend to be dominated by smaller diameter holdfast discs, implying smaller fronds (Grazhdankin and Ivantsov, 1996).

4.4 Ediacara Nama Assemblage (549–542 Mya)

The youngest Ediacara fossils are known from Namibia, where they range through the last 7 million years of the Ediacaran to just below the base of the Cambrian. The Nama Assemblage occurs primarily in very shallow marine delta plain to distributary mouth bar environments (Saylor *et al.*, 1995; Grazhdankin and Seilacher, 2002; Grazhdankin, 2004) and is preserved, often transported, as three-dimensional casts within event beds (typical "Nama-style" preservation; Narbonne, 2005). Compared to the earlier Avalon and White Sea Assemblages, the Nama Assemblage in its type area in Namibia has low diversity, represented primarily by 4 genera, *Pteridinium* (Fig. 5A), *Rangea, Swartpuntia* (Fig. 5B), and *Ernietta*. Examples of Nama-type fossil assemblages found in Russia, preserved in distributary mouth-bar sediments in the middle Verkhova Formation and lower Yorga Formation (Grazhdankin, 2004), are slightly more diverse, including the probable anemone-like *Nemiana* in addition to unusual forms such as *Ausia* and *Ventogyrus*. Although trace fossils are present, bilaterian body fossils are absent. The major evolutionary innovation found in the Nama Assemblage is the earliest calcified fossils, including *Cloudina* and *Namacalathus* (Fig. 5C).

The transported nature of Nama Assemblage fossil occurrences hinders ecological studies; however, in Namibia most fossil localities contain monospecific or nearly monospecific occurrences, for example, of *Pteridinium* (Grazhdankin and Seilacher, 2002) or *Swartpuntia* (Narbonne *et al.*, 1997). In contrast, Russian localities contain 4–6 taxa in typical Nama Assemblage occurrences (Grazhdankin, 2004). The tiering diorama presented in Fig. 6C is based on the qualitative data presented by Grazhdankin (2004) and shows a prominent upper tier of fronds such as *Rangea* and *Swartpuntia* with a lower tier possibly containing *Nemiana* (if it represents an anemone-like organism, not a holdfast like *Aspidella*). Based on presence-absence data, frondose fossils dominate the Nama Assemblage, implying a similar two-level tiering structure to the White Sea but with a dense upper tier of fronds and only a few species present near the sediment-water interface.

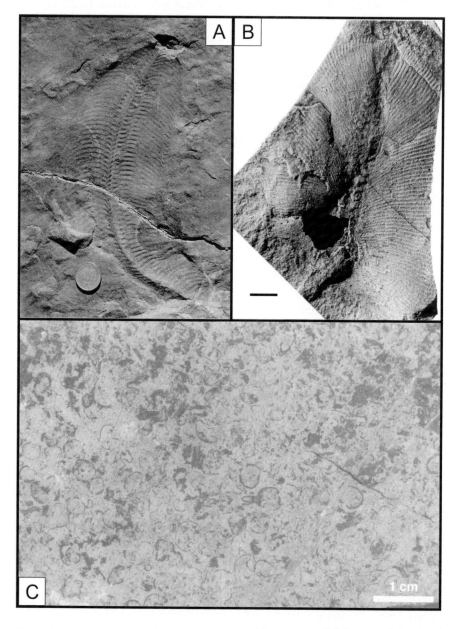

Figure 5. Representative members of the Ediacara Nama Assemblage. (A) *Pteridinium* from Namibia. Coin is 22.5 mm diameter. (B) *Swartpuntia* frond from Namibia. Scale bar 2 cm. From Narbonne *et al*. (1997). (C) *Namacalathus*, one of the earliest known skeletal fossils. From Grotzinger *et al*. (2000).

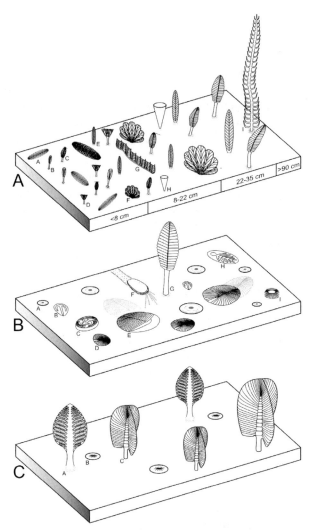

Figure 6. Tiering dioramas for the Avalon, White Sea, and Nama Assemblages. The dioramas are based on quantitative data from Clapham and Narbonne (2002) for the Avalon assemblage and presence-absence data from Grazhdankin (2004) for the White Sea and Nama assemblages but are composites of a number of individual communities and do not intend to show the proportion of organisms in each tier. (A) Avalon Assemblage. Fossils depicted: a. "Spindle" rangeomorph, b. "ostrich feather" rangeomorph, c. *Charniodiscus*, d. "feather duster" rangeomorph, e. *Charnia* rangeomorph, f. *Bradgatia* rangeomorph, g. "pectinate" rangeomorph, h. *Thectardis*, i. "xmas tree." After Clapham and Narbonne (2002). (B) White Sea Assemblage. Fossils depicted: a. Discoidal fossil (probable holdfast), b. *Parvancorina*, c. *Tribrachidium*, d. *Dickinsonia*, e. *Yorgia* with grazing impression, f. *Kimberella* with *Radulichnus* trace, g. *Charniodiscus*, h. Vendomiid (e.g., *Vendia*), i. *Eoporpita*. (C) Nama Assemblage. Fossils depicted: a. *Rangea*, b. *Nemiana*, c. *Swartpuntia*.

5. DISCUSSION

To understand benthic evolutionary paleoecology of Ediacaran animals we are totally dependent on the presence of Lagerstätten (e.g., Bottjer, 2002). More Lagerstätten clearly are needed and should be searched for. In particular, additional Lagerstätten with microscopic preservation for deposits younger than 570 Ma would be very valuable in order to evaluate whether the microscopic fauna of the Doushantuo biota co-existed with the Ediacara biota.

One of the biggest early breakthroughs for understanding the ecology of Ediacaran benthic animals has been the development of evidence showing that these organisms inhabited unique ecosystems that were based on the presence of microbial mats. Thus, many mat-based lifestyles existed that are not found in the Phanerozoic (Seilacher, 1999), and were expressed as morphologies that seem strange when compared with modern benthic invertebrates. The conservative interpretation is that the Ediacaran organisms discussed in this paper were all eukaryotes and most were likely metazoans. Trophically, that would mean a predominance of suspension-feeders and grazers. As the "garden of Ediacara," the prevailing interpretation is that, although some of the "suspension feeders" may be more accurately termed "micro-predators" (Lipps and Valentine, 2004), true macro-predators did not evolve until late in the Ediacaran or even in the Early Cambrian (McMenamin, 1986; Waggoner, 1998).

There is an enormous jump in size from the microscopic adult sponges and stem group cnidarians of the Doushantuo assemblage to fossils 1 meter or more in height such as *Charnia wardi* in the Avalon assemblage. This may be a taphonomic phenomenon, and larger fossils in the Doushantuo certainly need to be searched for. Recent discovery of large Ediacara biota in the Dengying Formation (Xiao *et al.*, 2005), which overlies the Doushantuo Formation phosphorites, indicates the possibility that if animal macrofossils exist in the Doushantuo Formation they might be found in the carbonate facies of this unit. If this size increase from Doushantuo to Avalon biotas is a primary signal, it is likely because of a rise in oceanic dissolved oxygen levels that affected and allowed the evolution of non-bilaterians to reach macroscopic size (Knoll, 2003).

White Sea assemblages contain the first evidence for large bilaterians at approximately 560 Ma, recording greater morphological disparity and an increase in ecosystem complexity (Droser *et al.*, 2006), and again this may be linked to a further increase in dissolved oceanic oxygen. White Sea assemblages also potentially show a decrease in both height of tiering and the abundance of epifaunal organisms inhabiting higher tiers compared to the earlier Avalonian assemblage and later Nama assemblage. Although this may be, in part, a taphonomic pattern, it is possible there is an underlying biotic cause. It is striking that both the Avalonian and Nama assemblages display abundant tall fronds (*Charniodiscus* and rangeomorphs in Avalon, *Pteridinium* and *Swartpuntia* in Nama) but lack mobile bilaterian grazers. In addition, the community data collected by Grazhdankin and Ivantsov (1996) and Droser *et al.* (2006) show that bilaterian-dominated assemblages usually contain no or few fronds whereas those assemblages with abundant discs (presumably reflecting the holdfasts of fronds) contain no or few bilaterians. Finally, Grazhdankin and Ivantsov (1996) also note that the average diameter of discoidal fossils (approximately proportional to the average height of the fronds) in those assemblages where they co-occur with bilaterians is smaller than the diameter in disc-only assemblages, implying a strong link between the activity of bilaterians and the tiering structure of frondose fossils. It is possible that this apparent exclusion of large frondose taxa resulted from the grazing actions of mobile bilaterians that may have disrupted the mat enough to prevent secure attachment of the fronds' holdfast discs. If true, then this drop in tiering foreshadowed the effects of the agronomic and Cambrian substrate revolutions that ushered in the beginning of the Phanerozoic.

The tiering structure displayed by many Neoproterozoic communities, with low-level bilaterian taxa living among tall non-bilaterian taxa, is similar to the pattern in Early Cambrian communities but quite distinct from later Phanerozoic assemblages where upper tiers contained abundant bilaterian groups (e.g., crinoids). For example, the upper tiers in Early Cambrian assemblages from both China and the Burgess Shale in British Columbia were dominated by sponges, whereas nearly all suspension-feeding bilaterians (echinoderms, molluscs, brachiopods) were less than 10 cm tall (Yuan *et al.*, 2002). The well-developed tiering structure among non-bilaterian taxa in the Neoproterozoic and the apparent rarity of suspension-feeding bilaterians in Ediacaran communities is consistent with Yuan *et al.*'s (2002) argument that the later restriction of suspension-feeding bilaterians to lower tiers in the Early Cambrian likely resulted from the strong incumbency of their non-bilaterian competitors. In addition, the abundance of non-bilaterian clades in upper tiers during the Neoproterozoic and Early Cambrian strongly suggests that they were well adapted for attachment to firm substrates but did not fare as well as bilaterian counterparts in the aftermath of the Cambrian substrate revolution (e.g., Yuan *et al.*, 2002).

The field of Ediacaran evolutionary paleoecology is still in its infancy, but the few studies conducted to date have yielded intriguing insights into the early evolution of animal communities. Analysis of the tiering structure has shown that Ediacaran communities were quite similar to Phanerozoic counterparts and may have implications for the early interplay between bioturbation, substrate, and adaptive strategies of benthic metazoans. The preservation of many Ediacaran-aged assemblages as *in situ* census populations is ideal for evolutionary paleoecology research, suggesting that additional study will further our knowledge of the early evolution of metazoans and their interactions with each other and their environment.

ACKNOWLEDGEMENTS

We thank W. I. Ausich, J.-Y. Chen, F. A. Corsetti, E. H. Davidson, S. Q. Dornbos, M. L. Droser, A. G. Fischer, J. G. Gehling, J. W. Hagadorn, M. Laflamme, C.-W. Li, G. M. Narbonne, B. Runnegar, A. Seilacher, B. M. Waggoner, S. Xiao, and M. Zhu for stimulating discussions through the course of this research. Reviews by M. Droser, G. Narbonne, S. Xiao, and an anonymous reviewer greatly improved this chapter.

REFERENCES

Allmon, W. D., and Bottjer, D. J., 2001, *Evolutionary Paleoecology*, Columbia University Press, New York.

Arnold, G. L., Anbar, A. D., Barling, J., and Lyons, T. W., 2004, Molybdenum isotope evidence for widespread anoxia in mid-Proterozoic oceans, *Science* **304**: 87–90.

Ausich, W. I., and Bottjer, D. J., 2001, Sessile Invertebrates, in: *Palaeobiology II* (D. E. G. Briggs and P. R. Crowther, eds.), Blackwell Scientific, Oxford, pp. 384–386.

Awramik, S. M., 1971, Precambrian columnar stromatolite diversity: reflection of metazoan appearance, *Science* **174**: 825–827.

Barfod, G. H., Albarède, F., Knoll, A. H., Xiao, S., Télouk, P., Frei, R., and Baker, J., 2002, New Lu–Hf and Pb–Pb age constraints on the earliest animal fossils, *Earth Planet. Sci. Lett.* **201**: 203–212.

Bengtson, S., and Budd, G. E., 2004, Comment on "Small Bilaterian Fossils from 40 to 55 Million Years Before the Cambrian", *Science* **306**: 1291.

Bottjer, D. J., 2002, Enigmatic Ediacara fossils: Ancestors or aliens? in: *Exceptional Fossil Preservation: A Unique View on the Evolution of Marine Life* (D. J. Bottjer, W. Etter, J. W. Hagadorn, and C. Tang, eds), Columbia University Press, New York, pp. 11–33.

Bottjer, D. J., 2005, The early evolution of animals, *Sci. Am.* **293**: 42–47.

Bottjer, D. J., and Ausich, W. I., 1986, Phanerozoic development of tiering in soft substrata suspension-feeding communities, *Paleobiology* **12**: 400–420.

Bottjer, D. J., Etter, W., Hagadorn, J. W., and Tang, C. M., 2002, *Exceptional Fossil Preservation: A Unique View on the Evolution of Marine Life*, Columbia University Press, New York.

Bottjer, D. J., Hagadorn, J. W., and Dornbos, S. Q., 2000, The Cambrian substrate revolution, *GSA Today* **10(9)**: 1–7.

Bowring, S. A., Myrow, P. M., Landing, E., Ramezani, J., and Grotzinger, J. P., 2003, Geochronological constraints on terminal Neoproterozoic events and the rise of metazoans, *Geophys. Res. Abstr.* **5**: 13219.

Boynton, H., and Ford, T. D., 1995, Ediacaran fossils from the Precambrian (Charnian Supergroup) of Charnwood Forest, Leicestershire, England, *Mercian Geol.* **13**: 165–182.

Brenchley, P. J., and Harper, D. A. T., 1998, *Palaeoecology: Ecosystems, Environments and Evolution*, Chapman & Hall, London.

Brocks, J. J., Logan, G. A., Buick, R., and Summons, R. E., 1999, Archean molecular fossils and the early rise of Eukaryotes, *Science* **285**: 1033–1036.

Canfield, D. E., 1998, A new model for Proterozoic ocean chemistry, *Nature* **396**: 450–453.

Chen, J.-Y., Bottjer, D. J., Oliveri, P., Dornbos, S. Q., Gao, F., Ruffins, S., Chi, H., Li, C.-W., and Davidson, E. H., 2004, Small bilaterian fossils from 40 to 55 million years before the Cambrian, *Science* **305**: 218–222.

Chen, J.-Y., Oliveri, P., Gao, F., Dornbos, S. Q., Li, C.-W., Bottjer, D. J., and Davidson, E. H., 2002, Precambrian animal life: Probable developmental and adult cnidarian forms from southwest China, *Developmental Biol.* **248**: 182–196.

Clapham, M. E., and Narbonne, G. M., 2002, Ediacaran epifaunal tiering, *Geology* **30**: 627–630.

Clapham, M. E., Narbonne, G. M., and Gehling, J. G., 2003, Paleoecology of the oldest-known animal communities: Ediacaran assemblages at Mistaken Point, Newfoundland, *Paleobiology* **29**: 527–544.

Clapham, M. E., Narbonne, G. M., Gehling, J. G., Greentree, C., and Anderson, M. M., 2004, *Thectardis avalonensis*: a new Ediacaran fossil from the Mistaken Point biota, Newfoundland, *J. Paleontol.* **78**: 1031–1036.

Condon, D. J., Zhu, M.-Y., Bowring, S. A., Wang, W., Yang, A., and Jin, Y., 2005, U–Pb ages from the Neoproterozoic Doushantuo Formation, China, *Science* **308**: 95–98.

Dalrymple, R. W., and Narbonne, G. M., 1996, Continental slope sedimentation in the Sheepbed Formation (Neoproterozoic, Windermere Supergroup), Mackenzie Mountains, N.W.T., *Can. J. Earth Sci.* **33**: 848–862.

Dornbos, S. Q., Bottjer, D. J., and Chen, J.-Y., 2004, Evidence for seafloor microbial mats and associated metazoan lifestyles in Lower Cambrian phosphorites of southwest China, *Lethaia* **37**: 127–137.

Dornbos, S. Q., Bottjer, D. J., Chen, J.-Y., Gao, F., Oliveri, P., and Li, C.-W., 2006, Environmental controls on the taphonomy of phosphatized animals and animal embryos from the Neoproterozoic Doushantuo Formation, southwest China, *Palaios* **21**: 3–14.

Dornbos, S. Q., Bottjer, D. J., Chen, J.-Y., Oliveri, P., Gao, F., and Li, C.-W., 2005, Precambrian animal life: taphonomy of phosphatized metazoan embryos from southwest China, *Lethaia* **38**: 101–109.

Droser, M. L., Gehling, J. G., and Jensen, S., 2005, Ediacaran trace fossils: true or false? in: *Evolving Form and Function: Fossils and Development* (D. E. G. Briggs, ed.), Peabody Museum of Natural History, New Haven, CT, pp. 125–138.

Droser, M. L., Gehling, J. G., and Jensen, S. R., 2006, Assemblage palaeoecology of the Ediacara biota: the unabridged edition? *Palaeogeogr. Palaeoclimatol. Palaeoecol.* **232**: 131–147.

Fedonkin, M. A., 2003, The origin of the Metazoa in the light of the Proterozoic fossil record, *Paleontol. Res.* **7**: 9–41.

Fedonkin, M. A., and Waggoner, B. M., 1997, The late Precambrian fossil *Kimberella* is a mollusc-like bilaterian organism, *Nature* **388**: 868–871.

Gehling, J. G., 1999, Microbial mats in terminal Proterozoic siliciclastics: Ediacaran death masks, *Palaios* **14**: 40–57.

Gehling, J. G., 2000, Environmental interpretation and a sequence stratigraphic framework for the terminal Proterozoic Ediacara Member within the Rawnsley Quartzite, South Australia, *Precambrian Res.* **100**: 65–95.

Gehling, J. G., Droser, M. L., Jensen, S. R., and Runnegar, B. N., 2005, Ediacara organisms: relating form to function, in: *Evolving Form and Function: Fossils and Development* (D. E. G. Briggs, ed.), Peabody Museum of Natural History, New Haven, CT, pp. 43–66.

Glaessner, M. F., and Wade, M., 1966, The late Precambrian fossils from Ediacara, South Australia, *Palaeontology* **9**: 599–628.

Grazhdankin, D. V., 2004, Patterns of distribution in the Ediacaran biotas: facies versus biogeography and evolution, *Paleobiology* **30**: 203–221.

Grazhdankin, D. V., and Ivantsov, A. Y., 1996, Reconstructions of biotopes of ancient Metazoan of the Late Vendian White Sea biota, *Palaeontol. J.* **30**: 674–678.

Grazhdankin, D. V., and Seilacher, A., 2002, Underground Vendobionta from Namibia, *Palaeontology* **45**: 57–78.

Grotzinger, J. P., and Knoll, A. H., 1999, Stromatolites in Precambrian carbonates: Evolutionary mileposts or environmental dipsticks, *Annu. Rev. Earth Planet. Sci.* **27**: 313–358.

Grotzinger, J. P., Watters, W. A., and Knoll, A. H., 2000, Calcified metazoans in thrombolite-stromatolite reefs of the terminal Proterozoic Nama Group, Namibia, *Paleobiology* **26**: 334–359.

Hagadorn, J. W., and Bottjer, D. J., 1997, Wrinkle structures: Microbially mediated sedimentary structures common in subtidal siliciclastic settings at the Proterozoic–Phanerozoic transition, *Geology* **25**: 1047–1050.

Hagadorn, J. W., and Bottjer, D. J., 1999, Restriction of a late Neoproterozoic biotope: Suspect-microbial structures and trace fossils at the Vendian–Cambrian transition, *Palaios* **14**: 73–85.

Hoffman, P. F., and Schrag, D. P., 2002, The snowball Earth hypothesis: Testing the limits of global change, *Terra Nova* **14**: 129–155.

Knoll, A. H., 2003, The geological consequences of evolution, *Geobiology* **1**: 3–14.

Knoll, A. H., Grotzinger, J. P., Kaufman, A. J., and Kolosov, P., 1995, Integrated approaches to terminal Neoproterozoic stratigraphy: An example from the Olenek Uplift, northeastern Siberia, *Precambrian Res.* **73**: 251–270.

Knoll, A. H., Walter, M. R., Narbonne, G. M., and Christie-Blick, N., 2004, A new period for the geological time scale, *Science* **305**: 621–622.

Laflamme, M., Narbonne, G. M., and Anderson, M. M., 2004, Morphometric analysis of the Ediacaran frond *Charniodiscus* from the Mistaken Point Formation, Newfoundland, *J. Paleontol.* **78**: 827–837.

Li, C.-W., Chen, J.-Y., and Hua, T.-E., 1998, Precambrian sponges with cellular structures, *Science* **279**: 879–882.

Lipps, J. H., and Valentine, J. W., 2004, Late Neoproterozoic metazoa: weird, wonderful, and ghostly, in: *Neoproterozoic–Cambrian Biological Revolutions* (J. H. Lipps and B. M. Waggoner, eds.), Paleontological Society Papers 10, New Haven, pp. 51–66.

Martin, M. W., Grazhdankin, D. V., Bowring, S. A., Evans, D. A. D., Fedonkin, M. A., and Kirschvink, J. L., 2000, Age of Neoproterozoic bilaterian body and trace fossils, White Sea, Russia: implications for metazoan evolution, *Science* **288**: 841–845.

McCaffrey, M. A., Moldowan, J. M., Lipton, P. A., Summons, R. E., Peters, K. E., Jeganathan, A., and Watt, D. S., 1994, Paleoenvironmental implications of novel C_{30} steranes in Precambrian to Cenozoic age petroleum and bitumen, *Geochim. Cosmochim. Acta* **58**: 529–532.

McMenamin, M. A. S., 1986, The Garden of Ediacara, *Palaios* **1**: 178–182.

Narbonne, G. M., 2004, Modular construction of early Ediacaran complex life forms, *Science* **305**: 1141–1144.

Narbonne, G. M., 2005, The Ediacara biota: Neoproterozoic origin of animals and their ecosystems, *Annu. Rev. Earth Planet. Sci.* **33**: 421–442.

Narbonne, G. M., and Gehling, J. G., 2003, Life after Snowball: The oldest complex Ediacaran fossils, *Geology* **31**: 27–30.

Narbonne, G. M., Saylor, B. Z., and Grotzinger, J. P., 1997, The youngest Ediacaran fossils from southern Africa, *J. Paleontol.* **71**: 953–967.

Olcott, A., Corsetti, F. A., and Awramik, S. M., 2002, A new look at stromatolite form diversity, *GSA Abstr. with Progr.* **34**: 271.

Olcott, A. N., Sessions, A. L., Corsetti, F. A., Kaufman, A. J., and Flavio de Oliviera, T., 2005, Biomarker evidence for photosynthesis during Neoproterozoic glaciation, *Science* **310**: 471–474.

Peterson, K. J., and Butterfield, N. J., 2005, Origin of the eumetazoa: testing ecological predictions of molecular clocks against the Proterozoic fossil record, *Proc. Nat. Acad. Sci. USA* **102**: 9547–9552.

Peterson, K. J., Lyons, J. B., Nowak, K. S., Takacs, C. M., Wargo, M. J., and McPeek, M. A., 2004, Estimating metazoan divergence times with a molecular clock, *Proc. Nat. Acad. Sci. USA* **101**: 6536–6541.

Peterson, K. J., Waggoner, B. M., and Hagadorn, J. W., 2003, A fungal analog for Newfoundland Ediacaran fossils? *Integr. Comp. Biol.* **43**: 127–136.

Runnegar, B., 2000, Loophole for snowball Earth, *Nature* **405**: 403–404.

Saylor, B. Z., Grotzinger, J. P., and Germs, G. J. B., 1995, Sequence stratigraphy and sedimentology of the Neoproterozoic Kuibis and Schwarzrand subgroups (Nama Group), southwestern Namibia, *Precambrian Res.* **73**: 153–171.

Seilacher, A., 1999, Biomat-related lifestyles in the Precambrian, *Palaios* **14**: 86–93.

Seilacher, A., Grazhdankin, D. V., and Legouta, A., 2003, Ediacaran biota: The dawn of animal life in the shadow of giant protists, *Paleontol. Res.* **7**: 43–54.

Selden, P., and Nudds, J., 2004, *Evolution of Fossil Ecosystems*, University of Chicago Press, Chicago.

Shields, G. A., Stille, P., Brasier, M. D., and Atudorei, N.-V., 1997, Stratified oceans and oxygenation of the late Precambrian environment: A post glacial geochemical record from the Neoproterozoic of W. Mongolia, *Terra Nova* **9**: 218–222.

Taylor, W. L., and Brett, C. E., 1996, Taphonomy and paleoecology of Echinoderm *Lagerstätten* from the Silurian (Wenlockian) Rochester Shale, *Palaios* **11**: 118–140.

Waggoner, B. M., 1998, Interpreting the earliest Metazoan fossils: What can we learn? *Am. Zool.* **38**: 975–982.

Waggoner, B. M., 1999, Biogeographic analyses of the Ediacara biota: a conflict with paleotectonic reconstructions, *Paleobiology* **24**: 440–458.

Watkins, R., 1991, Guild structure and tiering in a high-diversity Silurian community, Milwaukee County, Wisconsin, *Palaios* **6**: 465–478.

Wood, D. A., Dalrymple, R. W., Narbonne, G. M., Gehling, J. G., and Clapham, M. E., 2003, Paleoenvironmental analysis of the late Neoproterozoic Mistaken Point and Trepassey formations, southeastern Newfoundland, *Can. J. Earth Sci.* **40**: 1375–1391.

Xiao, S., and Knoll, A. H., 2000, Phosphatized animal embryos from the Neoproterozoic Doushantuo Formation at Weng'an, Guizhou, South China, *J. Paleontol.* **74**: 767–788.

Xiao, S., Shen, B., Zhou, C., Xie, G.-W., and Yuan, X., 2005, A uniquely preserved Ediacaran fossil with direct evidence for a quilted bodyplan, *Proc. Nat. Acad. Sci. USA* **102**: 10227–10232.

Xiao, S., Yuan, X., and Knoll, A. H., 2000, Eumetazoan fossils in terminal Proterozoic phosphorites? *Proc. Nat. Acad. Sci. USA* **97**: 13684–13689.

Xiao, S., Zhang, Y., and Knoll, A. H., 1998, Three-dimensional preservation of algae and animal embryos in a Neoproterozoic phosphorite, *Nature* **391**: 553–558.

Yuan, X., Xiao, S., Parsley, R. L., Zhou, C., Chen, Z., and Hu, J., 2002, Towering sponges in an Early Cambrian Lagerstätte: Disparity between nonbilaterian and bilaterial epifaunal tierers at the Neoproterozoic-Cambrian transition, *Geology* **30**: 363–366.

Chapter 5

A Critical Look at the Ediacaran Trace Fossil Record

SÖREN JENSEN, MARY L. DROSER and JAMES G. GEHLING

Área de Paleontología, Facultad de Ciencias, Universidad de Extremadura, E-06071 Badajoz, Spain; Department of Earth Sciences, University of California, Riverside, CA 92521, USA; South Australian Museum, South Terrace, 5000 South Australia, Australia.

1. Introduction	116
2. Problems in the Interpretation of Ediacaran Trace Fossils	117
2.1. Tubular Organisms	119
2.2. *Palaeopascichnus*-type Fossils	120
3. List of Ediacaran Trace Fossils	120
4. Discussion	135
4.1. True and False Ediacaran Trace Fossils	136
4.1.1 *Archaeonassa*-type trace fossils	136
4.1.2 *Beltanelliformis*-type fossils	136
4.1.3 *Bilinichnus*	137
4.1.4 *Chondrites*	137
4.1.5 *Cochlichnus*	137
4.1.6 *Didymaulichnus*	137
4.1.7 *Gyrolithes*	138
4.1.8 *Harlaniella*	138
4.1.9 *Helminthoidichnites*-type trace fossils	138
4.1.10 *Lockeia*	139
4.1.11 *Monomorphichnus*	139
4.1.12 *Neonereites*	139
4.1.13 *Palaeopascichnus*-type fossils	139
4.1.14 *Planolites-Palaeophycus*	140
4.1.15 "*Radulichnus*"	140

4.1.16. *Skolithos*.	141
4.1.17. *Torrowangea*	141
4.1.18. Dickinsonid trace fossils	142
4.1.19. Meniscate trace fossils	142
4.1.20. Star-shaped trace fossils	142
4.1.21. Treptichnids	143
4.2. Ediacaran Trace Fossil Diversity	143
4.3. Stratigraphic Distribution and Broader Implications of Ediacaran Trace Fossils	145
Acknowledgements	147
References	147

1. INTRODUCTION

It has long been known that there was an increase in the size, complexity and diversity of trace fossils around 555–535 Ma (e.g., Seilacher, 1956; Alpert, 1977; Crimes, 1987; Narbonne *et al.*, 1987). Crimes (1987, 1992, 1994) published important overviews of Ediacaran and Cambrian trace fossils, including tabulations of ichnotaxa in time blocks spanning the Ediacaran to the Ordovician. These showed a respectable diversity of Ediacaran trace fossils as measured in number of ichnotaxa, with one list (Crimes, 1994, Table 4.2) giving 36 ichnogenera for the Ediacaran, with an increase to 66 ichnogenera for pre-trilobite lower Cambrian. Among these Ediacaran trace fossils were forms that suggest moderately complex behavior, such as guided meander traces. Also, some subsequent studies (e.g., Jenkins, 1995) have suggested a great diversity of Ediacaran trace fossils including arthropod-type scratch marks. The actual diversity and complexity of Ediacaran trace fossils have, however, recently come under scrutiny as alternative interpretations for several of these Ediacaran trace fossils have emerged (Gehling *et al.*, 2000; Jensen, 2003; Seilacher *et al.*, 2003). It has become increasingly clear that Ediacaran strata are particularly rich in problematic structures and that these often have been mistaken for trace fossils (see Seilacher *et al.*, 2005; Droser *et al.*, 2005). It may not be entirely inappropriate to compare the changing view on Ediacaran trace fossils to the shift in the interpretation of Ediacaran body fossils over the last several decades. In the case of trace fossils, however, the problem is not in organismic affinities, but in recognizing what is a genuine trace fossil and what is not. A better understanding of the diversity of Ediacaran trace fossils is not merely an exercise in numbers, but has direct impact on questions of animal evolution and for understanding of Ediacaran communities. It can be argued that trace fossils provide a good indicator for the appearance of macroscopic bilaterian animals (e.g., Valentine, 1994; Budd and Jensen, 2000), and that conditions for the preservation of shallow tier trace fossils were particularly favorable in the pre-Phanerozoic (Droser *et al.*, 2002a,b). The record of pre-Ediacaran trace fossils is at present problematic for reasons concisely summarized by Conway Morris (2003, p. 506). Even if some of these pre-Ediacaran trace fossils turn out to be genuine, their relation to Ediacaran trace fossils or to the radiation of animals is doubtful (cf., Rasmussen *et al.*, 2002). A better understanding of the Ediacaran trace fossil record is therefore of great interest because the complexity of trace fossils may tell us something about the complexity of their producers. For example, Ediacaran imprints of arthropod-type limbs, if accepted, would provide anchoring points for likely crown-group bilaterians. An improved understanding of Ediacaran trace fossil diversity is also necessary in order to explore possible roles of trace fossils in the establishment of Ediacaran Period subdivisions.

The purpose of this paper is to provide a brief critical overview of the Ediacaran trace fossil record in order to establish a more solid foundation for assessing changing patterns in the diversity and complexity of trace fossils at the end of the Ediacaran.

2. PROBLEMS IN THE INTERPRETATION OF EDIACARAN TRACE FOSSILS

General principles on which to distinguish trace fossils from body fossils and structures of inorganic origin have been discussed by Ekdale *et al.* (1988). Jensen *et al.* (2005a) and, in particular, Droser *et al.* (2005), discuss the identification and preservation of Ediacaran trace fossils. For example, signs of sediment displacement, such as raised levees, provide convincing evidence for a trace fossil origin. Sediment-filled shrinkage cracks (Fig. 1A) are the inorganic structures most commonly reported as trace fossils. However, a more serious problem is that making the distinction between Edicaran body fossils and trace fossils is surprisingly difficult. We believe that expectations from the Phanerozoic, where a trace fossil-like bedding-plane structure generally is a trace fossil has played a part in causing an inflation of Ediacaran ichnotaxa. Ediacaran siliciclastic sediments appear to have had conditions particularly favorable to the preservation of non-mineralized organisms as casts and molds, including well-known forms such as *Aspidella* and *Dickinsonia* (Gehling, 1999). A contributing factor was the extensive development of biomats, which led to the formation of protective husks of mineral precipitation, and also protected the carcasses from physical disturbance (Gehling, 1999). The absence of deep and intensive bioturbation (e.g., Droser *et al.*, 1999) also meant that such structures were not destroyed when preserved close to the sediment-water interface. There were Ediacaran organisms with a morphology that, when preserved as casts and molds in siliciclastic sediments, are easily mistaken for trace fossils. It is important to note that these were as much part of the Ediacaran biotas as emblematic forms like *Dickinsonia* (see Droser *et al.*, 2006). The majority of these organisms are poorly understood and remain little studied, but in terms of gross morphology two types are particularly important for the study of Ediacaran trace fossils.

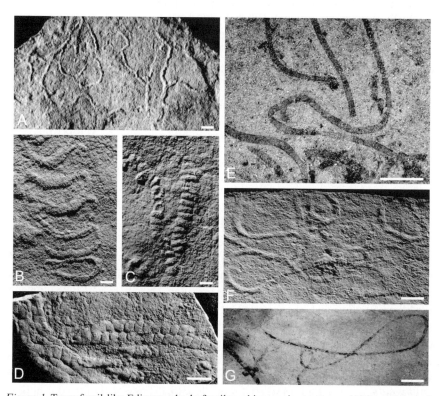

Figure 1. Trace fossil-like Ediacaran body fossils and inorganic structures. (A) Irregular sand-filled shrinkage cracks on sole of thin sandstone bed. Nama Group, Namibia. (B, C) Different styles of preservation of palaeopaschichnids (?algae or protists) resulting in similarity to meander trace fossils or rows of fecal pellets, Ust Pinegia Formation, White Sea area, northern Russia. (D) Branching tubular fossil with transverse grooves. Fossils of this type have been interpreted as fecal rows. Ediacara Member, South Australia. (E) Tubular organism (probably a sabelliditid) preserved as dark films. Note angular termination near center of image, and smooth curvature as well as folds with reduced tube width. Ibor group, central Spain. (F) Looping shallow furrows interpreted as trace fossil by Vidal *et al.* (1994), but probably representing a tubular organism. Light from lower right. Domo Extremeño group, central Spain. (G) Looping filaments preserved as dark film. Note uniform width of film. Ibor Group, central Spain, terminal Ediacaran or earliest Cambrian. Scale bars represent 10 mm in A; 2 mm in B and C; 20 mm in D; 5 mm in E, F and G.

2.1 Tubular Organisms

The three-dimensional preservation of relatively large tubular organisms of uncertain affinities has been a common source of mistaken identification of trace fossils. Examples of such structures have been discussed by Runnegar (1994) and Droser *et al.* (2005). Scalloped margins in strongly curved portions and varying degree of sand-fill along the length of the tubes suggest that these were tubular organisms that were deformed and filled with sediment during transport and entombment in sand. Some of these tubular organisms have a segmentation and complex branching (Fig. 1D) that is clearly incompatible with a trace fossil. However, where only fragments are preserved these have been interpreted as trace fossils (see Droser *et al.*, 2005).

There is also the three-dimensional preservation of smaller parallel-sided tubular organisms that may include, but are not restricted to, vendotaeniaceans and, in particular, sabelliditids. Such organisms are generally known from flattened carbonized specimens but they were originally tubular and it has recently been recognized that these can be preserved as grooves and ridges in siliciclastics (Jensen *et al.*, 2005b; Fig. 1F). The presence of angular terminations and abrupt changes in diameter, where the tube has been folded or twisted, are diagnostic features for a non-trace fossil in carbonized compressions (Fig. 1E). Such features, however, may not be readily identifiable in casts and molds because of the inevitable loss in morphological definition in these modes of preservations. Abrupt changes in tube diameter depend on its mechanical properties and may not be present (see Fig. 1G). Finally, the presence or absence of organic material is not a reliable criterion as it is a factor of preservational conditions as well as being sensitive to weathering. Weathered and bleached flattened tubular organisms have been a source of confusion with trace fossils. This is the most likely interpretation of *Planolites* and *Chondrites* of Wu and Li (1987) and *Helanoichnus helanensis* of Yang and Zheng (1985).

A related case is the preservation as casts and molds in siliciclastics of the early biomineralized organism *Cloudina*, which have been confounded with meniscate and other trace fossils (Germs, 1972; Chen *et al.*, 1981). Another interesting tubular form that may have been biomineralized is *Gaojiashania* from the terminal Ediacaran of China. It consists of a series of ring-like structures and is found preserved as flattened films as well as in three dimensions (Zhang, 1986; Chen *et al.*, 2002). Zhang's (1986) report of *Planolites annularis* in the same biota appears to be a *Gaojiashania*.

2.2 *Palaeopaschichnus*-type Fossils

These are bedding plane-parallel fossils consisting of numerous closely positioned relatively small round to kidney-shaped objects that have a superficial similarity to meandering trace fossils or rows of fecal pellets. These structures have been the basis for a range of "ichnogenera" such as *Palaeopaschichnus* and *Yelovichnus* (Fig. 1B–C). Their biological affinity is not resolved but clearly they are not trace fossils (Gehling *et al.*, 2000; Jensen, 2003; Seilacher *et al.*, 2003; Shen *et al.*, in press). These are discussed in more detail below under *Palaeopaschichnus*-type fossils.

3. LIST OF EDIACARAN TRACE FOSSILS

The list of Ediacaran trace fossils presented in Table 1 includes the main reports of Ediacaran trace fossils known to us. We are under no illusion that this constitutes a complete listing of Ediacaran trace fossils and it is certainly weaker in some geographic areas. The list does not include occurrences that have only been reported from abstracts. Taxa that are generally considered body fossils but for which a trace fossil interpretation has been discussed, such as *Beltanelliformis* and *Nemiana*, are not included in the list but are briefly discussed in the text below. The list is restricted to occurrences that with reasonable confidence are Ediacaran (ca. 635–542 Ma, e.g., Knoll *et al.*, 2004a,b; Condon *et al.*, 2005). There are numerous successions that are known as Precambrian–Cambrian but where more precise age constraints are lacking. For example, Webby (1970) described several types of trace fossils including *Planolites ballandus* and *Curvolithus? davidis* from the Lintiss Vale Beds of New South Wales, which he initially assigned a latest Proterozoic age, and later (Webby 1984) suggested to be older than the basal Cambrian Uratanna Formation. This unit contains trace fossils that potentially are of great interest to the discussion of latest Ediacaran trace fossil zonation, but pending further constraints we follow Walter *et al.* (1989) in considering it Cambrian. Another example is the Puncoviscana Formation of northern Argentina which is thought to straddle the Ediacaran–Cambrian boundary, and includes reports of Ediacaran trace fossils (e.g., Aceñolaza and Alonso, 2001). The Puncoviscana Formation is notable as one of relatively few units that yield both shallow and deep-water early Cambrian trace fossils (e.g., Buatois and Mángano, 2004; Aceñolaza, 2004). A considerable number of ichnotaxa have been reported and been the topic of ichnotaxonomic re-evaluations (see Buatois and Mángano, 2003, 2004, 2005). However, at present there exist no compelling evidence that any of the Puncoviscana trace fossils are Ediacaran (Buatois and Mángano, 2005).

A Critical Look at the Ediacaran Trace Fossil Record

Several reports from the former Soviet Union of latest Vendian (Rovno, Nemakit Daldynian) trace fossils (e.g., Palij *et al.*, 1979; Bekker and Kishka, 1991) are not included as they are likely Cambrian as formally defined.

Below we briefly explain the different posts and the construction of the table.

Ichnotaxa: Reports are listed alphabetically following the identification of a principal reference. For the sake of consistency, sp. (species) has been changed to isp. (ichnospecies). Where there have been substantial taxonomic reassignments, it is noted in interpretation and comments.

Reference: This gives a paper, or papers, that provides the fullest description and/or illustration of the specific structure. Because ichnotaxonomy is not a primary aim of this paper, the references do not necessarily include the paper where the taxon was erected.

Location: This provides basic information on the geographic and stratigraphic setting. An attempt has been made to give widely used and up-to-date lithostratigraphic nomenclature. For reasons of convenience, we have retained Ust Pinega Formation instead of the more detailed lithostratigraphic nomenclature of Grazhdankin (2003)

Age: The following main categories of age constraints are used. Usually only one age constraint is listed, even though more may be available.

t – Occurrence stratigraphically below the lowest local occurrence of Cambrian-type trace fossils. This evidence must be treated as tentative.

c – Occurrence with or stratigraphically below the weakly biomineralized tubular fossil *Cloudina*. *Cloudina* is currently considered to be terminal Ediacaran with a stratigraphic range of ca 550–542 Ma (Grant, 1990; Grotzinger *et al.*, 1995; Amthor *et al.*, 2003).

e – Occurrence with or in close stratigraphic proximity to core Ediacara-type body fossils for which there is no post-Ediacaran record, such as *Dickinsonia* and *Tribrachidium*. This stratigraphic relationship provides undisputed evidence for an Ediacaran age.

a – Occurrence constrained by disc-shaped fossils such as *Aspidella*, strongly suggesting an Ediacaran age.

p – Occurrence with or in close stratigraphic proximity to palaeopascichnids. These have a long Ediacaran stratigraphic range (Gehling *et al.*, 2000), and are not known from younger strata.

r – Occurrence constrained by radiometric ages.

Interpretation and comments: Wherever possible we have evaluated the published occurrences if they are credible trace fossils. Ideally such evaluations should be based on a first-hand examination of the material. Considering the geographic spread of the material this is hardly practical. Nevertheless, we believe that our experience in studying Ediacaran material from Australia, Namibia, and the White Sea region makes it possible to do meaningful evaluations for other regions, where this is permitted by detailed descriptions and high-quality illustrations. Our experience is that the diversity of trace fossils has been greatly inflated; therefore evaluation of all material is made with the same critical eyes. In so doing it is certainly possible that we have in some cases been in error in disputing previous interpretations. The following symbols are used.

T – Identification as trace fossil accepted. This only means that a trace fossil interpretation is likely and further studies may change the evaluation. This in particular applies to simple horizontal forms without unequivocal evidence for sediment displacement.

T? – Identification as trace fossil problematic

? – The published information does not allow for a critical evaluation.

* – An interesting form that requires further documentation.

- – Not accepted as a trace fossil.

Table 1. List of Ediacaran trace fossils.

Ichnotaxon	Reference	Location	Age	Interpretation and comments
Annulusichnus regularis	Zhang '86	Yangtze platform, Dengying Fm.	c	-	*Nimbia*: Hofmann *et al.* '90
Archaeichnium haughtoni	Glaessner '77	Namibia, Nasep Mbr.	c	-	Tubular fossil: Glaessner '77
Archaeonassa isp	Jensen '03	White Sea area, Ust Pinega Fm.	e	T	
Arenicolites isp.	Banks '70	N. Norway, Innerelv Mbr.	t	-	Pseudofossil: Farmer *et al.* '92
Arthrophycus isp.	Lin *et al.* '86	Yangtze platform, Dengying Fm.	c	-	Body fossil, *Gaojiashania* ?
Asterichnus vialovi	Gureev '84	Ukraine, Mogilev Fm.	e	*	*Hiemalora*?
cf. *Asterichnus* isp.	Jenkins '95	South Australia, Ediacara Mbr.	e	?	
Aulichnites isp.	Jenkins '95	South Australia, Ediacara Mbr.	e	T	
Aulichnites isp.	Narbonne and Aitken '90	NW Canada, Blueflower Fm.	t	T	
Aulichnites isp.	Fedonkin '80, '85	White Sea area, Ust Pinega Fm.	e	T	
Bergaueria isp.	Crimes and Germs '82	Namibia, Kuibis Subgr.	c	-	Dubiofossil; *Intrites*: McIlroy *et al.* '05
Bergaueria isp.	Fedonkin '81	Ukraine, Mogilev Fm.	e	-	*Gaojiashania* ?
Bilinichnus simplex	Palij *et al.* '79	White Sea area, Ust Pinega Fm.	e	?	Trace fossil or pseudofossil?
Bilinichnus isp	Shanker *et al.* '97	N. India, ?Krol Gr.	t?	?	Pseudofossil?
Brooksella isp.	Crimes and Germs '82	Namibia, Kuibis Subgr.	c	?	Pseudofossil?

124   S. JENSEN ET AL.

Table 1 (Continued).

Bucerusichnus octoideus	Li et al. '92	Yangtze platform, Dengying Fm.	c	T?	If trace fossil, not a new igen.
Bucerusichnus gaojiashanensis	Zhang '86	Yangtze platform, Dengying Fm.	c?	T?	If trace fossil, not a new igen.
Buchholzbrunnichnus kroeneri	Germs '73	Namibia, Kuibis Subgr.	c	-	Probably body fossil
Bunyerichnus dalgarnoi	Glaessner '69	South Australia, Bunyeroo Fm	e	-	Pseudofossil: Jenkins et al. '81
Bunyerichnus isp	Bekker '92	Urals area, Bakeevsko Fm.	p	?	Dubiofossil,
Catellichnus oktonarius	Bekker and Kishka '89	Urals area, Basinska Fm.	p	-	Palaeopascichnid
Chomatichnus loevcensis	Gureev '81, '84	Ukraine, Zhamovka Fm.	t,p	?	Water escape structures?
Chondrites isp	Jenkins '95	South Australia, Ediacara Mbr.	e	*	Body fossil or trace fossil?
Chondrites isp A, B	Wu and Li '87	Xingjiang?		-	Flattened tubular organism?
Chondrites isp.?	Crimes and Germs '82	Namibia, Nudaus Fm.	c	T	Coincidence of simple traces?
? *Chondrites* isp.	Li et al. '92	Yangtze platform, Dengying Fm.	c	?	No resemblance to *Chondrites*
Circulichnus montanus	Gureev '83, Gureev et al. '85	Ukraine, Danilovka Fm.	t,p	-	*Nimbia*: Hofmann et al. '90
Circulichnus isp	Sokolov '97	White sea area, Ust Pinega Fm.	e	-	*Nimbia*
Cochlichnus isp.	Cope '83	Wales	p	T	*Helminthoidichnites tenuis*: Runnegar '92
Cochlichnus isp.2	Palij et al. '79	Ukraine, Mogilev Fm.	e	?	
Codonichnus wujiensis	Ding & Xing '88	Yangtze platform, Dengying Fm.	c	?	
Cylindrichnus isp.	Glaessner '69	South Australia, Ediacara Mbr.	e	-	Tubular fossil
Dengvingella shibantanensis	Li et al. '92	Yangtze platform, Dengying Fm.	c	*	Tubular fossil or trace fossil

Table 1 (Continued).

Didymaulichmus cf. *miettensis*	Gureev '84	Ukraine, Mogilev Fm.	e	*	
Didymaulichmus isp.	Poire et al. '84	Argentina, Sierra Bayas Fm.	t	T?	Age constraint?
Diplocraterion isp.	Crimes and Germs '82	Namibia, Nudaus Fm.		T*	Documentation of spreite needed
Eilscaptichnus	Li and Ding '96	China, Doushantou Fm.	c	?	Pseudofossil: Xiao et al. '02
Funisichnus grammatus	Li et al. '92	Yangtze platform, Dengying Fm.	c	?	
Funisichnus wuheensis	Li et al. '92	Yangtze platform, Dengying Fm.	c	?	
Gordia arcuata?	Gibson '89	North Carolina, Floyd Church Mbr.	e	T?	*Circulichnus*: Seilacher et al. '05
Gordia aff. *arcuata*	Vidal et al. '94	Spain, Domo Extremeno Gp.	c	?	Trace fossil or tubular organism
Gordia marina	Narbonne and Hofmann '87	NW Canada, Blueflower Fm.	t,a	T	*H. tenuis*: Narbonne, Aitken '90
Gordia marina	Shanker et al. '97	N. India, ?Krol Gp.	?t	-	Dubiofossil
Gordia marina	Tangri et al. '03	Bhutan, Deschiling Fm.	t	T	
Gordia marina	Vidal et al. '94	Spain, Domo Extremeno Gp.	c	?	Trace fossil or tubular organism
Gordia isp. cf. *G. marina*	Mathur and Shanker '89	N. India, (upper) Krol Gp.	?t	-	Dubiofossil
Gordia isp.	Paześna '86	Poland, Lublin Fm.	t	T?	Not *Gordia*
Gordia isp.	Vidal et al. '94	Spain, Domo Extremeno Gp.	t,c	?	Trace fossil or tubular organism
Gordia isp.	Jenkins '95	South Australia, Ediacara Mbr.	e	T?	
Harlaniella podolica	Sokolov '72, Palij et al. '79	Ukraine, Studenitsa Fm.	t	-	Palaeopascichnid?
Harlaniella podolica	Narbonne et al. '87	Newfoundland, Chapel Island Fm.	t	-	Palaeopascichnid?

Table 1 (Continued).

cf. *Harlaniella* isp.	Jenkins '95	South Australia, Ediacara Mbr.	e	-	Tubular fossil?
Helanoichnus helanensis	Yang and Zheng '85	Ningxia, Zhengmuguan Fm.	?	-	Flattened tubular organism?
Helminthoida helanshanensis	Xing *et al.* '85	Ningxia, Zhengmuguan Fm.	?	-	Same as *Helanoichnus helanensis*
Helminthoida isp.	Fedonkin '85	White Sea area, Ust Pinega Fm.	t	?	Tubular organism?
Helminthoida isp.	Narbonne and Aitken '90	NW Canada, Blueflower Fm.	t	T	
Helminthoidichnites tenuis	Narbonne and Aitken '90	NW Canada, Blueflower Fm.	t	T	
Helminthoidichnites isp.	Hagadorn and Waggoner '00	Nevada Wood Canyon Fm.	t	T?	
Helminthoidichnites isp.	Corsetti and Hagadorn '03	California, Wyman Fm.	c	T?	Trace fossil or tubular organism?
Helminthopsis abeli	Narbonne and Aitken '90	NW Canada, Blueflower Fm.	t	T	
Helminthopsis irregularis	Narbonne and Aitken '90	NW Canada, Blueflower Fm.	t	T	
Helminthopsis quanjishanensis	Xing *et al.* '85, Zhang '86	Yangtze platform, Dengying Fm.	c	-	Similar to *Helanoichnus helanensis*
?*Helminthopsis* isp.	Gibson '89	North Carolina, Floyd Church Mbr.	e	T?	Trace fossil or tubular organism?
Hormosiroidea arumbera	Walter *et al.* '89	Cent.Australia, Central Mt. Stuart Fm.	t,e?	T	
Intrites punctatus	Fedonkin '80	White Sea area, Ust Pinega Fm.	e	-	Palaeopascichnid (pars?)
Intrites punctatus	Bekker '92	Urals area, Bakeevsko Fm.	p	-	Dubiofossil
Intrites cf. *punctatus*	Lin *et al.* '86	Yangtze platform, Dengying Fm..	c	?	Biogenic?
Linbotulichnus	Li and Ding '96	China, Doushantou Fm.	r	?	
Lockeia isp.	Narbonne and Aitken '90	NW Canada, Blueflower Fm.	t	T?	Assignment dubious

Table 1 (Continued).

Lockeia isp.	McMenamin '96	Mexico, Clemente Fm.	t	-	Dubiofossil
cf. *Lockeia* isp.	Jenkins '95	South Australia, Ediacara Mbr.	e	-	Body fossil
Margaritichnus linearis	Fedonkin '76	White Sea area, Ust Pinega Fm.	e	-	*Neonereites uniserialis*: Fedonkin '81
Medvezichnus pudicum	Fedonkin '85	White Sea area, Ust Pinega Fm.	e	*	Unique specimen
Microspirolithus tianquanensis	Yang et al. '82	Sichuan	?	?	
Monocraterion isp.	Gureev '84	Ukraine, Studenitsa Fm.	t,p	?	
?*Monocraterion* isp.	Gibson '89	North Carolina, Tillery Fm.	e	-	Probably inorganic: Seilacher et al. '05
cf. *Monomorphichnus* isp	Jenkins '95	South Australia, Ediacara Mbr.	e	T	See *Radulichnus*
cf. *Monomorphichnus* isp	Waggoner and Hagadorn '02	Nevada, Wood Canyon Fm.	t	?	Tool mark or trace fossil?
Muensteria isp	Germs '72	Namibia, Huns Mbr.	c	-	Cast of *Cloudina*
cf. *Muensteria* isp	Jenkins '95	South Australia, Ediacara Mbr.	e	-	Tubular fossil
Nenoxites curvus	Fedonkin '76	White Sea area, Ust Pinega Fm.	e	T?*	Seilacher et al '05 support trace fossil
Neonereites biserialis	Palij et.al. '79, Fedonkin'81	White Sea area, Ust Pinega Fm.	e		Palaeopascichnid
Neonereites biserialis	Gibson '89	North Carolina, Floyd Church Mbr.	e	T?	
Neonereites dengyingensis	Li et al. '92	Yangtze platform, Dengying Fm.	c	?	
Neonereites renarius	Fedonkin '80	White Sea area, Ust Pinega Fm.	e	-	Palaeopascichnid
Neonereites renarius	Jenkins '95	South Australia, Ediacara Mbr.	e	-	Palaeopascichnid

Table 1 (Continued).

Neonereites renarius	Gehling et al. '00	Newfoundland, Fermeuse Fm.	r	-	Palaeopascichnid
Neonereites uniserialis	Palij et al. '79, Fedonkin '81	White Sea area, Ust Pinega Fm.	e	-	Palaeopascichnid
Neonereites uniserialis	Jenkins '95	South Australia, Ediacara Mbr.	e	-	Palaeopascichnid
Neonereites uniserialis	Chistyakov et al. '84	White Sea area (Onega)	e	-	Palaeopascichnid
Neonereites uniserialis	Yang and Zheng '85	Ningxia, Zhengmuguan Fm.	?	-	Palaeopascichnid
Neonereites uniserialis	Gibson '89	North Carolina, Floyd Church Mbr	e	T?	
Neonereites uniserialis	Brasier and McIlroy '98	Scotland, Bonhaven Fm.	r	-	Dubiofossil: Brasier and Shields '00
Neonereites cf. uniserialis	Li et al. '92	Yangtze platform, Dengying Fm.	c	?	
Neonereites isp.	Lin et al. '86	Yangtze platform, Dengying Fm.	c	?	
Neonereites isp?	Narbonne and Hofmann '87	NW Canada, Blueflower Fm.	t	T?	Poorly preserved treptichnid?
Neonereites isp?	Narbonne and Aitken '90	NW Canada, Blueflower Fm.	t	?	
?Neonereites isp.	Gibson '89	North Carolina, Floyd Church Mbr.	e	T?	Treptichnus? isp: Seilacher et al. '05
Nereites isp.	Crimes and Germs '82	Namibia, Nudaus Fm.	c	T	Archeonassa-type trace fossil
cf.Nereites isp.	Jenkins '95	South Australia, Ediacara Mbr.	e	T	Archeonassa-type trace fossil
Ningxiaichnus minimum	Yang and Zheng '85	Ningxia, Zhengmuguan Fm.	?	-	Palaeopascichnid: Shen et al. in pres
Ningxiaichnus stuyukonensis	Yang and Zheng '85	Ningxia, Zhengmuguan Fm.	?	-	Palaeopascichnid: Shen et al., in pres
Oldhamia recta	Seilacher et al. '05	North Carolina, Floyd Church Mbr.	e	T?	Trace fossil or tubular fossil?

Table 1 (Continued).

Taxon	Reference	Location			Notes
Palaeohelminthoida isp.	Gehling '91, Jenkins '95	South Australia, Ediacara Mbr.	e	-	Palaeopascichnid
Palaeopascichnus delicatus	Palij '76	Ukraine, Studenitsa Fm.	t,p	-	Palaeopascichnid?
Palaeopascichnus delicatus	Pacześna '86	Poland, Lublin Fm.	t	-	Palaeopascichnid
Palaeopascichnus delicatus	Palij et al. '79, Fedonkin'81	White Sea area, Ust Pinega Fm.	e	-	Palaeopascichnid
Palaeopascichnus delicatus	Gehling et al. '00	Newfoundland, Fermeuse Fm.	e,r	-	Palaeopascichnid
Palaeopascichnus delicatus	Narbonne et al. '87	Newfoundland, Chapel Island Fm.	t	-	Palaeopascichnid
Palaeopascichnus cf. delicatus	Jenkins '95	South Australia, Ediacara Mbr.	e	-	Palaeopascichnid
Palaeopascichnus cf. delicatus	Lin et al. '86	Yangtze platform, Dengying Fm.	c?	?	Palaeopascichnid?
Palaeopascichnus sinuosus	Fedonkin '81	White Sea area, Ust Pinega Fm.	e	-	Palaeopascichnid
Palaeopascic. wangjiawanensis	Yin et al. '93	China, Yunnan	?	-	Palaeopascichnid
Palaeophycus tubularis	Jenkins '95	South Australia, Ediacara Mbr.	e	-	Tubular fossil?
Palaeophycus tubularis	Jenkins et al. '92	Centr. Austral., Grant Bluff Fm.	t,e?	T?	
Palaeophycus tubularis	Narbonne and Aitken '90	NW Canada, Blueflower Fm.	t	T	
Palaeophycus tubularis	McMenamin '96	Mexico, Clemente Fm.	t	-	Dubiofossil
Palaeophycus tubularis	Bartley et al. '98	Siberia, Platonovska Fm.	t	T	
cf. Palaeophycus tubularis	Waggoner and Hagadorn '02	Nevada, Wood Canyon Fm.	t	T	Could be Cambrian: Bartley et al. 98
Palaeophycus isp. A, B	Geyer and Uchman'95	Namibia, Nasep Mbr.	c	T	

Table 1 (Continued).

Parascalarituba ningxiaensis	Yang and Zheng '85	Yangtze platform, Dengying Fm.	c	-	Same as *Helminthopsis quanjishanensis*
?Pelecypodichnus isp.	Zhang '86	Yangtze platform, Dengying Fm.	c?	?	
Phycodes? n. isp.	Liñan and Palacios '87	Spain, Domo Extremeño Gp.	t	T	
Plagiogmus isp	Cloud and Nelson '66	California, Deep Springs Fm.	t	?	Probably body fossil: cf. Alpert '74
Planolites annularis	Zhang '86	Yangtze platform, Dengying Fm.	c?	-	Probably the tubular fossil *Gaojiashania*
Planolites ballandus	Walter *et al.* '89	Centr. Austral., Elkera Fm.	t,e?	T	
Planolites beverleyensis	Jenkins '95	South Australia, Ediacara Mbr.	e	?	
Planolites beverleyensis	Bartley *et al.* '98	Siberia, Platonovska Fm.	t	T	Could be Cambrian: Bartley *et al.* 98
Planolites beverleyensis	Gibson '89	North Carolina, Floyd Church Mbr.	e	T?	*H. tenuis*: Narbonne and Aitken '90
Planolites beverleyensis	Zhang '86	Yangtze platform, Dengying Fm.	c	?	
Planolites liantuoensis	Ding *et al.* '93	Yangtze platform, Dengying Fm.	c	?	
Planolites montanus	Jenkins *et al.* '92	Centr. Austral., Grant Bluff Fm.	t,e?	T?	
Planolites montanus	Narbonne and Aitken '90	NW Canada, Blueflower Fm.	t	T	
Planolites montanus	Alpert '75	California, Deep Springs Fm.	c	T	
Planolites montanus	Gibson '89	North Carolina, Floyd Church Mbr.	e	T?	
Planolites isp. cf. *P. montanus*	Geyer and Uchman '95	Namibia, Nasep Mbr.	c	T	
Planolites reticulatus	Isakov '90	Siberia, Aldan	?	-	Filled shrinkage cracks

Table 1 (Continued).

Planolites sinensis	Li et al. '92	Yangtze platform, Dengying Fm.	c	T?	*Gordia* or tubular fossil
Planolites taishanmiaoensis	Ding et al. '85	Yangtze platform, Dengying Fm.	c	T	*Planolites ballandus*?
Planolites isp.	Acenolaza et al. '98	Uruguay, Lavalleja Gp.	t	?	
Planolites isp.	Fedonkin '76	White Sea area, Ust Pinega Fm.	e	T?	
Planolites isp.	Lin et al. '86	Yangtze platform, Dengying Fm.	c	T?	
Planolites isp.	Sovetov and Komlev '05	Siberia, Marnya Fm.		-	Filled shrinkage cracks
Planolites isp.	Tangri et al. '03.	Bhutan, Deshichling Fm.	t	T	
Planolites isp. A, B	Wu and Li '87	China, Xingjiang?	?		Flattened tubular organism?
cf.*Planolites* isp.	Jenkins '95	South Australia, Ediacara Mbr.	e		
Planotaenichnus	Li and Ding '96	China, Doushantou Fm.	r	-	Pseudofossil: Xiao et al.'02
Psammichnites isp.	Jenkins '95	South Australia, Ediacara Mbr.	e	-	Tubular fossil?
Radulichnus isp.	Seilacher '95,'97	South Australia, Ediacara Mbr.	e	T	Awaiting description
Scolecocoprus isp	Germs '72	Namibia, Huns Mbr.	c	-	Cast of *Cloudina*
Scolicia isp?	Germs '72	Namibia, Huns Mbr.	c	-	Cast of *Cloudina*
cf.. *Scolicia* isp?	Hagadorn and Waggoner '00	Nevada, Wood, Canyon Fm.	t	T?	Trace fossil or body fossil?
Shaanxilithes ningqiangensis	Lin et al. '86	Yangtze platform, Dengying Fm.	c	-	Tubular fossil: Shen et al. in press
Shaanxilithes erodus	Zhang '86	Yangtze platform, Dengying Fm.	c	?	Tubular fossil: Shen et al. in press

Table 1 (Continued).

Taxon	Reference	Locality			Notes
Shipaitubulus baimatouensis	Chen et al. '81	China, Xilingxia	?	?	
Skolithos declinatus	Fedonkin '85	White Sea area, Ust Pinega Fm.	e	T?*	
Skolithos hubeiensis	Ding & Xing 88'; Li et al. '92	Yangtze platform, Dengying Fm.	c	?	
Skolithos isolatus	Zhang '86	Yangtze platform, Dengying Fm.	c	?	
Skolithos miaoheensis	Chen et al. '81	Yangtze platform, Dengying Fm.	c?	-	*Sinotubulites*: Ding et al. '93
Skolithos isp	Banks '70	Norway, Innerelv Mbr.		-	Pseudofossil: Farmer et al. '92
Skolithos isp	Sokolov '97	White sea area, Ust Pinega Fm.	e	T?	
Skolithos isp. B	Geyer and Uchman '95	Namibia, Nasep Mbr.	c	T	*Planolites*?
Skolithos isp. ?	Germs '72	Namibia, Kuibis Subgr	c	-	Body fossil: Crimes and Fedonkin, 96
cf.*Spirorhaphe involuta*	Jensen '03	South Australia, Ediacara Mbr.	e	T	Unique specimen
Stelloglyphus isp.	Gureev et al. '85	Ukraine, Nagoryany Fm.	t	-	Benthic medusoid
Suzmites volutatus	Fedonkin '76	White Sea area, Ust Pinega Fm.	e	-	Tool mark
Syringomorpha nilssoni?	Gibson '89	North Carolina, Floyd Church Mbr.	e	T?	see *Oldhamia recta*
Taenidium cf. *serpentinium*	Jenkins '95	South Australia, Ediacara Mbr.	e	-	Tubular fossil: Droser et al. '05
Taenidium isp.	Germs '72	Namibia, Huns Mbr.	c	-	Cast of *Cloudina*
Taenioichnus spiralis	Li and Yang '88	China, Yunnan	?	-	Flattened tubular organism?
Taenioichnus zhengmuguanensis	Yang&Zheng 85; Li&Yang'88	China, Yunnan	?	-	= *Shaanxilithes*: Shen et al. in press

Table 1 (Continued).

?*Tomaculum* isp.	Gibson '89	North Carolina, Floyd Church Mbr.	e	T?	
Torrowangea rosei	Narbonne and Aitken '90	NW Canada, Blueflower Fm.	t	T	
Torrowangea rosei	Geyer and Uchman '95	Namibia, Nasep Mbr.	c	T	
Torrowangea rosei	Paczesna '86	Poland, Lublin Fm.	c	T	
Torrowangea rosei	Liñan and Palacios '87	Spain, Domo Extremeño Gp.	t	T?	Not *Torrowangea*
Torrowangea aff. *rosei*	Liñan and Tejero '88	Spain, Saviñán Fm.	t	?	
Torrowangea cf. *rosei*	Lin *et al.* '86	Yangtze platform, Dengying Fm.	c	T	Igen. assignment?
Torrowangea isp.	Zhang '86	Yangtze platform, Dengying Fm.	c	T	Igen. assignment?
?*Torrowangea* isp.	Gibson '89	North Carolina, Floyd Church Mbr.	e	T?	
Trichophycus pedum	Geyer and Uchman '95	Namibia, Nasep Mbr.	c	T?	?*Treptichnus* isp, not figured
Trichophycus tripleurum	Geyer and Uchman '95	Namibia, Nasep Mbr.	c	T	*Treptichnus* isp: Jensen *et al.* '00
Vendichnus vendicus	Palij *et al.* '79	White Sea area, Ust Pinega Fm.	e	?	Indistinct
Vermiforma antiqua	Cloud *et al* '76	North Carolina, Carolina slate belt	e	-	Pseudofossil: Seilacher *et al.* '00
Vimenites bacillaris	Fedonkin '80	White Sea area, Ust Pinega Fm.	e	?	
Vimenites bacillaris	Acenolaza *et al.* '98	Uruguay, Lavalleja Gp.	t	?	
Yelovichnus gracilis	Fedonkin '85	White Sea area, Ust Pinega Fm.	e	-	Palaeopascichnid
Yelovichnus gracilis	Gehling *et al.* 00	Newfoundland, Fermeuse Fm.	e,r	-	Palaeopascichnid

Table 1 (Continued).

Curved burrows	Germs '72	Namibia, Nudaus Fm.	c	-	Dubiofossil
Discontinuous trail ...	Germs '72	Namibia, Huns Mbr.	c	T	*Treptichnus* isp: Jensen et al. '00
Form A	Glaessner '69	South Australia, Ediacara Mbr.	e	-	Tubular fossil: Droser et al. '05
Form B	Glaessner '69	South Australia, Ediacara Mbr.	e	T	*Helminthoidichnites* etc
Form C	Glaessner '69	South Australia, Ediacara Mbr.	e	-	Palaeopascichnids
Form D	Glaessner '69	South Australia, Ediacara Mbr.	e	-	Tubular fossil: Droser et al. '05
Form E	Glaessner '69	South Australia, Ediacara Mbr.	e	T?	
Form F	Glaessner '69	South Australia, Ediacara Mbr.	e	-	Tubular fossil
Horizontal back-fill burrow	Narbonne and Hofmann '87	NW Canada, Blueflower Fm.	t,a	-	Wrinkled tubular fossil?
Knotted burrow	Narbonne and Aitken '90	NW Canada, Blueflower Fm.	e	?	
Meandering feeding trail	Cope '83	Wales	p	-	Palaeopascichnid: Runnegar '92
Problematic fossil	Haines '00	South Australia. Wonoka Fm	e	-	Palaeopascichnid
Shallow branching burrow	Cope '83	Wales	p	T?	
Sinuous to meandering trails	Webby '70	New South Wales, Fowlers Gap beds	t	T	

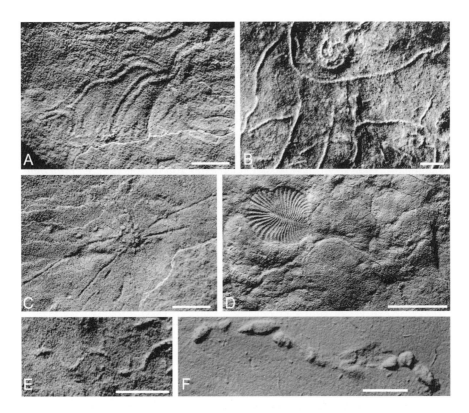

Figure 2. Ediacaran trace fossils. (A) *Archaeonassa* isp., on bed top with prominent marginal levees demonstrating displacement of sediment. Ust Pinegia Formation, White Sea area, north-west Russia. (B) Bed top with *Tribrachidium* and *Helminthoidichnites*-type trace fossils preserved as natural casts as sand moved up to fill negative features on the bed sole. Ediacara Member, South Australia. (C) Stellate structure on bed top interpreted as a trace fossil comparable to *Oldhamia*. Ediacara Member, Chace Range, South Australia. (D) Bed sole with *Dickinsonia* preserved as an external mold at the end of a series of overlapping resting traces, preserved as low casts. Ediacara Member, Heysen Range, South Australia. (E) Field photograph of sinusoidal trace fossil *Cochlichnus* isp., on bed sole from the Huns Member, Namibia. (F) *Treptichnus* isp., on bed sole from the Huns Member, Namibia. Scale bars represent 10 mm in A, B, C, E, F; 20 mm in D.

4. DISCUSSION

On the basis of Table 1 it is possible to discuss several aspects of the Ediacaran trace fossil record. First we discuss in more detail the interpretation and Ediacaran status of some taxa.

4.1 True and False Ediacaran Trace Fossils

4.1.1 *Archaeonassa*-type trace fossils

These are trace fossils with an upper surface that is either bilobed or consists of prominent marginal raised ridges flanking a central area (Fig. 2A). The nature of the base of the trace typically is not known. Similar Ediacaran trace fossils have been identified as *Aulichnites* (Fedonkin, 1981; Waggoner, 1998), *Archaeonassa* (Jensen, 2003), or as *Sellaulichnus* (Fedonkin and Runnegar, 1992). The assignment to *Sellaulichnus* for this type of trace is doubtful in view of Zhu's (1997) examination of the type material, which showed that the bilobed epireliefs might be a result of burrow collapse. Furthermore, *Sellaulichnus* forms branching burrow systems. The definitions of Phanerozoic *Archaeonassa* and *Aulichnites* are not without problems (see Buckman, 1994; Yochelson and Fendonkin, 1998; Mángano *et al.*, 2002) and it may turn out that neither is entirely suitable for the Ediacaran trace fossils. The clear evidence for sediment displacement makes these some of the least problematic Ediacaran trace fossils.

4.1.2 *Beltanelliformis*-type fossils

A reasonable case has been made to compare Ediacaran globular structures variously referred to as *Beltanelliformis, Beltanelloides, Nemiana,* and *Hagenetta* with Phanerozoic plug-shaped burrows such as *Bergaueria* (Fedonkin and Runnegar, 1992; Crimes, 1994; Crimes and Fedonkin, 1996). The distinction between a shallow plug-shaped trace fossil and the imprint or cast of a body fossil is indeed problematic. It is, however, more likely that these represent body fossils rather than trace fossils (see Jensen, 2003), and that these are examples of common and long-ranging Ediacaran body fossils, many probably of non-metazoan origin (Xiao *et al.*, 2002). A crucial feature is the absence of evidence for movement as well as transition to similar forms preserved as carbonized compressions. To this group should probably also be related *Nimbia occlusa* and Ediacaran reports of *Circulichnus*. Several specimens reported as *Intrites* may belong to this group (see McIlroy *et al.*, 2005 for discussion).

Of particular interest are reports of Ediacaran *Bergaueria* associated with lunates structures interpreted to suggest lateral movement (e.g. Fedonkin, 1981). The Ediacaran forms have been compared to *Bergaueria sucta,* a Cambrian trace fossils interpreted as recording lateral movement of an actinian type cnidarian (Seilacher, 1990). Leaving aside the interpretation of the Cambrian forms, we suggest that the signs of lateral movement in the

Ediacaran forms are a structural part of a globular structure, and that comparison should instead be made to the complex tubular fossil *Gaojiashania* from the terminal Ediacaran of China (see Zhang, 1986).

4.1.3   *Bilinichnus*

Bilinichnus consists of two parallel ridges on bed soles, and was interpreted as marginal imprints of a creeping organism (Fedonkin, 1981). There have also been reports of *Bilinichnus* in the Cambrian (e.g., Fedonkin et al., 1985; Pacześna, 1996), but these appear to be different. The biogenicity of the type material of *Bilinichnus* has been questioned by Runnegar (1992), who labelled it a pseudofossil, and doubts were raised also by Keighley and Pickerill (1996) and Buatois et al. (1998).

4.1.4   *Chondrites*

Reports of the the regularly and repeatedly branching *Chondrites* from the Ediacaran (e.g., Crimes and Germs, 1982; Jenkins, 1995) are unconvincing. They lack the orderly branching systems of Phanerozoic examples and likely are chance occurrences of simple unbranched structures.

4.1.5   *Cochlichnus*

Cochlichnus is a small, regularly sinuous, horizontal trace fossil. Though *Cochlichnus* has been reported from the Ediacaran (Cope, 1983; Palij et al., 1979), we know of no convincingly documented examples. As noted by Seilacher et al. (2005, p. 331), the absence of Ediacaran regular sinusoidal trace fossils may be significant. However, there are probable *Cochlichnus* in the terminal Ediacaran Huns Member of southern Namibia (Fig. 2E).

4.1.6   *Didymaulichnus*

Didymaulichnus is a trace fossil with a bilobed lower surface without obvious ornamentation. It occurs widely in the lower Cambrian, including *Didymaulichnus miettensis*, a large form restricted to the lower Cambrian characterized by prominent lateral bevels (e.g., Walter et al., 1989). The Ediacaran record of *Didymaulichnus* is dubious. Most occurrences (e.g., Poire et al., 1984) are in sections stratigraphically close to the Cambrian and with few geochronological constraints. The type material of *Didymaulichnus miettensis* occurs in strata that are terminal Ediacaran–early Cambrian in age (Young, 1972). However, all other known occurrences of this wide-spread and distinctive ichnospecies, as well as of *Didymaulichnus tirasensis*, are associated with diverse Cambrian-type trace fossils.

4.1.7 Gyrolithes

This small, vertical, spirally coiled trace fossil is common in lower Cambrian strata, and straddle the GSSP in Newfoundland (e.g., Pacześna, 1996, Jensen, 1997; Gehling *et al.*, 2001; Droser *et al.*, 2002a), but appears to be absent in the Ediacaran. The single report known to us (Jenkins, 1981) has never been substantiated by a photographic documentation.

4.1.8 Harlaniella

The small rope-like *Harlaniella podolica* is a candidate for reinterpretation as a body fossil (Jensen, 2003). It is principally known from the Ukraine and Newfoundland and is the name-bearer of a latest Ediacaran trace fossil zone, the *Harlaniella podolica* Zone (Narbonne *et al.*, 1987). As pointed out by Palij (1976), *Harlaniella* is often found together with and appears to grade into *Palaeopascichnus*. The re-interpretation of *Palaeopascichnus* casts doubt also on *Harlaniella*. The common reconstruction of this form as a spiral coil is geometrically problematic (see Jensen, 2003).

4.1.9 *Helminthoidichnites*-type trace fossils

These are without doubt the most common Ediacaran trace fossils (Fig. 2B). They appear to have been essentially horizontal and probably formed within the uppermost 10 mm of sediment (e.g. Narbonne and Aitken, 1990; Droser *et al.*, 2005). Depending on the type of meandering they have been assigned to various Phanerozoic ichnotaxa, such as *Helminthoidichnites*, *Gordia* and *Helminthopsis*. A distinctive feature of the Ediacaran examples is their common preservation as negative epireliefs or negative hyporeliefs. Sediment may have been displaced to form narrow marginal raised ridges or levees. These combined features suggest that these trace fossils were formed by animals displacing sediment and leaving behind a tunnel that stayed open for some time after the animal's passage. Some specimens appear to show guided meanders similar to *Helminthoida* (or perhaps more properly *Helminthorhaphe*) (Narbonne and Aitken, 1990) and there is also a specimen bearing comparison to *Spirorhaphe* (Jensen, 2003).

While it is not in doubt that these include genuine trace fossils, there are good reasons to be exceptionally cautious in the interpretation of individual cases. They are essentially horizontal and therefore preservationally similar to tubular organisms. As noted by Fedonkin and Runnegar (1992, p. 391)

A Critical Look at the Ediacaran Trace Fossil Record 139

they often show abrupt bends. While this could indeed suggest that the producing animal was equi-dimensional (Fedonkin and Runnegar, 1992, p. 391), it need be considered that some of the abrupt bends are better explained as those of tubular organisms.

4.1.10 *Lockeia*

Lockeia (junior synonym *Pelecypodichnus*), is a trace fossil consisting of almond-shaped casts, generally attributed to infaunal bivalves. Reports of Ediacaran *Lockeia* (Zhang, 1986; Narbonne and Aitken, 1990; Jenkins, 1995; McMenamin 1996) are not convincing as they lack the distinctive shape and mode of occurrence of their Phanerozoic counterparts.

4.1.11 *Monomorphichnus*

Monomorphichnus consists of series of ridges that may be repeated laterally, and which have been explained as leg imprints of swimming or grazing arthropods. It appears close to the base of the Cambrian (e.g., Narbonne *et al.*, 1987) and represents the earliest arthropod-type trace fossil. There have been reports of Ediacaran *Monomorphichnus* but none are convincing. Sets of paired ridges from the Ediacaran of South Australia, once interpreted as arthropod-type scratch marks (e.g., Jenkins, 1995), are now interpreted as rasping structures (see "*Radulichnus*") comparable to radular marks (Seilacher *et al.*, 2005; Gehling *et al.*, 2005).

4.1.12 *Neonereites*

Neonereites consists of irregular chains, typically uniserial or biserial, of dimples or knobs. It has long been a topic of discussion if it represents a separate ichnogenus or a preservational variant of the ichnogenera *Scalarituba, Phyllodicites, Nereites* and *Helminthoida* (e.g., Uchman, 1995). The vast majority of Ediacaran reports of this ichnogenus (e.g., Fedonkin, 1981; Chistyakov *et al.*, 1984) are palaeopascichnid-type fossils and therefore not trace fossils (Gehling *et al.*, 2000; Jensen, 2003; Seilacher *et al.*, 2003, 2005). Some reports of Ediacaran *Neonereites* may represent treptichnids; the *Neonereites* isp.? of Narbonne and Hofmann (1987) compares to a preservational variety of *Treptichnus* isp. from Nambia (cf., Jensen *et al.*, 2000, Fig. 2B).

4.1.13 *Palaeopascichnus*-type fossils

Palaeopascichnus and numerous other similar forms are some of the more familiar structures that should be removed from the tally of Ediacaran trace fossils (Haines, 1990; Gehling *et al.*, 2000; Jensen, 2003; Seilacher *et al.*, 2003; Shen *et al.*, in press). The exact morphology and affinities of these organisms are still unclear. Gehling *et al.* (2000) found tantalizing if inconclusive connections to *Aspidella*, whereas Seilacher *et al.*, (2003) linked them to xenophyophoran protists. *Palaeopascichnus* preserved in carbonates have been interpreted as stratiform stromatolites (Runnegar, 1995), or compared to algae (Haines, 1990). What is clear is that these were organisms growing strictly in two dimensions (though some show spiral twisting) and possibly in direct connection to microbial mats. *Palaeopascichnus* and *Yelovichnus* are clearly identical. Other Ediacaran structures described as *Neonereites renarius*, *Neonereites biserialis*, *Neonereites uniserialis* and *Catellichnus* also are related structures. *Orbisiana* represents a further preservational variation in shales (Jensen, 2003), and this may apply also to *Catenasphaerophyton* and *Serisphaera* from the Ediacaran of South China (see Xiao and Dong, this volume). The same is probably the case for some but not all reports of *Intrites* (see McIlroy *et al.*, 2005).

4.1.14 *Planolites-Palaeophycus*

These are short horizontal or oblique trace fossils without a distinct meandering pattern and represent some of the more widely reported Phanerozoic trace fossils. This type of trace fossil presents problems in terms of interpretation and naming. The morphology is sensitive to preservation, often being preserved as only a small fraction of the original burrow. Many papers have dealt with features by which these two ichnotaxa can be distinguished, mainly with *Palaeophycus* being lined open burrows with a passive fill and *Planolites* a lined burrow with an active fill (e.g., Pemberton and Frey, 1982, Keighley and Pickerill, 1995), but the practical application of these characters is problematic (e.g., Jensen, 1997). The presence or absence of a lining often is a subtle distinction and one strongly controlled by the nature and state of preservation of the trace fossil. Although there have been numerous reports of Ediacaran *Planolites,* we are unaware of any examples where it can be demonstrated that these were formed by reworking of sediment. It is doubtful that the distinction of Ediacaran *Planolites* and *Palaeophycus* is meaningful. As with all such simple structures care is needed in distinguishing these from casts and molds of tubular organisms.

4.1.15 "*Radulichnus*"

Paired fine ridges, arranged in fans and occasionally associated with *Kimberella* in South Australia and the White Sea region represent structures analogous to mollusk radula-type grazing (Gehling, 1996; Seilacher, 1995, 1997; Fedonkin, 2003; Seilacher *et al.*, 2003, 2005; Gehling *et al.*, 2005). The animal was located at the apex of the fan from which it appears to have scraped biomats with an extensible proboscis (Gehling *et al.*, 2005). This trace fossil is broadly analogues to the rasping trace fossils of mollusks (*Radulichnus*) and echinoids (*Gnatichnus*). The Ediacaran raspings have been assigned to *Radulichnus* (Seilacher 1995) but are sufficiently different in morphological details and preservation that they should be assigned to a new ichnogenus (cf., Gehling, 1996).

4.1.16   *Skolithos*

Skolithos consists of vertical or inclined unbranched burrows that are typically cylindrical. *Skolithos*-type trace fossils may represent a range of behaviors and widely different producers. They may be a structure built by a mobile organism of small dimension relative to the tube or a protective dwelling structure closely corresponding to the animal's size. A further possibility, which in particular applies to short vertical knobs, is that they may represent an anchoring structure of an organism that otherwise protruded out of the sediment. Such vertical structures are sometimes surrounded by concentric ridges formed by the rotation of the anchored organism (see Jensen *et al.*, 2002). Ediacaran reports of *Skolithos* are problematic. An interpretation as a basal attachment should be considered for short knobs such as the *Skolithos* isp. from the Ediacaran of the White Sea area figured by Sokolov (1997, pl. 24:3). Some are short fragments that may as well be the vertical portion of a *"Planolites"*-type trace (Geyer and Uchman, 1995). The most convincing Ediacaran *Skolithos* so far described is *S. declinatus* from the White Sea Area (Fedonkin, 1985). Precise information on its occurrence and a more detailed presentation would be of great interest.

4.1.17   *Torrowangea*

Torrowangea rosei is a sinuous to meandering trace fossil with regularly spaced constrictions, which may reflect peristalsis (Narbonne and Aitken, 1990). The type material from the Lintiss Vale Beds (Webby, 1970) is of disputed terminal Ediacaran–earliest Cambrian age (see above). Ediacaran material has been figured by among others Narbonne and Aitken (1990). Several other reported occurrences (Liñan and Palacios, 1987; Lin *et al.*, 1986) do not show these constrictions and are better assigned to other ichnogenera or tubular fossils. Also, in each individual case it is necessary to consider the possible body fossil origin of *Torrowangea*.

4.1.18 Dickinsonid trace fossils

One of the most remarkable recent additions to the record of Ediacaran trace fossils is the discovery of serial "foot prints" in direct continuation with *Dickinsonia* and *Yorgia* in the White Sea area and South Australia (Ivantsov and Malakhovskaya, 2002; Fedonkin, 2003; Gehling *et al.*, 2005). The producer is preserved in negative relief on the bed sole with sharply defined imprints of the body. The associated trace fossils have the general size and outline of the producer but are preserved in low positive relief with diffuse impressions of the body (Fig. 2D). In addition to providing solid evidence that dickinsonids were mobile, these associations show that dickinsonids lived in close association to biomats, which likely provided a food source (Fedonkin, 2003; Gehling *et al.*, 2005).

4.1.19 Meniscate trace fossils

There have been a number of reports of Ediacaran meniscate trace fossils; some were identified as *Taenidium* and *Muensteria*, suggesting manipulation of material packed behind the animal in crescent shaped packages of sediment. Alternative explanations as body fossils seem to better explain the majority of the Ediacaran occurrences. Some are probably casts of *Cloudina*, for example the *Muensteria* and *Taenidium* of Germs (1972). Other reports appear to owe their meniscate appearance to deformational wrinkles in tubular organisms. For example, we suggest that this better explains a structure figured by Narbonne and Hofmann (1987, their Fig. 10E).

4.1.20 Star-shaped trace fossils

By this we refer to trace fossils that consist of elements that radiate from a central area. This is clearly a heterogeneous group with a problematic Ediacaran record. One possible source of confusion is with algal rhizoids. These are known from the Doushantou Formation (Steiner, 1994; Xiao *et al.*, 2002), and likely are present also in younger Ediacaran rocks. However, the most common source of confusion is with Ediacara-type fossils with prominent radial elements. For example, *Hiemalora* and *Eoporpita* have radial elements that in themselves are indistinguishable from a simple trace fossil or a ray of a star-shaped trace fossil (e.g., Dzik, 2003; Sokolov, 1997, pl. 18;2,3). In general the radial elements meet in a prominent central structure, which is more consistent with a body fossils, probably an attachment structure, but the distinction is not without problems. A trace fossil interpretation for the medusoid *Mawsonites* (Seilacher *et al.*, 2003) is unlikely (e.g. Runnegar, 1991). *Stelloglyphus* sp. reported from Ukraine may be a further example of a *Mawsonites* type structure. Martin *et al.* (2000) reported that Ediacaran sections in the White Sea area had radial trace fossils at various levels but these remain to be published. Certainly, extending a trace fossil interpretation for *Hiemalora* is hard to accept.

The delicate, radiating to fan-shaped *Oldhamia* has a lower stratigraphic range extending to the Ediacaran-Cambrian boundary, but reported Ediacaran occurrences are generally in successions that are without precise stratigraphic control (see Lindholm and Casey, 1990). The Ediacara Member of South Australia contains finely radiating structures (Fig. 2C) that appear to be trace fossils possibly akin to *Oldhamia*. There are possible, simple *Oldhamia* from the Ediacaran of North Carolina (Seilacher *et al.*, 2005), but it remains debatable if these should be included in *Oldhamia* (cf., Hofmann *et al.*, 1994).

4.1.21 Treptichnids

These are trace fossils consisting of repeatedly branching curved elements, representing three-dimensional burrow systems (Fig. 2F). In bedding-plane expression this type of trace fossil may show isolated vertical pipes. All Ediacaran reports of this type of trace fossil are in strata that are close to the Ediacaran–Cambrian boundary. The best-constrained occurrence is in Namibia where treptichnids are ca. 548–545 Ma and occur within the stratigraphic range of *Cloudina* (Jensen *et al.*, 2000). Seilacher *et al.* (2005, Fig. 6), figured Ediacaran treptichnids from an unspecified formation of the Vanrhynsdorp Group in a subsidiary basin of the Nama basin in South Africa. Other possible Ediacaran treptichnids occur in Spain (see Jensen *et al.*, 2000). Dzik (2005) has suggested that early treptichnids and, indeed, much of the early infaunal activity can be attributed to priapulids.

4.2 Ediacaran Trace Fossil Diversity

Trace fossils are given binomial names using the Linnean nomenclature system. No better alternative has been suggested and some form of label is needed for communication. Ichnotaxa have been repeatedly used as a measure in search of evolutionary or ecological trends, such as the colonization of the deep sea (Orr, 2001; Uchman, 2003). It is, however, important to understand what is entailed in an ichnotaxon. There is a broad consensus among trace fossil workers that trace fossil taxa—ichnogenera and ichnospecies—as far as possible be based on morphology as an expression of behavior (Pickerill, 1994; Bromley, 1996). Most traces do not carry enough morphological information to allow identifying the producer even at the phylum level. The inclusion of two traces fossils in the same ichnotaxon does not imply commonality of producer. Trace fossil names therefore bear no phylogenetic information, even though, for historical reasons, their names, such as *Helminthopsis* or *Archaeonassa*, may suggest so. It also follows that ichnotaxa are not between themselves comparable entities that can be ranked. This is one reason why contrasting numbers of ichnotaxa in itself may have little meaning. It is also important to realize that the inclusion of two trace fossils in the same ichnotaxon does not mean that they necessarily represent the same behavior or feeding strategy. This is particularly true in considering the typical preservation of simple Ediacaran trace fossils as negative reliefs on bed bases and bed tops, a style of preservation not common in the Phanerozoic.

From Table 1, and earlier papers (Jensen, 2003; Seilacher *et al.*, 2003; 2005; Droser *et al.*, 2005), it is clear that the number of Ediacaran ichnotaxa has been inflated. The above examination reinforces the thrust of recent reports (Jensen, 2003; Seilacher *et al.*, 2003, 2005; Droser *et al.*, 2005) that a substantial proportion of the reported Ediacaran trace fossils are better interpreted as body fossils or remain doubtful. In some cases a final verdict must await detailed examination, with particular attention to accompanying biogenic and non-biogenic structures. For reasons discussed above we do not find it useful to provide a numerical count of what we consider genuine Ediacaran ichnogenera. It may, however, be instructive to give an indication of the extent to which the list of Ediacaran ichnogenera need be reduced. Of the 36 ichnogenera in Crimes' (1994) list, we consider that the following 22 either lack a convincing Ediacaran record or they are body fossils: *Asterichnus*, *Beltanelliformis*, *Bergaueria*, *Brooksella*, *Buchholzbrunnichnus*, *Corophioides*, *Gyrolithes*, *Intrites*, *Lockeia*, *Monomorphichnus*, *Neonereites*, *Nereites*, *Nimbia*, *Palaeopascichnus*, *Stelloglyphus*, *Suzmites*, *Syringomorpha*, *Vendichnus*, *Vimenites*, and *Yelovichnus*. Also on Crimes' list *Sellaulichnus* refers to forms better assigned to a different ichnogenus (see *Archaeonassa*-like trace fossils). Furthermore, the following are forms that require additional documentation: *Bilinichnus*, *Medvezichnus*, and *Nenoxites*.

In earlier tabulations several ichnotaxa were restricted to the Ediacaran, and of these several had morphologies that were described as unusual. This led to the suggestion of an end-Ediacaran extinction of ichnogenera (Crimes, 1994). Recent revisions (Gehling *et al.*, 2000; Jensen, 2003; Seilacher *et al.*, 2003; 2005; this paper) suggest that these taxa are all body fossils. There still exist a number of ichnotaxa reported only from the Ediacaran, mainly from China. The majority of these appear to be either body fossils of palaepascichnid type, or could be included in existing ichnotaxa (this paper; Shen *et al.*, in press).

4.3 Stratigraphic Distribution and Broader Implications of Ediacaran Trace Fossils

Trace fossils appear towards the end of the Ediacaran and probably are younger than 560 Ma (Martin *et al.*, 2000; Droser *et al.*, 2002b; Knoll *et al.*, 2004a). The Ediacaran record is strongly dominated by essentially horizontal unbranched forms such as *Helminthoidichnites* and *Helminthopsis* (e.g., Narbonne and Aitken, 1990; Droser *et al.*, 1999). These can be assigned to various ichnogenera but all occupy the same position in the sediment, close to the sediment water interface. Among these are forms with a more complex pattern including open meanders and spirals, though these are very rare. There are Ediacaran trace fossil with a pronounced bilobed upper surface (*Archaeonassa*). Also these represent movement close to the sediment-water interface and are likely transitional to the *Helminthoidichnites*-type trace fossil (Jensen, 2003). In addition there are probable rare radial trace fossils as well as trace fossils from the activity of dickinsonids and *Kimberella*. All of these trace fossils have been interpreted as representing feeding strategies closely related to biomats (e.g., Seilacher, 1999; Gehling *et al.*, 2005). A somewhat greater diversity of trace fossils, including simple branching burrow systems overlap with the upper stratigraphic range of *Cloudina*, close to the Ediacaran–Cambrian boundary (see Jensen *et al.*, 2000; Narbonne *et al.*, 1997; Jensen and Runnegar, 2005). The pattern outlined above suggests two zones of Ediacaran trace fossils (Table 2; see also Jensen, 2003). The temporal and geographic distribution of these zones broadly correspond to, respectively, the White Sea and Nama assemblages of the Ediacara biota (see Waggoner, 1999, 2003; Gehling, 2004; Gehling and Narbonne, 2002; Narbonne, 2005). The presence of trace fossils preceding diverse assemblages of Ediacara-type fossils (Assemblage I of Walter *et al.*, 1989; ?Prot I of Jensen, 2003) is questionable. This largely hinges on the correlation of the Elkera Formation of the Georgina Basin as preceding the Ediacara-biota bearing units of the Adelaide syncline (see Walter *et al.*, 1989; Walter *et al.*, 2000). Further tests of the validity of these trace fossil zones will not only depend on improved precision in the correlation of sections (including new information from chemostratigraphy), and the understanding of facies control on fossil distribution (Grazhdankin, 2004), but also depend on correct identification of trace fossils as well as additional data. For example, the younger Ediacaran trace fossil zone is so far confidently recognized only in Namibia, with central Spain providing a probable additional case (Jensen *et al.*, 2000). Additional and updated information on trace fossils from the *Cloudina*-bearing Dengying Formation of south China will be of great interest. This younger zone is also contingent on the assumption that the upper range of *Cloudina* and a pronounced negative excursion in carbon isotopes roughly corresponds to the basal Cambrian GSSP. The vast majority of the Ediacaran trace fossil record is from shallow-marine strata. Relatively scant data suggests that Ediacaran deep-sea bottoms were also colonized by mobile benthos (see Seilacher *et al.*, 2005). It should be noted that no trace fossils have been found associated with the oldest (ca. 575–560 Ma) assemblage of Ediacara-type fossils, the Avalon assemblage, which occur in deep-water volcanoclastic settings (Narbonne, 2005).

Table 2. Stratigraphic distribution of Ediacaran trace fossils and trace fossil-like fossils with respect to Ediacaran assemblages.

Assemblages	Characteristic trace fossils	Trace fossil-like fossils
Nama/Namibian: ca 550–542 Ma. Low diversity of Ediacara-type fossils. *Cloudina* and other biomineralized tubes. E.g. Schwarzrand Group, Dengying Formation.	Upper age range with treptichnids and trace fossils with three-lobed lower surface.	Palaeopascichnid-type fossils. Tubular organisms including cloudinids and sabelliditids.
White Sea/Vendian: ca 560–550 Ma. Diverse Ediacara-type fossils including bilaterally symmetrical forms. E.g., Ediacara Member, Ust Pinega Formation.	Unbranched horizontal trace fossils (*Helminthoidichnites* etc). Dickinsonid imprints. Raspings.	Palaeopascichnid-type fossils. Tubular organisms.
Avalon: ca 575–560 Ma. Fronds and fractal forms. E.g. Mistaken Point Fm.	No trace fossils.	Palaeopascichnid-type fossils.

Ediacaran trace fossils provide important constraints on animal evolution. The absence of trace fossils prior to about 560 Ma makes the existence of mobile benthic animals in deep time unlikely (e.g., Budd and Jensen, 2000, 2003; Jensen *et al.*, 2005a). The simple trace fossils of the late Ediacaran are evidence of mobile organisms, probably small bilaterians, but the diversity and complexity of these trace fossils is low. The evidence for a through-gut in the form of rows of supposed fecal material is no longer convincing (but see Seilacher *et al.*, 2005, for a new interpretation of *Nenoxites*). It is quite possible that the producers of Ediacaran trace fossils have left no fossil record. Priapulids, for example, may have been among the earliest burrowers (e.g., Valentine, 1994) and more specifically have been linked to treptichnid-type trace fossils (Dzik, 2005). Nevertheless, and in spite of the much needed rethinking of Ediacara-type fossils triggered by the Vendobiota hypothesis (e.g., Seilacher 1999), the Ediacara biota was an assemblage diverse in size and biological affinities and it remains an obvious place to look for early bilaterians and trace makers. The apparently unusual morphologies may have their explanation in unusual ecologies (e.g., Gehling *et al.*, 2005; Dornbos *et al.*, 2005). The association of members of the Ediacara biota, notably *Dickinsonia* and *Kimberella,* with trace fossils shows that at least some of these were mobile. Furthermore, taphonomic features of *Dickinsonia* suggest that it possessed muscular tissue (see Gehling *et al.*, 2005). The identity of the producers of the *Helminthoidichnites*-type trace fossils is unknown but could include some of the smaller bilaterally symmetrical Ediacara-type fossils. The increase in depth and intensity of bioturbation through the Ediacaran–Cambrian transition led to a restriction of biomats and to a change from firm to increasingly soupy surface sediments (Seilacher and Pflüger, 1994; McIlroy and Logan, 1999; Bottjer *et al.*, 2000; Droser *et al.*, 2002a,b; Dornbos *et al.*, 2004, 2005). Biomat-related life-styles continued also into the Cambrian (e.g., Dornbos *et al.*, 2004), and may be seen in *Psammichnites*-type trace fossils (Hagadorn *et al.*, 2000). As observed by Seilacher *et al.* (2005, p. 331), the absence of Ediacaran arthropod-type trace fossils and regular sinusoidal trace fossils is notable. Probable *Cochlichnus* (see above) in strata with treptichnids is further evidence of a modest increase in trace fossil diversity immediately preceeding the more dramatic Cambrian diversification of trace fossils.

ACKNOWLEDGEMENTS

SJ acknowledges funding from the Spanish Ministry of Education and Science (MEC) through Programa Ramon y Cajal and project CGL-2004-02967 (co-financed by FEDER). Aspects of this research was supported by a National Science Foundation Grant to MLD (EAR-0074021) and an Australian Research Council Grant to JGG. This paper benefitted greatly from the reviews of Luis A. Buatois, Stephen Q. Dornbos and Robert B. MacNaughton, as well as from the insights and editorial assistance of Shuhai Xiao.

REFERENCES

Aceñolaza, F. G., and Alonso, R. N., 2001, Icno-asociaciones de la transición Precambrico/Cámbrico en el noroeste de Argentina, *J. Iberian Geol.* **27**: 11–22.

Aceñolaza, F. G., Bettucci, L. S., and Fernicola, J. C., 1998, Icnofósiles del Grupo Lavalleja Neoproterozoico de Uruguay, *Coloquios de Paleontologia* **49**: 9–21.

Aceñolaza, G. F., 2004, Precambrian–Cambrian ichnofossils, and enigmatic "annelid tube" and microbial activity in the Puncoviscana Formation (La Hilguera; Tucumán Province, NW Argentina), *Geobios* **37**: 127–133.

Alpert, S. P., 1974, Trace fossils of the Precambrian–Cambrian succession, Whitey-Inyo mountains, California, Ph D. Thesis, Univeristy of California, Los Angeles.

Alpert, S. P., 1975, *Planolites* and *Skolithos* from the upper Precambrian–Lower Cambrian, White-Inyo mountains, California, *J. Paleontol.* **49**: 508–521.

Alpert, S. P., 1977, Trace fossils and the basal Cambrian, in: *Trace fossils 2* (T. P. Crimes and J. C. Harper, eds.), pp. 1–8, Geological Journal Special Issue 9, Seel House Press, Liverpool

Amthor, J. E., Grotzinger, J. P., Schröder, S., Bowring, S. A., Ramezani, J., Martin, M. W., and Matter, A., 2003, Extinction of *Cloudina* and *Namacalathus* at the Precambrian–Cambrian boundary in Oman, *Geology* **31**: 431–434.

Banks, N. L., 1970, Trace fossils from the late Precambrian and Lower Cambrian of Finnmark, Norway, in: *Trace fossils* (T. P. Crimes and J. C. Harper, eds.), pp. 91–138, Geological Journal Special Issue 3, Seel House Press, Liverpool.

Bartley, J. K., Pope, M., Knoll, A. H., Semikhatov, M.A., and Petrov, P. Yu., 1998, A Vendian–Cambrian boundary succession from the northwestern margin of the Siberian Platform: stratigraphy, paleontology, chemostratigraphy and correlation, *Geol. Mag.* **135**: 473–494.

Bekker, Yu. R., 1992, Drevnejshaya ediakarskaya biota urala, *Izvestia Akademiya Nauk, Seria Geologicheskaya* **1992**(6): 16–24.

Bekker, Yu. R., and Kishka, N. V., 1989, Otkrytie ediakarskoj bioty na juzhnom Urale, in: *Teoreticheskie i prikladnye aspekty sovremennoj paleontologii* (T .N. Bogdanov and L. I. Khozatsky, eds.), pp. 109–120, Nauka, Leningrad.

Bekker, Yu. R., and Kishka, N. V., 1991, Iskopaemye sledy v verkhnevendskikh otlozheniya juzhnogo urala, *Izvestia Akademiya Nauk, Seria Geologicheskaya* **1991**(6): 66–78.

Bottjer, D. J., Hagadorn, J. W., and Dornbos, S. Q., 2000, The Cambrian substrate revolution, *GSA Today* **10**(9): 1–7.

Brasier, M. D., and McIlroy, D., 1998, *Neonereites uniserialis* from c. 600 Ma year old rocks in western Scotland and the emergence of animals, *J. Geol. Soc. London* **155**: 5–12.

Brasier, M. D., and Shields, G., 2000, Neoproterozoic chemostratigraphy and correlation of the Port Askaig glaciation, Dalradian Supergroup of Scotland, *J. Geol. Soc. London* **157**: 909–914.

Bromley, R. G., 1996, *Trace fossils, Biology, Taphonomy and Applications*, 2nd ed., Chapman and Hall, London.

Buatois, L. A., and Mángano, M. G., 2003, La icnofauna de la Formacción Puncoviscana en le noroestes argentino: Implicancias en la colonización de fondos oceánicos y reconstrucción de paleoambientes y paleoecosistemas de la transición precámbrica-cámbrica, *Ameghiniana* **40**: 103–117.

Buatois, L. A., and Mángano, M. G., 2004, Terminal Proterozoic–Early Cambrian ecosystems: ichnology of the Puncoviscana Formation, northwest Argentina, *Fossils and Strata* **51**, 1–16.

A Critical Look at the Ediacaran Trace Fossil Record

Buatois, L. A., and Mángano, M. G., 2005, The Cambrian system in northwestern Argentina: stratigraphical and palaeontological framework. Discussion, *Geologica Acta* **3**: 65–72.

Buatois, L. A., Mángano, M. G., Maples, C. G., and Lanier, W. P., 1998, Taxonomic reasessment of the ichnogenus *Beaconichnus* and additional examples from the Carboniferous of Texas, *Ichnos* **5**: 287–302.

Buckman, J. O., 1994, *Archaeonassa* Fenton and Fenton 1937 reviewed, *Ichnos* **3**: 185–192.

Budd, G. E., and Jensen, S., 2000, A critical reappraisal of the fossil record of the bilaterian phyla, *Biol. Rev.* **75**: 253–295.

Budd, G. E., and Jensen, S. 2003., The limitations of the fossil record and the dating of the origin of the Bilateria, in: *Telling the evolutionary time: molecular clocks and the fossil record* (P.C.J. Donoghue, and M. P. Paul, eds.), pp. 166–189, CRC Press, Boca Raton.

Chen, M., Chen, Y., and Qian, Y., 1981, Some tubular fossils from Sinian–Lower Cambrian boundary sequences, Yangtze Gorge, *Bull. Tianjin Inst. Geol. Mineral Resources* **1981**(3): 117–124.

Chen, Z., Sun, W., and Hua, H., 2002, Preservation and morphologic interpretation of late Sinian *Gaojiashania* from southern Shaanxi, *Acta Palaeontol. Sinica* **41**: 448–454.

Chistyakov, B. G., Kalmykova, N. A., Nesov, L. A., and Suslov, G. A., 1984, O nalichii vendskikh otlozhenij v srednem techenii r. Onegi i vozmozhnom syshchestvovanii obolochnikov (Tunicata: Chordata) v dokembrii, *Vestnik Leningradskogo gosudarstvennogo Universiteta* **1984**(6): 11–19.

Cloud, P. E., and Nelson, C. A., 1966, Phanerozoic–Cryptozoic and related transitions: new evidence, *Science* **154**, 1–5.

Cloud, P. E., Wright, J., Glover, L., 1976, Traces of animal from 620 million-year old rocks on North Carolina, *Am. Sci.* **64**: 396–406.

Condon, D., Zhu, M., Bowring, S., Wang, W., Yang, A., and Jin, Y., 2005, U–Pb ages from tthe Neoproterozoic Doushantou formation, China, *Science* **308**: 95–99.

Conway Morris, S., 2003, The Cambrian "explosion" of metazoans and molecular biology: would Darwin be satisfied? *Int. J. Developmental Biol.* **47**: 505–515.

Cope, J. C. W., 1983, Precambrian fossils of the Carmarthen area, Dyfed, *Natur in Wales* **1**(2): 11–16.

Corsetti, F. A., and Hagadorn, J. W., 2003, The Precambrian–Cambrian transition in the southern Great Basin, USA. *The Sedimentary Rec.* **1**: 4–8.

Crimes, T. P., 1987, Trace fossils and correlation of late Precambrian and early Cambrian strata, *Geol. Mag.* **124**: 97–119.

Crimes, T. P., 1992, Changes in the trace fossil biota across the Proterozoic–Phanerozoic boundary, *J. Geol. Soc. London* **149**: 637–646.

Crimes, T. P., 1994, The period of early evolutionary failure and the dawn of evolutionary success: the record of biotic changes across the Precambrian–Cambrian boundary, in: *The Palaeobiology of Trace fossils* (S. K. Donovan, ed.), pp. 105–133, John Wiley, Chichester.

Crimes, T. P., and Fedonkin, M. A., 1996, Biotic changes in platform communities across the Precambrian–Phanerozoic boundary, *Rivista Italiana de Paleontologia i Stratigrafia* **102**: 317–332.

Crimes, T. P., and Germs, G. J. B., 1982, Trace fossils from the Nama Group (Precambrian–Cambrian) of southwest Africa (Namibia), *J. Paleontol.* **56**: 890–907.

Ding, Q., and Xing, Y., 1988, Trace fossils, in: *The Sinian System of Hubei* (Z. Zhao et al., eds.), pp. 182–185, China University of Geosciences Press, Wuhan.

Ding, Q., Xing, Y., and Chen, Y., 1985, Metazoa and trace fossils, in: *Biostratigraphy of the Yangtze Gorge Area 1. Sinian* (Z. Zhao, ed.), pp. 115–117, Geological Publishing House, Beijing.

Ding, Q., Xing, Y., Wang, Z., Yin, C., and Gao, L., 1993, Tubular and trace fossils from the Sinian Dengying Formation in the Miaohe–Liantuo area, Hubei Province, *Geol. Rev.* **39**, 118–123

Dornbos, S. Q., Bottjer, D. J., and Chen, J., 2004, Evidence for seafloor microbial mats and associated metazoan lifestyles in Lower Cambrian phosphorites of Southwest China, *Lethaia* **37**: 127–137.

Dornbos, S. Q., Bottjer, D. J., and Chen, J., 2005, Paleoecology of benthic metazoans in the Early Cambrian Maotianshan Shale biota and the Middle Cambrian Burgess Shale biota: evidence for the Cambrian substrate revolution, *Palaeogeogr. Palaeoclimatol. Palaeoecol.* **220**: 47–67.

Droser, M. L., Gehling, J. G., and Jensen, S., 1999, When the worm turned; concordance of Early Cambrian ichnofabric and trace fossil record in siliciclastic rocks of South Australia, *Geology* **27**: 625–629.

Droser, M. L., Jensen, S., Gehling, J. G., Myrow, P., and Narbonne, G.M., 2002a, Lowermost Cambrian Ichnofabrics from the Chapel Island Formation, Newfoundland: implications for Cambrian substrates, *Palaios* **17**: 3–15.

Droser, M. L., Jensen, S., and Gehling, J.G. 2002b, Trace fossils and substrates of the terminal Proterozoic–Cambrian transition: implications for the record of early bilaterians and sediment mixing, *Proc. Nat. Acad. Sci. USA* **99**: 12572–12576.

Droser, M. L., Gehling, J.G., and Jensen, S., 2005, Ediacaran trace fossils: true and false, in: *Evolving Form and Function: Fossils and Development* (D. E. G. Briggs, ed.), pp. 125–138, Peabody Museum of Natural History, New Haven.

Droser, M. L., Gehling, J.G., and Jensen, S., 2006, Assemblage palaeoecology of the Ediacara biota: the unabridged edition? *Palaeogeogr. Palaeoclimatol. Palaeoecol.* **232**: 131–147.

Dzik, J., 2003, Anatomical information content in the Ediacaran fossils and their possible zoological affinities, *Integr. Comparative Biol.* **32**: 114–126.

Dzik, J., 2005, Behavioral and anatomical unity of the earliest burrowing animals and the cause of the "Cambrian explosion", *Paleobiology* **31**: 503–521.

Ekdale, A. A., Bromley, R. G., Pemberton, S. G., 1984, *Ichnology: The use of Trace Fossils in Sedimentology and Stratigraphy.* SEPM Short Course 15, 317 pp.

Farmer, J., Vidal, G., Moczydłowska, M., Strauss, H., Ahlberg, P., and Siedlecka, A., 1992, Ediacaran fossils from the Innerelv Member (late Proterozoic) of the Tanafjorden area, northeastern Finnmark, *Geol. Mag.* **129**: 181–195.

Fedonkin, M. A., 1976, Sledy mnogokletochnykh iz Valdajskoj serii, *Izvestia Akademiya Nauk, Seriya Geologicheskaya* **1976**(4): 129–132.

Fedonkin, M. A., 1980, Iskopaemye sledy dokembrijskikh metazoa, *Izvestia Akademiya Nauk SSSR, Seriya Geologicheskaya* **1980**(1): 39–46.

Fedonkin, M. A., 1981, *Belomorskaya biota venda,* Trudy Akademii Nauk SSSR 342, pp. 1–100.

Fedonkin, M. A, 1985, Paleoichnology of Vendian metazoa, in: *The Vendian system: historic-geological and paleontological basis* (B. S. Sokolov and M. A. Ivanovskiy, eds.), pp. 112–116, Nauka, Moscow.

Fedonkin, M. A., 2003, The origin of the Metazoa in the light of the Proterozoic fossil record, *Palaeontol. Res.* **7**: 9–41.

Fedonkin, M. A., and Runnegar, B. N., 1992, Proterozoic metazoan trace fossils, in: *The Proterozoic Biosphere: a multidisciplinary study* (J. W. Schopf, and C. Klein, eds.), pp. 389–395. Cambridge University Press, Cambridge.

Fedonkin, M. A., Liñán, E., and Perejón, A., 1985, Icnofósiles de las rocas precámbrico-cámbricas de la Sierra de Cçordoba. España, *Boletín de la Real Sociedad Española de Historia Natural, Sección Geológica* **81**: 125–138.

Gehling, J. G., 1991, The case for Ediacaran fossil roots to the Metazoan tree, *Geol. Soc. India Mem.* **20**: 181–224.

Gehling, J. G., 1996, The stratigraphy and sedimentology of the late Precambrian Pound Sungroup. [dissertation]. University of California, Los Angeles, 234 pp.

Gehling, J. G., 1999, Microbial mats in Terminal Proterozoic siliciclastics: Ediacaran death masks, *Palaios* **14**: 40–57.

Gehling, J. G., 2004, Fleshing out the Ediacaran period, in: Abstract Volume for the Workshop of Meeting 2 of IGCP493 (UNESCO) (P. Komarower and P. Vickers-Rich eds.), 4 pages (unpaginated).

Gehling, J. G., and Narbonne, G. M., 2002, Zonation of the terminal Proterozoic (Ediacaran), *Geol. Soc. Australia Abstr.* **68**, 63–64.

Gehling, J. G., Narbonne, G. M., and Anderson, M. M., 2000, The first named Ediacaran body fossil, *Aspidella terranovica*, *Palaeontology* **43**: 427–456.

Gehling, J. G., Jensen, S. Droser, M. L., Myrow, P. M., and Narbonne, G. M., 2001, Burrowing below the basal Cambrian GSSP, Fortune Head, Newfoundland. *Geol. Mag.* **138**: 213–218.

Gehling, J. G., Droser, M. L., Jensen, S., and Runnegar, B. N., 2005, Ediacaran organisms: relating form to function, in: *Evolving Form and Function: Fossils and Development* (D. E. G. Briggs, ed.), pp. 43–66, Peabody Museum of Natural History, New Haven.

Germs, G. J. B., 1972, Trace fossils from the Nama Group, south-west Africa, *Journal of Paleontology* **46**: 864–870.

Germs, G. J. B., 1973, Possible sprigginid worm and a new trace fossil from the Nama Group, South West Africa, *Geology* **1**: 69–70.

Geyer, G., and Uchman, A., 1995, Ichnofossil assemblages from the Nama Group (Neoproterozoic–Lower Cambrian) in Namibia and the Proterozoic–Cambrian boundary problem revisited, in: *Morocco '95. The Lower-Middle Cambrian standard of western Gondwana* (G. Geyer and E. Landing, eds.), pp. 175–202. Beringeria Special Issue 2

Gibson, G. G., 1989, Trace fossils from late Precambrian Carolina Slate Belt, south-central North Carolina, *J. Paleontol.* **63**: 1–10.

Glaessner, M. F., 1969, Trace fossils from the Precambrian and basal Cambrian, *Lethaia* **2**: 369–393.

Glaessner, M. F., 1977, Re-examination of *Archaeichnium*, a fossil from the Nama Group, *Ann. South African Museum* **74**: 335–342.

Grant, S. W. F., 1990, Shell structure and distribution of *Cloudina*, a potential index fossil for the terminal Proterozoic, *Am. J. Sci.* **290A**, 261–294.

Grazhdankin, D., 2003, Structure and depositional environment of the Vendian complex in the southeastern White Sea area, *Stratigr. Geol. Correlation* **11**: 313–331.

Grazhdankin, D., 2004, Patterns of distribution in the Ediacaran biotas: facies versus biogeography and evolution, *Paleobiology* **30**: 203–221.

Grotzinger, J., Bowring, S. A., Saylor, B. Z., and Kaufman, A. J., 1995, Biostratigraphic and geochronologic constraints on early animal evolution, *Science* **270**: 598–604.

Gureev, Yu. A., 1981, Nova znakhidka zhittediyalnosti o vidkladakh Vendu severnogo Pridnistrovya, *Doklady Akademija Nauk USSR, Serija B* **1981**(12): 5–12.

Gureev, Yu. A, 1983, Koltsevye bioglify iz otlozhenij kanilovskoj serii Venda Pridnestrovya, *Geologicheskij Zhurnal* **43**(1):130–132.

Gureev, Yu. A., 1984, Bioglifi fanerozojskogo vidu o vendi podillya ta ikh stratigrafichne znacheniya, *Doklady Akademija Nauk USSR, Serija B* **1984**(4): 5–8.

Gureev, Yu, A., Velikanov, V. A., and Ivantjenko, V. Ya, 1985, Besskeletnaya fauna v otlozheniyakh baltiskoj i berezhkovskoj serij podolii, *Doklady Akademija Nauk USSR, Serija B* **1985**(6): 10–14.

Hagadorn, J. W., and Waggoner, B. M., 2000, Ediacaran fossils from the southwestern Great Basin, United States, *J. Paleontol.* **74**: 349–359.

Hagadorn, J. W., Schellenberg, S. A., and Bottjer, D. B., 2000, Paleoecology of a large early Cambrian bioturbator, *Lethaia* **33**: 142–156.

Haines, P. W., 2000, Problematic fossils in the late Neoproterozoic Wonoka Formation, South Australia, *Precambrian Res.* **100**: 97–108.

Hofmann, H. J., Narbonne, G. M., and Aitken, J. D., 1990, Ediacaran remains from intertillite beds in northwestern Canada, *Geology* **18**: 1199–1202.

Hofmann, H. J., Cecile, M. P., and Lane, L. S., 1994, New occurrences of *Oldhamia* and other trace fossils in the Cambrian of the Yukon and Ellesmere Island, arctic Canada, *Can. J. Earth Sci.* **31**: 767–782.

Isakov, A. V., 1990, Ikhnofauna i drugie teksturnye snaki v pozdnem dokembrem uchuro-aldanskogo vodorazdela, in: *Pozdnij dokembriij i rannij paleozoj sibiri* (V. V. Khomentovsky, *et al.* eds.), pp. 147–155, Novosibirsk.

Ivantsov, A.Yu., and Malakovskaya, Ya. E., 2002, Gigantskie sledy vendskikh zhivotnykh, *Doklady Akademii Nauk* **385**: 382–386.

Jenkins, R. J. F., 1981, The concept of an "Ediacaran Period" and its stratigraphic significance in Australia, *Trans. Roy. Soc. South Australia* **105**: 179–194.

Jenkins, R. J. F., 1995, The problems and potential of using animal fossils and trace fossils in terminal Proterozoic biostratigraphy, *Precambrian Res.* **73**: 51–69.

Jenkins, R. J. F., Plummer, P. S., and Moriarty, K., C., 1981, Late Precambrian pseudofossils from the Flinders Ranges, South Australia, *Trans. Roy. Soc. South Australia* **105**: 67–83.

Jenkins, R. J. F., McKirdy, D. M., Foster, C. B., O'Leary, T., and Pell, S. D., 1992, The record and stratigraphic implications of organic-waller microfossils from the Ediacaran (terminal Proterozoic) of South Australia, *Geol. Mag.* **129**: 401–410.

Jensen, S., 1997, Trace fossils from the Lower Cambrian Mickwitzia sandstone, south-central Sweden, *Fossils and Strata* **42**: 1–111.

Jensen, S., 2003, The Proterozoic and earliest Cambrian trace fossil record; patterns, problems and perspectives, *Integr. Comparative Biol.* **43**: 219–228.

Jensen, S., and Runnegar, B. N., 2005. A complex trace fossil from the Spitskop Member (Ediacaran -?Lower Cambrian) of south Namibia, *Geol. Mag.* **142**: 561–569.

Jensen, S., Saylor, B. Z., Gehling, J. G., and Germs, G. J. B., 2000, Complex trace fossils from the terminal Proterozoic of Namibia, *Geology* **28**: 143–146.

Jensen, S., Gehling, J. G., Droser, M. L., and Grant, S. W. F., 2002, A scratch circle origin for the medusoid fossil *Kullingia*, *Lethaia* **35**: 291–299.

Jensen, S., Droser, M. L., and Gehling, J. G, 2005a, Trace fossil preservation and the early evolution of animals, *Palaeogeogr. Palaeoclimatol. Palaeoecol.* **220**: 19–29.

Jensen, S., Palacios, T., and Martí Mus, M., 2005b, Megascopic filamentous organisms preserved as grooves and ridges in Ediacaran siliciclastics, *Paleobios* **25 Suppl. to No. 2**, 65–66.

Keighley, D. G., and Pickerill, R. K., 1995, Commentary: the ichnotaxa *Palaeophycus* and *Planolites*, historical perspectives and recommendations, *Ichnos* **3**: 301–309.

Keighley, D. G., and Pickerill, R. K., 1996, Small *Cruziana*, *Rusophycus*, and related ichnotaxa from eastern Canada: the nomenclatural debate and systematic ichnology, *Ichnos* **4**: 261–285.

Knoll, A. H., Walter, M. R., Narbonne, G. M., and Christie-Blick, N., 2004a, A new period for the geological time scale, *Science* **305**: 621–622.

Knoll, A. H., Walter, M. R., Narbonne, G. M., and Christie-Blick, N., 2004b, Three "first places" for Ediacaran Period, *Episodes* **27**: 222.

Li, Y., and Ding, L., 1996, Trace fossils, in: *Sinian Miaohe Biota* (Ding L. *et al.*, eds.), pp. 120–127, Geological Publishing House, Beijing.

Li, R., and Yang, S., 1988, Trace fossils near the Sinian–Cambrian boundary in eastern Yunnan and central Sichuan, China, *Geoscience* **2**: 158–174.

Li, Y., Ding, L., Zhang, L., Dong, J., and Chen, H., 1992, Metazoans and trace fossils, in: *The Study of the late Sinian – Early Cambrian biotas from the northern Margin of the Yangtze platform* (Ding, L., Zhang, L., Li, Y., and Dong, J., eds.), pp. 91–106, Scientific and Technical Documents Publishing House, Beijing.

Lin, S., Zhang, Y., and Zhang, L., 1986, Body and trace fossils of metazoa and algal macrofossils from the upper Sinian Gaojiashan Formation, southern Shaanxi, *Shaanxi Geol.* **4**(6): 9–17.

Liñan, E., and Palacios, T., 1987, Asociaciones de pistas fósiles y microorganismos de pared orgánica del Proterozoico, en las facies esquisto-grauváqicas del norte de Cáceres. Consecuensias regionales, *Boletín de la Real Sociedad Española de Historia Natural, Sección Geológica* **82**: 211–232.

Liñan, E., and Tejero, R., 1988, Las formaciones précambricas del antiform de Paracuellos (Cadenas ibéricas), *Boletín de la Real Sociedad Española de Historia Natural, Sección Geológica* **84**: 39–49.

Lindholm, R. M., and Casey, J. F., 1990, The distribution and possible biostratigraphic significance of the ichnogenus *Oldhamia* in the shales of the Blow Me Down Brook Formation, western Newfoundland, *Can. J. Earth Sci.* **27**: 1270–1287.

Mángano, M. G., Buatois, L. A., and Rindsberg, A. K., 2002, Carboniferous *Psammichnites*: systematic re-evaluation, taphonomy and autecology, *Ichnos* **9**: 1–22.

Martin, M. W., Grazhdankin, D. V., Bowring, S. A., Evans, D. A. D., Fedonkin, M. A., and Kirschvink, J. L., 2000, Age of Neoproterozoic Bilaterian Body and Trace Fossils, White Sea, Russia: Implications for Metazoan Evolution, *Science* **288**: 841–845.

Mathur, V. K., and Shanker, R., 1989, First record of Ediacaran Fossils from the Krol Formation of Naini Tal Syncline, *J. Geol. Soc. India* **34**: 245–254.

McIlroy, D, and Logan, G. A., 1999, The impact of bioturbation on infaunal ecology and evolution during the Proterozoic-Cambrian transition, *Palaios* **14**: 58–72.

McIlroy, D., Crimes, T. P., and Pauley, J. C., 2005, Fossils and matgrounds from the Neoproterozoic Longmyndian Suopergroup, Shropshire, UK, *Geol. Mag.* **142**: 441–455.

McMenamin, M., 1996, Ediacaran biota from Sonora, Mexico, *Proc. Nat. Acad. Sci. USA* **93**: 4990–4993.

Narbonne, G. M., 2005, The Ediacara biota: Neoproterozoic origin of animals and their ecosystems, *Annu. Rev. Earth Planet. Sci.* **33**: 421–442.

Narbonne, G. M., and Aitken, J. D., 1990, Ediacaran fossils from the Sekwi Brook area, Mackenzie mountains, northwestern Canada, *Palaeontology* **33**: 945–980.

Narbonne, G. M., and Hofmann, H. J., 1987, Ediacaran biota of the Wernecke Mountains, Yukon Territory, *Palaeontology* **30**: 647–676.

Narbonne, G. M., Myrow, P. M., Landing, E., and Anderson, M. M., 1987, A candidate stratotype for the Precambrian–Cambrian boundary, Fortune Head, Burin Peninsula, southeastern Newfoundland, *Can. J. Earth Sci.* **24**: 1277–1293.

Narbonne, G. M., Saylor, B. Z., and Grotzinger, J. P., 1997, The youngest Ediacaran fossils from southern Africa. *J. Paleontol.* **71**: 953–967.

Orr, P., 2001, Colonization of the deep-marine environment during the early Phanerozoic: the ichnofaunal record. *Geol. J.* **36**: 265–278.

Pacześna, J., 1986, Upper Vendian and Lower Cambrian ichnocoenoses of Lublin region, *Biuletyn Instytutu Geologicznego* **355**: 31–47.

Pacześna, J., 1996, The Vendian and Cambrian ichnocoenoses from the Polish part of the East-European platform, *Prace Państwowego Instytutu Geologicznego* **152**: 1–77.

Palij, V., M. 1976, Ostatki besskeletnoj fauny i sledy zhiznedeyatelnosti iz otlozhenij verkhnego dokembriya i nizhnego Kembriya Podolii, in: *Paleontologiya i stratigrafiya verkhnego dokembriya i nizhnego paleozoya jugo-zapadna vostochno-evropejskoj platformy*, pp. 63–77, Naukova Dumka, Kiev.

Palij, V. M., Posti, E., and Fedonkin, M. A., 1979, Myagkotelye metazoa i iskopaemye sledy zhivotnykh venda i rannego kembriya, in: *Paleontologiya verkhnedokembrijskikh i kembrijskikh otlozhenij Vostochno-Evropejskoj platformy* (B. M. Keller and A. Yu. Rozanov, eds.), pp. 49–82, Nauka, Moscow.

Pemberton, S. G.,and Frey, R. W., 1982, Trace fossil nomenclature and the *Planolites-Palaeophycus* dilemma, *J. Paleontol.* **56**: 843–881.

Pickerill, R. K., 1994, Nomenclature and taxonomy of invertebrate trace fossils, in: *The Palaeobiology of Trace fossils* (S. K. Donovan, ed.), pp. 3–42, John Wiley, Chichester etc.

Poire, D. G., del Valle, A., and Regalia, G. M., 1984, Trazas fosiles en cuartcitas de la formacion Sierras Bayas (Precambrico) y su comparacion con las de la Formacion Balcarce (Cambro–Ordovicico), sierras septentrionales de la provincia de Buenos Aires. *Noveno Congreso Geologico Argentino, Actas* **4**: 249–266.

Rasmussen, B., Bengtson, S., Fletcher, I. R., and McNaughton, N., 2002, Discoidal impressions and trace-like fossils more than 1200 million years old, *Science* **296**: 1112–1115.

Runnegar, B., 1991, Oxygen and the early evolution of the Metazoa, in: *Metazoan life without oxygen* (C. Bryant ed.), pp. 65–87, Chapman and Hall, London.

Runnegar, B. N., 1992, Proterozoic Metazoan trace fossils, in: *The Proterozoic Biosphere: a multidisciplinary study* (J. W. Schopf, and C. Klein, eds.), pp. 1009–1015, Cambridge University Press, Cambridge.
Runnegar, B., 1994, Proterozoic eukaryotes: evidence from biology and geology, in: *Early Life on Earth* (S. Bengtson, ed.), pp. 287–297, Columbia University Press, New York.
Runnegar, B., 1995, Vendobionta or metazoa? Developments in understanding the Ediacara "fauna", *Neues Jahrbuch für Geologie und Paläontologie Abhandlungen* **195**: 303–318.
Seilacher, A., 1956, Der Beginn des Kambriums als biologische Wende, *Neues Jahrbuch für Geologie und Paläontologie Abhandlungen* **103**: 155–180.
Seilacher, A., 1990, Paleozoic trace fossils, in: *The Geology of Egypt* (R. Said ed.), pp. 649–670, Balkema, Rotterdam.
Seilacher, A., 1995, *Fossile kunst*, Goldschneck, Korb
Seilacher, A., 1997, *Fossil Art*, The Royal Tyrrell Museum of Palaeontology, Drumheller.
Seilacher, A., 1999, Biomat-related lifestyles in the Precambrian, *Palaios* **14**: 86–93.
Seilacher, A., and Pflüger, F., 1994, From biomats to benthic agriculture: a biohistoric revolution, in: *Biostabilization of Sediments* (W. E. Krumbein ed.), pp. 97–105, Bibliotheks- und Informationssystem der Carl von Ossietzky Universität, Oldenburg.
Seilacher, A., Buatois, L. A., and Mángano, M. G., 2005, Trace fossils in the Ediacaran–Cambrian transition: behavioral diversification, ecological turnover and environmental shift, *Palaeogeogr. Palaeoclimatol. Palaeoecol.* **227**: 323–356.
Seilacher, A., Grazhdankin, D., and Legouta, A., 2003, Ediacaran biota: The dawn of animal life in the shadow of giant protists, *Palaeontol. Res.* **7**: 43–54.
Seilacher, A., Meschede, M., Bolton, E. W., Luginsland, H. 2000, Precambrian "fossil" *Vermiforma* is a tectograph, *Geology* **28**: 235–238.
Shanker, R., Mathur, V. K., Kumar, O., and Srivastava, M. C., 1997, Additional Ediacaran biota from the Krol Group, lesser Himalaya, Indian and their significance, *Geosci. J.* **18**: 79–94.
Shen, B., Xiao, S., Dong, L., Zhou, C., and Liu, J., in press, Problematic macrofossils from Ediacaran successions in north China and Chaidam blocks: implications for their evolutionary root and biostratigraphic significance, *J. Paleontol.*
Sokolov, B. S., 1972, Vendskij etap v istorii zemli, in: *Mezhdunarodnyj geologicheskij kongress XXIV sessiya. Dokladov sovetskikh geologov. Problema 7. Paleontologiya*, pp. 114–124, Nauka, Moscow.
Sokolov, B. S., 1997, *Ocherki stanovleniya Venda*, KMK Scientic Press, Moscow.
Sovetov, Yu. K., and Komlev, D. A., 2005, Tillites at the base of the Oselok Group, footfills of the Sayan Mountains, and the Vendian lower boundary in the southwestern Siberian Platform, *Stratigraphy and Geological Correlation* **14**, 337–366.
Steiner, M., 1994, Die neoproterozoischen Megaalgen Südchinas, *Berliner Geowissenschaftliche Abhandlungen, Series E* **15**: 1–146.
Tangri, S. K., Bhargava, O. N., and Pande, A. C., 2003, Late Precambrian–Early Cambrian trace fossils from the Tethyan Himalaya, Bhutan and their bearing on the Precambrian–Cambrian boundary, *J. Geol. Soc. India* **62**: 708–716.
Uchman, A., 1995, Taxonomy and palaeoecology of flysch trace fossils: the Marnoso-Arenacea Formation and associated facies (Miocene, northern Apennines, Italy), *Beringeria* **15**, 1–115.

Uchman, A., 2003, Trends in diversity, frequency and complexity of graphoglyptid trace fossils: evolutionary and palaeoenvironmental aspects, *Palaeogeogr. Palaeoclimatol. Palaeoecol.* **192**: 123–142.

Valentine, J. W., 1994, Late Precambrian bilaterians: grades and clades, *Proc. Nat. Acad. Sci. USA* **91**: 6751–6757

Vidal, G., Jensen, S., Palacios, T., 1994, Neoproterozoic (Vendian) ichnofossils from the Lower Alcudian strata in central Spain, *Geol. Mag.* **131**: 169–179.

Waggoner, B., 1998, Interpreting the earliest metazoan fossils: what can we learn? *Am. Zool.* **38**: 975–982.

Waggoner, B., 1999, Biogeographic analyses of the Ediacara biota: a conflict with paleotectonic reconstructions, *Paleobiology* **25**, 440–458.

Waggoner, B., 2003, The Ediacaran Biotas in space and time, *Integr. Comparative Biol.* **43**: 104–113.

Waggoner, B., and Hagadorn, J. W., 2002, New fossils from terminal Neoproterozoic strata of Southern Nye County, Nevada, in: *Proterozoic–Cambrian of the Great Basin and Beyond* (F. A. Corsetti, ed.), pp. 87–96, SEPM Volume and Guidebook 93.

Walter, M. R., Elphinstone, R., and Heys, G,R., 1989, Proterozoic and Early cambrian trace fossils from the Amadeus and Georgina Basins, *Alcheringa* **13**: 209–256.

Walter, M. R., Veevers, J. J., Calver, C. R., Gorjan, P., and Hill, A. C., 2000, Dating the 840–544 Ma Neoproterozoic interval by isotopes of strontium, carbon, and sulfur in seawater, and some interpretative models, *Precambrian Res.* **100**: 371–433.

Webby, B. D., 1970, Late Precambrian trace fossils from New South Wales, *Lethaia* **3**: 79–109.

Webby, B. D., 1984, Precambrian–Cambrian trace fossils from western New South Wales, *Australian J. Earth Sci.* **31**: 427–427.

Wu, X., and Li, Y., 1987, The discovery and significance of trace fossils from Sinian marine sediments in Xinjiang, *Scientia Geol. Sinica* **1987**: 239–245.

Xiao, S., and Dong, L., 2006, On the morphological and ecological history of Proterozoic macroalgae, in: *Neoproterozoic Geobiology and Paleobiology* (S. Xiao and A. J. Kaufman, eds), pp. 56–90, Kluwer.

Xiao, S., Yuan, X., Steiner, M., Knoll, A. H., 2002, Macroscopic carbonaceous compressions in a terminal Proterozoic shale: a systematic reassessment of the Miahoe biota, south China, *J. Paleontol.* **76**: 347–376.

Xing, Y., Ding, Q., Lin, W., Yan, Y., and Zhang, L., 1985, Metazoans and trace fossils, in: *Late Precambrian Palaeontology of China* (Xing Y. *et al.*, eds.), pp. 182–192, Geological Publishing House, Beijing.

Yang, S., and Zheng, Z., 1985, The Sinian traces fossils from Zhenmuguan Formation of Helashan Mountain, Ningxia, *Earth Science (Wuhan)* **10**: 9–18.

Yang, Z., Yin, J., and He, T., 1982, Early Cambrian trace fossils from the Emei-Ganluo region, Sichuan, and other localities, *Geol. Rev.* **28**: 291–298.

Yin, J., Li, D., and He, T., 1993, New discovery of trace fossils from the Sinian–Cambrian boundary beds in eastern Yunnan and its significance for global correlation, *Acta Geol. Sinica* **67**: 146–157.

Yochelson, E., and Fedonkin, M. A., 1997, The type specimens (Middle Cambrian) of the trace fossil *Archaeonassa* Fenton and Fenton, *Can. J. Earth Sci.* **34**: 1210–1219.

Young, F. G., 1972, Early Cambrian and older trace fossils from the Southern Cordillera of Canada, *Can. J. Earth Sci.* **9**: 1–17.

Zhang, L., 1986, A discovery and preliminary study of the late stage of late Gaojiashan biota from Sinian in Ningqiang county, Shaanxi, *Bull. Xian Inst. Geol. Mineral Resources* **13**: 67–88.

Zhu, M., 1997, Precambrian–Cambrian trace fossils from eastern Yunnan, China: implications for Cambrian explosion, *Bull. Nat. Museum Natural Science (Taiwan)* **10**: 275–312.

Chapter 6

The Developmental Origins of Animal Bodyplans

DOUGLAS H. ERWIN

Department of Paleobiology, MRC-121, National Museum of Natural History, Smithsonian Institution, Washington, DC 20013, USA; and Santa Fe Institute, 1399 Hyde Park Rd, Santa Fe, NM 87501, USA.

1. Introduction	160
2. Pre-Bilaterian Developmental Evolution	163
2.1 Phylogenetic Framework	163
2.2 Unicellular Development	165
2.3 Poriferan Development	166
2.4 Cnidarian Development	167
2.5 The Acoel Conundrum	171
3. Development of the Urbilateria	172
3.1 Arterior-Posterior Patterning and Hox and ParaHox Clusters	172
3.2 Head Formation and the Evolution of the Central Nervous System	174
3.3 Eye Formation	176
3.4 Dorsal-Ventral Patterning	178
3.5 Gut and Endoderm Formation	178
3.6 Segmentation	179
3.7 Heart Formation	181
3.8 Appendage Formation	182
3.9 Other Conserved Elements	183
4. Constructing Ancestors	184
4.1 Maximally Complex Ancestor	184
4.2 An Alternative View	186
5. Conclusions	188
Acknowledgements	189
References	189

1. INTRODUCTION

There are a variety of questions one might like to answer about the origin of animal bodyplans: When did these bodyplans arise? What was the rate of developmental and morphological innovation associated with these events? How reliably does the fossil record reflect the pattern of metazoan divergences and the timing of origin of bodyplans? How do these events relate to environmental and ecological changes? And more broadly, what, if anything, does this evolutionary episode tell us about the nature of the evolutionary process? Each of these questions has been the subject of learned discourse and even summarizing the history of these discussions would exhaust both the available space and the reader's attention (for an excellent recent and comprehensive review see Valentine, 2004).

Here I will focus largely on the developmental aspects of the origin of animal bodyplans, particularly as revealed over the past decade or so by studies of recent organisms. Contrary to all expectation, such comparative studies have revealed remarkable conservation of regulatory elements across considerable phylogenetic distance. Placed in a phylogenetic framework, these studies have permitted inferences about the nature of many nodes on the phylogenetic tree, and from this we can develop and evaluate models of the processes of developmental evolution. As will become evident, my own view is that evidence of conservation of sequence and even regulatory relationships are not a guarantee of functional conservation. Consequently inferring the morphologic attributes of early metazoa is much more problematic than some have argued (see also Erwin and Davidson, 2002).

Understanding these developmental innovations is important for another reason: identifying the complexity of various nodes during the early history of animals is critical to constraining the dates of these nodes and, more importantly, distinguishing between alternative forcing functions for the radiation of the bilaterian metazoans. If early animals, and in particular the last common ancestor of all bilaterians, already possessed high developmental complexity, then developmental innovations alone would seem to be an unlikely cause of the metazoan radiation (see Valentine and Erwin, 1987). Alternatively, it could be that we can identify a suite of developmental innovations both necessary and sufficient for some or all of the new bodyplans that appear during the Ediacaran–Cambrian metazoan radiation. If, however, we find that the necessary genetic and developmental toolkit for building the panoply of bodyplans pre-dates the metazoan radiation, this is strong evidence that we must search instead for either changes in the physical environment or in the dynamics of ecological interactions. These latter two issues are considered in more detail in Erwin (2005).

The Developmental Origins of Animal Bodyplans 161

A host of genomic and developmental changes are associated with the origin and radiation of early metazoa ranging from gene duplications and possibly whole-genome duplication (e.g., Lundin, 1999, but see Hughes, 2003) to enhanced gene complexity and post-translational modification. New metazoan genes constructed by splicing domains have created new signal transduction and cell communication abilities (Cohen-Gihon et al., 2005), for example. As important as these are as mechanisms of change, they are less clearly associated with body plan evolution. Consequently, here I will concentrate on these highly conserved genes that have been linked to particular aspects of body plan evolution.

The first recognition of highly conserved developmental and regulatory modules between various model organisms (initially *Drosophila* and various vertebrates but now including a broader range of organisms) led to a burst of speculation about the last common bilaterian ancestor. Variously known as the 'Urbilateria' or the protostome-deuterostome ancestor, such commentaries attempted to identify the shared features of the great bilaterian clades.

The recognition that the Hox cluster, involved in anterior-posterior patterning of the body, was highly conserved led Slack *et al.* (1993) to define animals as "organism[s] that displays a particular spatial pattern of gene expression..." (p. 490), that they defined as the zootype. Critical to this idea was the recognition that there are a number of patterning genes shared between *Drosophila* and vertebrates, including the Hox clusters, *orthodenticle (otd)*, *empty spiracles (ems)*, and *even-skipped*. Slack and colleagues emphasized the role of Hox genes in specifying relative position rather than specific structures, and based on the identification of a hox gene (*cnox-2*) in *Hydra* they proposed this as a synapomorphy of the Metazoa, and the zootype. They suggested that the zootype was expressed at the phylotypic stage of development, when the precursor of the individual bodyplans first becomes evident and the major elements of the body plan are present as undifferentiated cellular forms. Although the zootype played some role in later discussions, principally through depiction of an hourglass figure, with diverse early and late developmental patterns but the greater similarity of the phylotypic stage denoted by the neck of the glass, it is overly typological (e.g., Schierwater and Kuhn, 1998) and did little to define the early stages of metazoan evolution.

A more concrete step in 1993 came from Shenk and Steele in "A molecular snapshot of the metazoan 'Eve'." They identified a series of conserved elements within a phylogenetic framework. These included such transcription factors as the Hox cluster, *eve*, *engrailed*, *msh* and *NK*, a variety of cell-cell communication molecules and such architectural elements as extra-cellular matrix proteins like type IV collagen. They did not attempt to describe the nature of the earliest metazoans, but emphasized the importance of comparative studies to identify the nodes at which critical innovations had occurred.

Scott (1994) was more daring, employing conservation of the Hox cluster, *Nkx-2.5* and *tinman* as well as *Pax6* to suggest that the ancestral bilaterian had anterior-posterior (A/P) patterning with at least four Hox genes, head and brain formation controlled by *Otd* and *ems*, heart formation produced from *tinman* and at least simple photoreceptive capability.

Two years latter much additional information had appeared, leading to various discussions of developmental aspects of early metazoan evolution. The most provocative was from de Robertis and Sasai (1996) who revived Geoffroy Saint-Hillaire's suggestions that the dorsoventral body axis had been inverted between protostomes and deuterostomes, with the ventral region of arthropods homologous to the dorsal side of vertebrates. This proposal was stimulated by the discovery that dorsal-ventral patterning in *Drosophila*, including the genes *sog*, *dpp* and others, are also present in the African clawed toad *Xenopus* and other vertebrates as *chd* and *Bmp-4*. Indeed the entire regulatory circuit appears to be conserved, but in an inverted fashion. Thus *sog* is expressed ventrally in *Drosophila* where it antagonizes expression of *dpp*, which is thus restricted to the ventral region. The situation is reversed in *Xenopus*, with *chd* expressed dorsally and antagonizing the homolog of *dpp*, *Bmp-4*. De Robertis and Sasai went on to christen the "Urbilateria" as an organism possessing A/P and dorsal-ventral (D/P) patterning, a subdermal longitudinal central nervous system, primitive photoreception, and a circulatory system with a contractile organ. They also suggested that segmentation and appendages might have been present.

A proliferation of speculation soon followed concerning the nature of early metazoans, based on the surge of developmental information (e.g., Holland, 2000; Shankland and Seaver, 2000). Kimmel (1996) suggest that segmentation between arthropods and vertebrates was homologous based on the apparent similarities in expression patterns between the *Drosophila* pair-rule gene *hairy* and the zebrafish gene *her1* (Müller *et al.*, 1996).

Paleontologists soon became interested in these discussions as well, for the information from development promised to reveal much about evolutionary events of the latest Neoproterozoic and Cambrian. In particular, a number of paleontologists have addressed the issue of how the integration of developmental data with data from trace and body fossils may constrain the timing and even processes involved in the Cambrian metazoan radiation (e.g., Conway Morris, 1994, 1998; Erwin *et al.*, 1997; Knoll and Carroll, 1999; Valentine, 2004; Valentine and Jablonski, 2003; Valentine *et al.*, 1999)

These earlier discussions of the role comparative developmental information can play in elucidating the nature of the developmental innovations leading to animal bodyplans set the stage for the remainder of this contribution. I will focus first on what is known and can be inferred of pre-bilaterian developmental patterning before turning to a more exhaustive treatment of the conserved developmental features among the Bilateria. I then evaluate different models for how to interpret this developmental information, distinguishing between a high degree of functional conservation, leading to a maximally complex Urbilateria, from the alternative view that the ancestral role of many of these highly conserved elements was much simpler, more akin to a developmental toolkit than fully realized morphogenetic patterning. I then turn briefly to molecular and developmental information on the timing of the origins of these bodyplans and to the ecological context in which they occur.

2. PRE-BILATERIAN DEVELOPMENTAL EVOLUTION

2.1 Phylogenetic Framework

Understanding the pattern of developmental and morphological change leading to the diversity of existing animal bodyplans and others documented only from the fossil record requires a well-developed phylogenetic framework. Fortunately, combined molecular and morphological data sets have revolutionized our views of metazoan relationships over the past several decades (see recent reviews by Eernisse and Peterson, 2004; Halanych, 2004; Giribet, 2003; Valentine, 2004). The growing number of workers in this area and the steady development of both analytical techniques and growing data sets will probably provide further surprises in the years ahead.

A number of nodes on the metazoan tree remain uncertain, but consensus between molecular and morphological analyses has been achieved in others. Several critical issues in metazoan phylogeny remain in dispute (contrast Fig. 1A and Fig. 1B). Areas of agreement include: 1) Choanoflagellates are the closest sister group to metazoans; 2) The siliceous and calcareous sponges arose independently (e.g., Botting and Butterfield, 2005 and references therein); 3) Ctenophores are the most basal Eumetazoan clade, with cnidarians the next most basal branch; 4) The Ecdysozoa (Arthropoda,

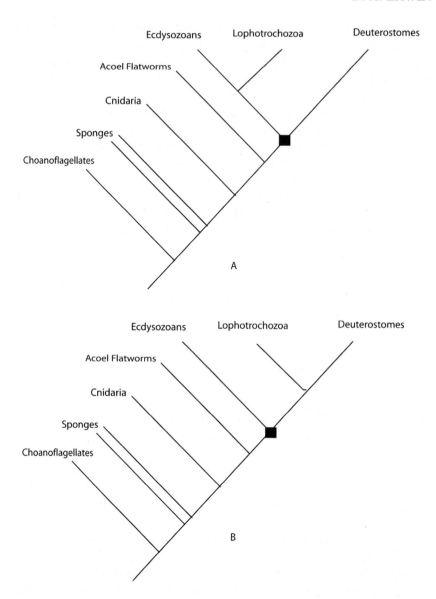

Figure 1. Phylogenetic framework for the metazoa used in this paper, based on recent molecular and morphological analyses. This topology largely follows Eernisse and Peterson (2004). Fig. 1A shows the topology accepted by many, uniting the Ecdysozoa and the Lophotrochozoa into the classic protostomes. Fig. 1B shows Eernisse and Peterson's preferred topology with the Ecdysozoa the sister clade to the deuterostomes, to the exclusion of the Lophotrochozoa. The square represents the position of the Urbilaterian node in the two topologies.

tardigrades, nematodes and priapulids plus others) are a monophyletic clade (Giribet, 2003). Areas of continuing uncertainty involve: 1) The position of the acoel flatworms, which have been separated from the remaining playhelminthes and appear to be the most basal bilaterians (Ruiz-Trillo *et al.*, 1999); 2) The relationships among the remaining major bilaterian clades. Since Aguinaldo *et al.* (1997), many have accepted the division between three large bilaterians subclades, the Ecdysozoa (arthropods, priapulids and allies), the deuterostomes (chordates, echinoderms and hemichordates) and the lophotrochozoans (annelids, molluscs, lophophorates and others). Although the Lophotrochozoa and Ecdysozoa have generally been united in the classic protostomes (Fig. 1A) Eernisse and Peterson note that there is a lack of support for this claim, and their analysis shows the Lophotrochozoa and deuterostomes as sister taxa (Fig. 1B) while Philip *et al.* (2005) claim support for the old coelomata hypothesis of arthropoda + chordata based on their molecular phylogeny. Halanych (2004), although cognizant of the difficulties identified by Eernisse and Peterson favors the Ecdysozoan + Lophotrochozoan topology based on the purported lophotrochozoan signatures in five hox genes (Balvoine *et al.*, 2002) as does Phillippe *et al.*'s (2005) reanalysis of molecular data. Note that the classic protostome-deuterostome ancestor does not exist in topology 1B where the critical node becomes the origin of the Bilateria and thus the critical hypothetical ancestor is that of the Urbilateria.

2.2 Unicellular Development

Multicellularity arose multiple times across a variety of eukaryotic lineages (Buss, 1987; Kaiser, 2001; King, 2004). The asymmetric pattern of these appearances suggests that some clades possessed more of the requirements for multicellularity than others (King, 2004). It has long been apparent that many features once considered as defining elements of the Metazoa are shared with a range of unicellular ancestors (see discussions in Wolpert, 1990, Erwin, 1993). On a molecular level, the specific cell-cell signalling pathways are also highly conserved (e.g., Gerhart, 1999).

The similarities between choanoflagellates and the collar cells of sponges have fueled views that they were the closest relatives of metazoa, a view now amply supported by molecular evidence (reviewed by King, 2004). The antecedents of cell adhesion, signal transduction and cellular differentiation are all found among the choanoflagellates. King *et al.* (2003) analyzed more than 5000 expressed sequence tags (ESTs) to identify representatives of a number of cell signalling and adhesion protein families in two choanoflagellate species. They found a variety of elements involved in cell-cell interactions in Metazoa including cadherins, C-type lectins, tyrosine kinases, and discovered that cell proliferation is controlled by tyrosine kinase inhibitors. Their presence in choanoflagellates demonstrates that they are exaptations co-opted for their role in animals. Much of metazoan diversity of tyrosine kinases, a critical component in cell proliferation and differentiation, apparently evolved between choanoflagellates and the base of Metazoa (Suga *et al.*, 2001), perhaps via rapid shuffling of protein domains (King, 2004).

Thus by the time extant metazoan lineages appeared, the earliest metazoa had acquired an extracellular matrix for cell support, differentiation and movement (as has long been apparent from microscopy); differentiated cell types produced by linking signalling pathways and the multitude of metazoan-specific transcription factors (Degnan *et al.*, 2005); cell junctions to facilitate communication between cells and the extra-cellular communication mediated by the tyrosine kinases.

2.3 Poriferan Development

In a recent review of sponge development Müller *et al.* (2004) described them as "complex and simple but by far not primitive" (p. 54). Müller and his group in Mainz coined the term "Urmetazoan" for the ancestral metazoan and for the past decade have been applying a range of molecular techniques to understanding the novelties that lie at the base of the metazoa. The urmetazoan appears to have had a suite of cell adhesion molecules with intracellular signal transduction pathways, the ability to produce morphogenic gradients, an immune system and a simple ability to pass messages between nerve cells (Müller, 2001; Müller *et al.*, 2004: this is the basis for the following review). Sponge morphogenesis is facilitated by extracellular morphogens and several transcription factors. Two T-box transcription factors have been recovered from the demosponge *Suberites douncula*, one a *Brachyury* homologue and the other related to Tbx3-4-5 from chordates; the former appears to be involved in axis formation. A *Forkhead* homologue has also been recovered from sponges and is apparently active in early morphogenetic cell movements. Among the homeodomain genes, a paired-class gene (*Pax-2/5/8*) and LIM and *Iroquois* transcription factors have been isolated and the available information suggests they are expressed in specific tissue regions. The identification of a *frizzled* gene, a receptor in the Wnt pathway, and other components has demonstrated that the Wnt signalling pathways is involved in cell specification and morphogenesis. The cell-cell and cell-matrix adhesion molecules include receptor tyrosine kinases, but cell adhesion is a prerequisite for immunity. The sponge immune system contains Ig-like molecules and pathways similar to deuterostomes, but not protostomes. (This

The Developmental Origins of Animal Bodyplans

is an interesting pattern that we will see repeatedly, with closer affinities between pre-bilaterians and deuterostomes than with protostomes.) Apopotosis (programmed cell death) also occurs among sponges, with molecules identified that are similar to tumor necrosis factor-α and caspases.

Müller *et al.* (2004) proposed a model for the appearance of the urmetazoan in which the critical evolutionary innovation was the construction of cell-cell and cell-matrix adhesion systems. This allowed cell aggregates to form and signal transduction facilitated cell differentiation and specialization. The addition of an immune system, apopototic machinery and the initial transcription factors permitted homeostasis and furthered differentiation of a body axis. Müller *et al.* do not consider the developmental data from choanoflagellates, but the presence of cell adhesion factors and the diversity of tyrosine kinases (King *et al.*, 2003) is generally consistent with the Müller hypothesis.

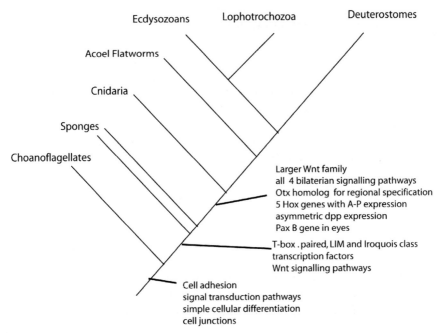

Figure 2. Major developmental innovations leading to the origin of bilateria, emphasizing features shared with sponges, cnidarians and acoel flatworms. See text for discussion.

2.4 Cnidarian Development

In contrast to the situation with sponges, there is a wealth of new developmental data on cnidarians and this greatly aids in defining the patterns of metazoan innovation. Four taxa have received the bulk of the attention from developmental biologists: the sea anemone *Nematostella vectensis*; the coral *Acropora millepora*; the freshwater hydrozoan *Hydra* (for which there is the least information); and the colonial marine hydroid *Podocoryne carnea* (see Ball *et al.*, 2004 for discussion of all four model organisms). The two anthozoans (*Nematostella* and *Acropora*) are of the most interest as representatives of the phylogenetically basal class. There is a surprising diversity of highly conserved bilaterian developmental genes among the Cnidaria. This has led to controversy over whether cnidarians are more complex than they appear, and perhaps even secondarily simplified from a bilaterian ancestor (although 18S rRNA analysis provide no support for such simplification: Collins, 2002). The more realistic alternative is that in many cases these conserved developmental elements serve a more primitive function in cnidarians, and new or enhanced functions have appeared among the Bilateria (Ball *et al.*, 2004). Examination of cnidarian development thus serves an important cautionary role for later discussions on the extent to which true functional conservation applies among the bilaterians.

Among the most important bilaterian cell signalling factors are those of the *Wnt* family, which control cell fate. Bilaterians have twelve known subfamilies, and Kusserow *et al.* (2005) have now reported the presence of all twelve from the sea anemone *N. vectensis*. Gene expression studies reveal a pattern of overlapping expression along the oral-aboral axis of the cnidarian planula, with five genes expressed in the ectodermal cells and another three in the endoderm. Two other *Wnt* genes are expressed only in particular cells, which Kusserow *et al.* suggest indicates a role in cell-type specification. *Wnt* expression near the blastopore may indicate pre-bilaterian evolution of this function. Taken together, this suggests an ancestral role of the *Wnt* genes in gastrulation and axis differentiation, surprisingly similar to patterns of *Hox* gene expression in bilaterians. (See also Wikramanayake *et al.*, 2003 on the role of β-catenin in *Wnt* signalling of *Nematostella* and Steele, 2002 for a review of the role of *Wnt* in *Hydra* development.) Caution must be used in reaching such a conclusion as *Wnt* genes have multiple roles in different animals.

Other signalling pathways present in cnidarians include elements of the TGF-β superfamily, Notch, and Hedgehog and many of the downstream receptors and other components (e.g. Galliot, 2000; Steele, 2002; see summary in Technau *et al.*, 2005). Thus all four of the major bilaterian developmental signalling pathways are present in cnidarians although the extent to which their functions are similar remains incompletely explored. The extent of genetic complexity of cnidarians is also illustrated by a recent expressed sequence tag (EST) study that showed that between 1.3% and

2.7% (depending on the criteria used) of *Acropora* and *Nematostella* genes were shared with fungi, plants, protists and other non-metazoan clades (Technau *et al.*, 2005). Assuming that these are not false positives due to contamination, these results suggest that bilaterians actually lost many genes present in ancestral metazoans.

This apparent loss of genes is a point worth emphasizing. Most biologists have tended to assume that genomic and developmental complexity increased in concert with the increases in metazoan morphologic complexity. The molecular studies discussed here suggest that there was an increase in regulatory specialization and a diversification of particular regulatory pathways to produce the additional morphologic complexity, there was also a loss of other regulatory systems found in other eukaryotic lineages.

Understanding the axial patterning systems of cnidarians and their relationship to axial patterning among bilaterians is critical to reconstructing the early evolution of animal body plans. A homologue of the homeobox gene *Otx*, which is involved in head formation in bilaterians, has been recovered from jellyfish (Müller *et al.*, 1999) and hydra (Smith *et al.*, 1999; see also Galliot and Miller, 2000). Cnidarians of course do not have a head, and in Cnidaria the genes seem to be involved in regional specification and cell movement, providing an important example of a setting where function does not appear to have been conserved from cnidarians. Cnidarians possess simple Hox and ParaHox clusters (Yanze *et al.*, 2001), with a single anterior-class and a single posterior-class gene in each cluster. In *Nematostella* the five Hox genes are expressed in an overlapping, staggered pattern along the oral-aboral axis, reminiscent of bilaterians and supporting suggestions that *Hox* genes are involved in anterior-posterior patterning (Finnerty, 2003; Finnerty *et al.*, 2004). In addition, Finnerty *et al.* found that *dpp* was initially expressed asymmetrically near the blastopore before encircling it. *Dpp* is also widely but asymmetrically expressed in ectoderm. These *dpp* expression patterns are similar to those in bilaterians where it specifies dorsal-ventral axis formation. Taken together, the *Hox* and *dpp* expression patterns suggest that some degree of axis specification was present in the ancestor of bilaterians and cnidarians. Finnerty *et al.*, (2004) suggest that this animal may itself have been bilaterally symmetrical. Interesting supporting results come from a report by Groger and Schmid (2001) describing the nerve net of *Podocoryne* which develops from anterior to posterior in a serially repeated fashion. This also suggests that at least some elements of A/P development were present in the cnidarian-bilaterian ancestor. The difficulty, as many authors have pointed out, is that it is far from clear that the oral-aboral axis of Cnidaria is truly homologous to the A/P axis of bilaterians (see discussion in Finnerty, 2003). The best evidence in support of this claim comes from the *Hox* and *dpp* expression patterns, but the issue remains unresolved.

One of the key characteristics of bilaterians is the presence of mesoderm, which arguably allowed far greater architectural diversity among triploblasts than is possible with only two tissue layers. In *Podocoryne,* Spring *et al.* (2002) studied the expression of homologues of *Brachyury*, *Mef2* and *snail*, all genes involved in bilaterian mesoderm formation. Cnidarian smooth and striated muscle cells in the medusa stage derive from the entocodon, and Spring *et al.*'s results are consistent with the entocodon being the evoluionary source for mesoderm. Martindale *et al.*, (2004) examined the expression in *Nematostella* of seven genes whose bilaterian homologues are involved in mesoderm formation and in the specification of cell types associated with mesoderm. Six genes (*twist, snailA, snailB, forkhead,* and GATA and LIM transcription factors) are restricted to endoderm; *mef2* is expressed in ectoderm. This suggests the genes are involved in germ-layer specification and that bilaterian endoderm and mesoderm are derived from diploblastic endoderm. From these results we can infer that the cnidarian-bilaterian ancestor at least possessed smooth and striated muscles derived from diploblastic ectoderm endoderm; these likely were the evolutionary precursor for bilaterian mesoderm. However, molecular evidence that anthozoans are the oldest clade within the cnidarians (e.g., Collins, 2002) raises difficulties for interpreting the evidence from *Podocoryne*, and suggests the similarities to bilaterian mesoderm could be due to convergence (Ball *et al.*, 2004). Technau and Scholz (2003), writing before publication of the data from *Nematostella,* argued that the role of these genes in the urmetazoan was in cell proliferation, adhesion and motility.

The cubozoan jellyfish *Tripedalia cytosphora* has both lens-containing eyes and simple photoreceptors on stalks beneath the bell, raising interesting questions about developmental similarities to bilaterian eyes. As will be discussed in greater detail in Section 3.4, comparative studies between *Drosophila* and vertebrates have shown that a member of the paired homeobox family of transcription factors, *Pax6*, appears to be responsible for eye formation. Piatigorsky and Kozmik (2004) were not able to isolate *Pax6* from *T. cytosphora*, but did recover *PaxB* (one of four *Pax* genes known to occur in cnidarians: see Miller *et al.*, 2000). *PaxB* appears to represent the ancestral metazoan representative of the *Pax* genes, and has been linked to regulation of lens crystallin, the proteins responsible for the optical nature of the lens in cnidarians. The *Pax* genes of bilaterians evidently evolved from *PaxB* via gene duplication and subsequent divergence of function. Piatigorsky and Kozmik (2004) also suggest that *PaxB* is more generally related to control of formation of mechanoreceptors, including the ancestor of the ear. *Pax2/5/8*, along with *Pax6* a descendent of

PaxB, is expressed in a wide range of bilaterian mechanosensory cells. For a discussion of the relationship between cnidarian nervous system and later bilaterians, see Holland (2003).

Martindale, Finnerty and their colleagues have concluded that the eumetazoan ancestor may not have been a simple, radially symmetrical organism, akin to a cnidarian planula. Instead they suggest it may have been bilaterially symmetrical with both A/P and D/V polarity and muscle cells of mesodermal affinities (see Martindale *et al.*, 2002, 2004; Ball *et al.*, 2004). The work of Piatigorsky and Kozmik (2004) implies that eyes encoded by *PaxB* could have been present as well.

2.5 The Acoel Conundrum

One of the most surprising results of the new studies of metazoan phylogeny over the past two decades has been the movement of the platyhelminthes into the Lophotrochozoa and the evaporation of the pseudocoelomates, long a staple of scenarios of metazoan evolution. The platyhelminthes appear to be polyphyletic, however, with the acoel flatworms and the nemertodermatida being basal-most bilaterians, lying below the divergence leading to the Ecdysozoa, Lophotrochozoa and Deuterostoma (Fig. 1) (Ruiz-Trillo *et al.*, 1999; Baguñà *et al.*, 2001; Baguñà and Riutort, 2004). This suggests there was an interval during the late Neoproterozoic with primitive bilaterian lineages (Knoll and Carroll, 1999), possibly including a number of now extinct lineages.

The acoelomorph flatworms are bilaterians with a simple brain and a sack-like gut. Several acoels have now been searched for *Hox* genes, and show members belonging to each of the four paralogy groups, but evidently without multiple members within any paralogy group. The genes lack the charateristic signatures of lophotrochozoan or ecdysozoan genes but are more similar to bilaterians than cnidarians. Thus the acoelomorph-eubilatieran ancestor likely had only four *Hox* genes. In addition, two ParaHox genes have been detected; the anterior ParaHox class appears to be missing (Cook *et al.*, 2004; Baguñà and Riutort, 2004). Baguñà and Riutort (2004) suggest that the most primitive bilaterians were thus simple, acoelomate, and unsegmented forms, probably with direct development and presumably benthic. Baguñà *et al.*, (2004) argued that the basal position of acoelomorphs supports the planuloid-acoeloid hypothesis for the origin of bilateria, with the first bilaterian evolving from a cnidarian planuloid-like form. While this would argue against a complex primitive bilaterian, because in the new metazoan phylogeny the origin of the bilateria is (again) distinct from the last common ancestor of the protostomes and deuterostomes (PDA), this still allows for a complex PDA.

3. DEVELOPMENT OF THE URBILATERIA

For a brief period of a year or so the basis of organismic complexity seemed fairly clear. After the *C. elegans* genome was released in 1998 (C. elegans sequencing consortium, 1998) with over 19,000 genes, it seemed clear that a substantially more complex organism like *Drosophila* should have perhaps double the number of genes and humans, near the apex of developmental sophistication, must have near 100,000 genes. This neat story began to crumble in 2000 with the announcement that *Drosophila melanogaster* had but 14,000 genes (Adams *et al.*, 2000) and then collapsed when the human genome came in at some 20,000 to 24,000 genes (International Human Genome Sequencing Consortium, 2004). Even then the additional genes were generally due to gene duplication not the origin of new genes through domain reshuffling or some related mechanism.

The roots of metazoan complexity clearly do not lie in a greater diversity of protein coding genes. Rather the key lies in more elaborate control of gene regulation, in particular by *cis*-regulatory transcriptional control and in diversification of multiple protein transcription complexes (Levine and Tjian, 2003; Davidson, 2001; but see also True and Carroll, 2002).

We turn now to the extensive highly conserved developmental elements among the Bilateria. In many cases little information is available from lophotrochozoan clades so the comparison is made between ecdysozoans and vertebrates. Here the relative topology between the ecdysozoan, lophotorochozoan and deuterostome clades becomes critical, but as mentioned in Section 2.1 it is currently unresolved. As Gerhart (1999) emphasized, there are seven major cell-cell signalling pathways that control most cell fate decisions across Bilateria (*Wnt*, TGF-β, hedgehog, receptor tyrosine kinase, nuclear receptor, Jak/STAT and *Notch*). Several of these have already appeared in the discussion of pre-bilaterians. Critically, although these pathways and the ways they operate (e.g., Barolo and Posakony, 2002) are highly conserved, they are used in many different developmental roles. In this Section, I will concentrate on those conserved elements that are relevant to the early evolution of body plans.

3.1 Anterior-Posterior Patterning and Hox and ParaHox Clusters

The *Hox* genes play a primary role in body patterning. The multiple genes within what is normally a single cluster among bilaterians (although it is two clusters in *Drosophila*) pattern the same regions of the body as their order on the chromosome, so that the anterior-most part of the developing larvae is controlled by the genes at the 3' end of the cluster, the middle part is pattern by genes in the middle and posterior regions are controlled by the 5' genes, a pattern described as temporal colinearity (see McGinnis and Krumlauf, 1992).

After the discovery of the role of *Hox* genes in body-patterning of arthropods and vertebrates many developmental and evolutionary biologists assumed that the sequential duplication of genes within the complex would prove to be a significant driver of morphological evolution. It was commonly assumed that as the *Hox* complex was sequenced in a greater diversity of bilaterians the number of genes would roughly correspond to the morphological complexity of the clade, and even that increased morphological complexity within the clade might be accompanied by within-clade gene duplications (e.g., Carroll, 1995; Valentine *et al.*, 1999). Resolving the evolutionary history of the *Hox* gene cluster was also complicated by the fact that in *Drosophila melanogaster* the cluster had been split into two, the Antennapedia and Bithorax clusters named for distinctive mutations to the body plan. In *Caenorahabditis elegans*, a nematode worm commonly used by developmental biologists, the *Hox* genes are found in three different groups. And in both *Drosophila* and *C. elegans* some genes have been lost. (The disruption of the gene cluster and loss of temporal colinearity may be associated with a change to more determinative development: see Seo *et al.*, 2004 and Ferrier and Holland, 2002).

It is now clear, however, that the sequential duplications producing the *Hox* cluster occurred relatively early in animal evolution and that the ancestral bilaterian had a suite of at least eight genes (de Rosa *et al.*, 1999; Balavoine *et al.*, 2002). These include at least five anterior-class genes, two central-class genes and at least one posterior class gene (Balavoine *et al.*, 2002). (The classes correspond to the general parts of the body where the genes are active.) Subsequent clade-specific deletions and duplications occurred to modify the number of central and posterior class genes, and two sequential duplications of the entire cluster led to at least four clusters (with some gene loss) within vertebrates (see Ferrier and Minguillón, 2003 for discussion of cluster patterns in a variety of clades and Hoegg and Meyer, 2005 for a discussion of the pattern of evolution among vertebrate *Hox* clusters). Control of body patterning is thus more an issue of regulation of *Hox* gene expression than gene duplication, at least among the vertebrates (see Gellon and McGinnis, 1998 for discussion). The multiple gene clusters seen among vertebrates may indicate a prominent role for gene duplication in this clade (Prince, 2002; Holland *et al.*, 1994).

The duplications of the entire *Hox* cluster in early vertebrates are an echo of earlier events in animal evolution. A protohox cluster of four genes appears to have duplicated to produce the ancestral *Hox* cluster as well as a second cluster; the four genes correspond to members of the anterior, group

3 (dispersed *Hox* genes in *Drosophila*) and posterior classes. Brooke *et al.*, (1998) termed this the ParaHox cluster. The ParaHox cluster appears to be widely conserved across bilaterians and the genes are expressed in a spatially collinear pattern in the developing gut. Castro and Holland (2003) have suggested that the establishment of the ParaHox cluster was just part of a larger pattern of gene duplication and dispersal across chromosomes in early animal evolution.

From these analyses it is clear that the urbilaterian contained at least seven to eight *Hox* genes, with the number of posterior class genes still uncertain (de Rosa *et al.*, 1999; Kourakis and Martindale, 2000), as well as at least four genes in the ParaHox cluster, one in each of the anterior, *Xlox*, central and posterior classes (Kourakis and Martindale, 2000).

While *Hox* genes pattern most of the body, patterning of anterior-most region is controlled by *Otx/Otd* (Finkelstein and Boncinelli, 1994; Hirth *et al.*, 2003). The posterior structures are controlled by *caudal* in *Drosophila*, *C. elegans* and vertebrates, suggesting that both elongation of the central body axis and elements of segmentation are conserved from the urbilaterian (Copf *et al.*, 2004).

3.2 Head Formation and the Evolution of the Central Nervous System

Formation of the anterior aspect of developing bilaterian embryos is closely linked to the formation of the brain and central nervous system, and the extensive evidence of genetic and regulatory homology implies that substantial components of this system were present in the urbilaterian. Apparently conserved elements include brain patterning, connectivity and neural cell specification, involving a large number of separate regulatory pathways. A related problem is the apparent conservation of a variety of sensory inputs to the nervous system, including both the eye and various mechano-sensory inputs. These will be discussed in the following section.

The brain and nervous system of vertebrates and arthropods is composed of several discrete domains arrayed along an anterior-posterior axis; in many invertebrates the nervous system is more diffuse, with a network, paired nerve cords or a ladder-like structure. Vertebrates posses a brain and dorsal spinal cord, with the brain subdivided into a fore-, mid- and hindbrain. In *Drosophila* the brain or cerebral ganglion is separated from a ventral nerve cord by a subesophageal ganglion. Patterning of the anterior-most components, including the cerebral ganglion and the fore- and mid-brain is controlled by *orthodenticle(otd)* in the fly and *Otx* in vertebrates. The midbrain-hindbrain boundary and the subesophoageal domain are characterized by expression of the *Pax 2/5/8* transcription factors. In both vertebrates and flies the posterior regions are controlled by the anterior-most *Hox* genes described earlier (see Hirth and Reichert, 1999). This has led to a widespread view that the *Otx/Pax/Hox* subdivisions of the brain are descended from the urbilateria. The urbilaterian brain would thus have comprised an anterior protobrain, the midbrain-hindbrain boundary region, a segmented hindbrain and a nerve cord (Arendt and Nübler-Jung, 1999; Reichert and Simeone, 2001; Ghysen, 2003; Hirth *et al.*, 2003; Lichtnechert and Reichert, 2005).

A challenge to this consensus in favour of a centralized brain in the urbilaterian was posed by Lowe *et al.*'s (2003) work on patterning of the brain in the hemichordate *Saccoglossus kowalevskii*. The ectodermal patterning of some 22 genes, including *otx, emx,* and *hox* genes matches that of chordates. *Saccoglossus kowalevskii* has a diffuse nerve net, however, and Lowe *et al.* argue that the most parsimonious solution is for ancestral deuterostomes to have also had a diffuse nerve net rather than a centralized system as suggested by the conservation of the *Otx/Pax/Hox* system described above. Critically, while the *Otx/Pax/Hox* systems has been viewed as defining a central nervous system (CNS), Lowe *et al.* have found a similar pattern of expression in a nerve net that some would have equated with the peripheral nervous system of other bilaterians. While Lowe *et al*'s analysis does show that the neural patterning system does not require centralization in a brain but could be associated with a diffuse nerve net, the critical issue is the polarization of this change across a phylogeny. In other words, are hemichordates and *Saccoglossus kowalevskii* likely to reflect the basal condition for deuterostomes? Looking across invertebrate nervous systems, arthropods and vertebrates provide a remarkably biased perspective, and phylogenetic considerations would favour the view that a CNS was not present in the urbilateria (see Holland [2003] for a perceptive review of these issues) but a simpler network controlled, as in *Saccoglossus kowalevskii*, by a suite of regulatory genes that were independently co-opted in arthropods and vertebrates as the nervous system became more centralized into a brain.

Beyond the formation of the brain, there are other elements of the developing nervous system that appear to be highly conserved across Bilateria. Axon guidance, the specialized movement of the growth cone in nerve cells within the CNS, reflects the activity of several signalling pathways which show strong functional conservation across bilaterians (Chisholm and Tessier-Lavigne, 1999). The conserved signalling factors include ephrins and their receptors the Eph tyrosine kinases, netrins and the slit/robo signalling system. The role of the sematophorins in repulsing axon guidance has been conserved, although the underlying mechanisms may differ substantially and the family is considerably expanded in vertebrates relative to protostomes.

The chordate nervous system is highly specialized relative to most invertebrate nervous systems and becomes even more so with the evolution of the neural crest in vertebrates. Although these events are unquestionably a component of the early evolution of body plans, the topic is sufficiently complex that it will not be treated here. Recent discussions and reviews include Lacalli (2001), Nielsen (1999), Lowe *et al.* (2003) and Holland (2003).

3.3 Eye Formation

Animal eyes have long played a pivotal role in evolutionary thought, from debates over the variety of morphological patterns to claims by creationists that they cast doubt on the primacy of evolution and natural selection (an utterly spurious claim neatly eviscerated by Nilsson and Pelger, 1994). The discovery of deep homologies in eye patterning mechanisms across the Bilateria has had an impact far beyond comparative developmental biology, and not simply because on morphological grounds eyes seem to have evolved many times (Salwini-Pawen and Mayr, 1977; Land and Fernald, 1992). There was thus considerable surprise at the discovery that *Pax6* appeared to control eye development across a wide variety of animals (Quiring *et al.*, 1994; Halder *et al.*, 1995a; Gehring and Ikeo, 1999; reviews in Callaerts *et al.*, 1997; Gehring, 2004). This claim was based on the early expression of *Pax6* in the developing eye (as well as other neural structures), by mutational studies and by ectopic expression experiments (gain-of function mutations) where eyes were induced on legs, wings and other parts of developing flies. Gehring and colleagues argued that this reflected conservation of eye formation pathways from a single ancestral photoreceptor, although there was some variability in the inferred complexity of this photoreceptor (Halder *et al.*, 1995b; Gehring, 1996). Gehring and Ikeo (1999) inferred this involved a simple photoreceptive cell and shading cell but without a lens, while Land and Fernald (1992) in their review of the morphological aspects of eye evolution emphasize eyes capable of forming an image (see also Treisman, 2004).

Over the past ten years considerably more information has been developed about eye development, and this system elucidates the difficulties in establishing the nature of homology across Bilatera, and the features that may have been present in the urbilateria. The transcription factor *Pax6* has been recovered from a wide variety of bilaterians where it is active in eye differentiation. *Pax6* is part of a dense network of genes required for early eye differentiation in *Drosophila*, including *eyeless (ey), twin of eyeless (toy), sine oculis (so), eyes absent (eya), daschund (dac)* with contributions from *eye gone (eyg), teashirt (tsh), optix* (Kumar and Moses, 2001; Gehring,

2004) and *wingless* (Baonza and Freeman, 2002). While homologues of these genes have been identified in mammals it is less clear that the regulatory relationships between them have been conserved (Hanson, 2001). (See also discussion in Treisman (2004) of the exceptions to *Pax6* as the master regulator of eye development.)

Fernald (2000) emphasized that despite the conservation of opsins and *Pax6*, lens proteins and eye structures show no such conservation. An additional complexity to interpreting the evolution of eyes is provided by the discovery of a ciliary photoreceptor in the brain of the polychaete annelid *Platyneris* (Arendt *et al.*, 2004) and the recognition of a rhapdomeric photoreceptor in a vertebrate retinal ganglion (Berson *et al.*, 2002). Ciliary photoreceptors normally occur in vertebrates while rhabdomeric photoreceptors occur in protostomes (Arendt and Wittbrodt, 2001; flatworm eyes, curiously, have both types). The two types of photoreceptors differ in their photoreceptive opsins: *Platyneris* has a normal rhabdomeric photoreceptor in its eyes, but a ciliary, vertebrate-like photoreceptor in its brain which Arendt *et al.* suggest is associated with photoperiodic behaviour (Arendt, 2003 contains an excellent discussion of homology issues related to the different photoreceptor types). In contrast, Panda *et al.* (2005) report that retinal ganglionic cells in vertebrates are derived from rhabdomeric photoreceptors. Apparently, both opsin types occurred in primitive bilateria and were likely derived from a common ancestral cell (Plachetzki *et al.*, 2005). But the two cell types followed distinct evolutionary trajectories after the divergence of early Bilateria, with the ciliary receptors co-opted for eye formation in the vertebrate lineage and rhabdomeric photoreceptors among the invertebrates.

The combination of the conservation of *toy, Pax6* and the related eye differentiation genes, the role of *Pax6* in control of eye differentiation across bilateria, and the probable presence of both ciliary and rhabdomeric photoreceptors in the urbilateria supports Gehring's early claims that at least differentiated cells must have been present in the urbilaterian phoreceptor (Arendt [2003] notes that such eyes occur in polychaete trochophores and similar ciliary larvae). The variety of opsins (see discussion in Fernald 2000) and lens types, however, demonstrates that the instantiation of the morphological variety of eyes across Bilateria followed unique pathways within different clades.

Gehring and Ikeo (1999) proposed a variant of Horowitz's (1945) model of retrograde evolution, in which progressive exhaustion of a necessary component of a biochemical pathway led to the construction of the current pathway in a reverse fashion. Gehring and Ikeo proposed a similar mechanism for eye development, in which control of rhodopsin by *Pax6* in early animals led to a more complex eyes by intercalation of new genes between the initial transcription factor and the final product. A similar but more thoroughly developed model in the language of *cis*-regulation was proposed by Erwin and Davidson (2002) and will be discussed below.

Finally, *Pax6* is active in the establishment of other head structures, including the nose and ear, which has led to suggestions of a conserved program for the production of sensory organs which was subsequently elaborated into separate control systems (Gehring, 2004; Niwa *et al.*, 2004). Niwa *et al.* (2004) document the role of the proneural gene *atonal (ato)* in the development of several segment specific sensory organs in *Drosophila* including the eye, auditory organ (Johnston's organ) and stretch receptor. They suggest that all of these evolved from a common protosensory organ (see also Ghysen, 2003).

3.4 Dorsal-Ventral Patterning

As discussed in the introduction, another mechanism which appears to be conserved from ancestral Bilateria involves dorsal-ventral patterning, mediated by a gradient of activity in genes known as *decapentaplegic (dpp)* in *Drosophila* and *bone morphogenic protein-4 (BMP-4)* in vertebrates. *Dpp* and *BMP-4* are homologous, but more importantly they form a critical part of a conserved axial patterning system which provides positional information to cells along a dorsal-ventral axis. *Dpp/BMP-4* is antagonized by *Sog/Chordin* and by *twisted gastrulation*, and *Sog/Chordin* in turn is cleaved by *Tolloid*. The result of this activity is a gradient in which *BMP-4* is high in the ventral aspect of vertebrates and is retarded dorsally by *Chordin*. In *Drosophila* on the other hand, the pattern is reversed, with *dpp* expressed dorsally and antagonized ventrally by *Sog* (Holley *et al.*, 1995; Oelgeschläger *et al.*, 2000). This pattern led de Robertis and Sasi (1996) to resurrect the suggestion that the vertebrate and arthropod D/V axis have been inverted relative to each other.

3.5 Gut and Endoderm Formation

The developmental origin of the mouth and anus is the fundamental divide between the protostomes, where the embryonic blastopore forms the mouth and anus, and the deuterostomes, in which it forms the anus and the mouth arises secondarily. This diference provides little reason to expect underlying similarities in developmental mechanisms of gut and endoderm formation, yet surprising similarities have been identified.

GATA transcription factors and *forkhead* have been identified in endoderm in cnidarians (Section 2.4) with some interpreting these as fundamental for germ-layer specification. GATA factors and zinc-finger transcription factors are widely distributed among eukaryotes and have significant roles in cell specification, differentiation, proliferation and movement. This family of transcription factors has not undergone extensive diversification within bilaterians, unlike some other gene families. Only six GATA factors are known in vertebrates, three or four in *Drosophila* and 11 in *C. elegans* (Patient and McGhee, 2002). In *Drosophila* ABF/Serpent and in *C. elegans end-1* is the GATA first factor expressed and appears to trigger a regulatory cascade leading to endoderm formation. Shoichet *et al*., (2000) demonstrated that when *end-1* was expressed in the African clawed toad *Xenopus* it also initiated vertebrate endoderm. This establishes that endoderm differentiation is initiated by homologous GATA factors across the bilaterians (see also Zaret, 1999). Studies of the polychaete annelid *Platyneris,* various arthropods, enteropneusts and vertebrates have suggested that the tripartite, tubular gut was present in the urbilaterian larvae. This is based on similarities in expression of *brachyury* and *goosecoid* in the developing foregut and expression of *otx* in the pre- and postoral ciliary bands of a variety of protostome and deuterostome larvae (Arendt *et al*., 2001).

The gene regulatory network, including all the *cis*-regulatory interactions, specifying endomesoderm formation in the sea urchin *Strongelocentrotus purpuratus* has been established (Davidson *et al*., 2002). This details the interactions between the various GATA genes, *brachyury*, and others. We can look forward to the elucidation of the regulatory architecture of other bilaterian components in the future, and this will allow us to identify the specific wiring changes associated with developmental innovations. The factors controlling gut and endoderm formation among bilaterians are highly conserved and this is among the morphological features which, on developmental grounds, seems likely to have been present in the urbilaterian.

3.6 Segmentation

Invertebrate biologists have historically viewed segmentation in vertebrates, annelids and arthropods as independent events. Thus one of the more contentious issues in deciphering the nature of the ancestral bilaterians is the possibility that segmentation arose once at the base of the bilaterian (see recent discussions by Balavoine and Adoutte, 2003; Seaver, 2003; Tautz, 2004; Minelli and Fusco, 2004). What constitutes segmentation is itself contentious. Willmer (1990) for example, continues the tradition of distinguishing between serial repetition and true segmentation, while Budd (2001) suggests that only organs can be segmented rather than entire body plans which thus allows partial segmentation.

Holland *et al.* (1997) provided the initial developmental argument for the conserved nature of segmentation. In *Drosophila*, *engrailed* controls formation of posterior compartment of each segment and this is widely conserved across arthropods. Holland *et al.* found that a homologue of *engrailed* in amphioxus, *AmphiEn*, is expressed in the posterior portion of the first eight somites. In addition, *Hairy/her-1* are pair-rule genes specifying the formation of alternating segments in both *Drosophila* and zebrafish, respectively (Müller *et al.*, 1996), although whether they are homologous genes has been unclear (Davis and Patel, 1999). Because other pair-rule genes have divergent expression patterns, even within arthropods, there was initially considerable doubt about the Holland *et al.* hypothesis. Furthermore, *Engrailed* expression in arthropods is normally driven by *wingless*, but this regulatory couple is not present in amphioxus (Holland 2000), raising doubts about the conserved role of *Engrailed* in segmentation.

There is good reason for doubting a single origin of segmentation for the developmental processes are quite different in each group. In *Drosophila* pair-rule and segmentation genes subdivide a broad region, what Tautz (2004) has called "top-down segmentation". In contrast, a Delta-Notch signalling cascade cyclically subdivides chicken embryos to form somites ("bottom-up segmentation"). Annelids follow yet another path, in which new segments form by budding from a growth zone (see discussion of all three in Tautz [2004]).

The *Delta-Notch* signalling system had been seen as unique to vertebrates, but work on spiders by Stollewerk *et al.* (2003) is a reminder of how often our views may be clouded by reliance on the relatively few (and often developmentally unusual) model animals studied by most developmental biologists. Expression of *Notch* in the developing spider is required for segmental patterning and establishing segment boundaries. In contrast to vertebrates where the segmental patterning encompasses the mesoderm, *Notch* expression in spiders patterns the ectoderm. Stollewerk *et al.* conclude that the *Delta-Notch* patterning system was present in the common ancestor of arthropods and vertebrates, and Tautz (2004) describes this as the best current evidence for segmentation in the Urbilateria. Further evidence in support of this comes from the *caudal(cad)* homeobox genes which are involved in the early phase of segmentation in a variety of arthropods. In the crustacean *Artemia* and the beetle *Tribolium caudal* is required for early segmentation, axis formation and Hox expression. The activity of *cad* is similar to its vertebrate homologue, *Cdx*, which Copf *et al.* (2004) suggest reflects an ancestral function in formation of body segments and elongation of the A/P axis.

In his review of segmentation Tautz (2004) notes that despite the great differences in segment formation between arthropods, annelids and vertebrates, there are 'intriguing similarities' as well: the activity of *hairy* related genes in early specification, *engrailed* in segmental boundary formation, *Delta-Notch* cycling in some arthropods and in vertebrates upstream of boundary formation, and possibly the conservation of head patterning systems across Bilateria (this depends on one's view of the relationship between anterior segmentation and segmentation along the remainder of the body). These similarities lead Tautz to resurrect the enterocoele theory of coelom formation, in which pouches in the endoderm give rise to the coelomic spaces and the triploblastic bilateria arose from radially symmetrical diploblasts (see Valentine [2004] for discussion of the enterocoely hypothesis). Finnerty *et al*'s (2004) recent results on the sea anemone *Nematostella*, described above, are also consistent with this hypothesis.

In their examination of the morphologic similarities between segmentation in different bilaterian groups Balavoine and Adoutte (2003) emphasize the variety of ways of forming segments. But underneath this they identify an underlying similarity in the formation of somites that supports Tautz's invocation of the enterocoely theory. Balavoine and Adoutte claim that in most phyla for which sufficient information is available, seriated, paired coelomic cavities or somites are present in the mesoderm, suggesting a close and ancestral relationship between segmentation and coelomic spaces. While these spaces eventually form the coelom of annelids they are transient in arthropods, being replaced by a haemocoel. In arthropods, annelids and vertebrates the somites form in a posterior growth zone, thus uniting the three segmentation mechanisms identified by Tautz (2004). Balavoine and Adoutte also examine a number of clades traditionally considered unsegmented, particularly those with trimeric organization such as brachiopods, phoronids, echinoderms, hemichordates and chaetognaths. In each case they identify features which they view as being consistent with an ancestral, segmented coelomic condition, and suggest that the the variety of trimeric forms are all derived states.

3.7 Heart Formation

The discovery that *Drosophila tinman*, and its probable vertebrate homologue *Nkx2.5* were both responsible for heart formation suggested a common developmental basis for cardiogenesis (Harvey, 1996) although other conserved factors are required as well (Bodmer and Venkatesh, 1998; Holland *et al.*, 2003 and references therein).

An important component of testing highly conserved developmental roles for transcription and signalling factors is introducing the factor into an individual of distantly related group that is defective in that factor. In a successful test the introduced factor will rescue the mutant and produce the appropriate phenotype. Ranganayakulu *et al.* (1998) introduced the mouse *Nkx2.5* gene into *Drosophila* and showed that it rescues visceral mesoderm function but not heart mesoderm. Thus while *tinman* and *Nkx2.5* are homologues, their function has diverged and the role of the ancestral gene may have been involved in visceral mesoderm specification. Further support for this view comes from studies of the *C. elegans* homologue, *ceh-22*, which is responsible for pharyngeal muscle development (Haun *et al.*, 1998). This suggests an ancestral function of producing a contractile muscular tube, rather than a heart *per se* (Harvey, 1996; Tanaka *et al.*, 1998; Ranganayakulu *et al.*, 1998).

3.8 Appendage Formation

The number and morphologic diversity of arthropod appendages has made them a prime target for developmental study. More importantly, however, along with eyes and possibly segmentation, appendages are one of the more provocative claims for morphological conservation from the urbilaterian. Prior to the advent of developmental genomics few morphologists had suggested the possibility of appendages in the earliest bilaterians. Yet the homeodomain transcription factor *Distal-less* (*Dll*/Dlx) exhibits a conserved expression pattern across bilaterian appendages. The earliest evidence for this pattern came from Panganiban *et al.* (1995) who showed that the great diversity of arthropod limbs, from the unbranched limbs of insects to the uniramous, biramous and even phyllopodous limbs of crustaceans, myriapods and chelicerates, all involved expression of *Dll* along the distal portion of the limb axis and its various branches.

Subsequent research has shown that the *Drosophila Dll* and the vertebrate Dlx genes have a variety of similar expression patterns (reviewed in Panganiban and Rubenstein, 2002; see also Panganiban *et al.*, 1997; Popadic *et al.*, 1996). *Dlx* plays a role in ear and nose development, while *Dll* defines the *Drosophila* antenna, which serves as both ear and nose. The genes are involved in formation of the peripheral nervous system as well as the mouthparts in both arthropods and vertebrates. Moreover, limb primordia in *Drosophila* are induced by *Wnt,* develop at the lateral margin of the neural ectoderm where they express *Dll* and then migrate before differentiation. This pattern is remarkably similar to the induction of neural crest in vertebrates, which begins with a Wingless signal (a *Wnt* homolog), followed by *Dlx* expression in the neural crest precursors adjacent to the neural plate before they migrate. There is also some evidence for *Dll*/Dlx involvement in central nervous system formation. Panganiban and Rubenstein suggest that the ancestral, urbilaterian role of *Dll/Dlx* involved formation of a primitive sensory system as well as specifying appendage formation. It is worth emphasizing that by appendage here we just mean "sticky-outy-bits" rather than the jointed appendage of an arthropod. Any outgrowth from the body wall would fulfil this definition. Thus the developmental program for appendage formation is largely homologous but the appendages themselves are not.

The view that *Dll* expression patterns reflect conservation of neural patterning and especially appendage formation from an urbilaterian ancestor has been challenged. Shubin *et al.* (1997) reviewed the similarities in axial systems of both arthropod and vertebrate appendages: The posterior compartment of the developing anterior-posterior axis is controlled by *hedgehog (Hh)* in *Drosophila,* and the wonderfully named *sonic hedgehog (Shh)* in mice. *Hh* induces expression of *decapentaplegic (dpp)* along the boundary between the anterior and posterior compartments. *Bmp-2*, a *dpp* homolog, does the same in vertebrates. A series of other genetic cascades covering the proximo-distal and dorsal-ventral axes are similarly preserved between the two groups. While this appears to be strong evidence of conservation of pathways from an urbilaterian ancestor with appendages, Shubin *et al.* consider several alternative explanations, however and concluded that the formation of arthropod and vertebrate limbs reflected "the cooption and redeployment of signals established in primitive metazoans" (p. 639). Vertebrate limbs are a secondary outgrowth derived from the branchial arches of fish that have co-opted, in the view of Shubin *et al.*, a preexisting patterning program.

Minelli (2003) advances another alternative, in which limbs involve duplication of the mechanisms involved in forming the central body axis. As evidence for what he terms axis peramorphosis, Minelli points to the fact that the only segmented appendages are found in clades where the main body is also segmented. In Minelli's view there is no requirement that an urbilaterian ancestor shares some form of outgrowth, only that the axial patterning mechanisms can be co-opted as a unit when appendages arise separately in a variety of clades.

3.9   Other Conserved Elements

At this point there should be little surprise that a variety of other features of bilaterians are also preserved from the common bilaterian ancestor. Many of these are only incompletely studied at this point and often understanding them does not impinge upon understanding the Metazoan radiation, so my discussion of them will be brief.

In vertebrates, olfaction requires GTP-binding protein coupled receptors. Interaction with an odor molecule stimulates a G-protein cascade. Sequencing of *C. elegans* has revealed about 500 possible genes of this type, although it is not yet clear how many are involved in chemoreception (Krieger and Breer, 1999; Prasad and Reed, 1999). Similar genes have been found in *Drosophila* as well where the sequences are highly divergent from vertebrates and *C. elegans* but the protein structure has been preserved. As Strausfeld and Hildebrand (1999) described, patterns of cellular arrangement and physiological response in olfactory cells are very similar between insects and vertebrates, as is the development of the glomeruli, the region of the brain stimulated by the olfactory cells.

The deployment of certain photoreceptors in non-eye tissues and their probable involvement in controlling photoperiodic behaviour has already been noted. It therefore comes as little surprise that the timekeepers of circadian rhythms also appear to be conserved from flies to mammals, at least in broad outline (reviewed in Panda *et al.*, 2002). Other features charaterizing the urbilaterian may include innate immunity (Hoffmann, 2003); the formation of branching structures such as those used in respiration and controlled by a firbroblast growth factor system (although the ancestral role remains unclear) (Metzger and Krasnow, 1999); muscle formation via *mef2* and *twist* (e.g., Baylies and Michelson, 2001) and the myoD family of bHLH transcription factors (Zhang *et al.*, 1999).

4. CONSTRUCTING ANCESTORS

4.1 Maximally Complex Ancestor

In the preceding section I have highlighted the apparent developmental homologies implied by highly conserved transcription factors, signally elements and other aspects of the regulatory machinery. In a number of cases, particularly the nature of the CNS, heart formation and eye development I have highlighted alternative explanations of the developmental data as they exist in the developmental literature.

This body of work provides a view of a maximally complex urbilaterian ancestor (Table 1), if we assume the maximal permissible functional homology based on the developmental data. Such an organism would possess anterior-posterior and dorsal-ventral differentiation, a differentiated head with a tripartite brain and nerve cord as part of the CNS and with sophisticated signalling systems, sensory systems including at least a primitive eye with both ciliary and rhabdomeric photoreceptors, and mechanoreceptors, a differentiated gut and probable mesoderm, segmentation, a heart and appendages. Other conserved elements support the presence of chemoreception, innate immunity, and photoperiodicity.

Table 1. Two alternative views of the developmental and morphological complexity of the urbilaterian (or protostome-deuterostome) ancestor. The maximally complex urbilaterian assumes that the highly conserved developmental elements are also functionally conserved. The alternative urbilaterian is a structurally less complex form with less morphogenesis, and differentiated cell types (see also Erwin and Davidson, 2002).

Developmental componenets	Maximally complex Urbilaterian	Alternative Urbilaterian
Anterior-posterior differentiation	Present	Present
Dorsal-ventral differentiation	Present	Present
Anterior differentiation (*otx/Otd*)	Present	Differentiation of specific neuronal cells
Nervous system	CNS	Nerve net
Eye (*Pax6*)	eye	Visual pigments
Gut (*caudal*)	Differentiated gut	CTS2 intestinal cells
Segmentation	Present	absent
Heart	Present	Contractile muscle cells
Appendages	Present	?

Such an organism possesses all of the developmental requirements of most non-vertebrate animals, and while there has been considerable elaboration of the developmental control machinery in specific clades, much of which has yet to be worked out, there would not appear to be any critical developmental innovations required for the rapid diversification of bilaterian bodyplans during the Ediacaran and Cambrian. If this urbilaterian ancestor immediately predated the Cambrian explosion, (say *Kimberella* in the late Ediacaran [555 Ma]) one could argue that the acquisition of this suite of novelties was both necessary and sufficient for the formation of animal body plans. Molecular clock evidence (e.g. Peterson *et al.*, 2004; Peterson and Butterfield 2005), however, suggests that ancestral bilaterians preceded the diversification of bilaterians as seen on the fossil record by at least tens of millions of years. If this is so, then our search for explanations of the triggering events of the diversification must turn toward changes in the physical environment or the nature of ecological relationships (Erwin, 2005).

The real significance of these patterns of developmental and regulatory complexity for paleontologists come from inferring the likely morphological complexity of animals through the late Neoproterozoic and the plausibility of finding records of them, as either trace or body fossils, in the geological record. The greater the morphological sophistication of the urbilaterian, particularly if it possessed appendages, segmentation or a through gut, the greater the likelihood of it producing trace fossils of various types or fecal pellets.

4.2   An Alternative View

There is, however, an alternative view to this highly complex urbilaterian ancestor, a view which takes a more nuanced view of homology and recognizes that the existence of highly conserved sequences does not necessarily imply that function has been as highly conserved. This issue of determining developmental homologies has been discussed at length elsewhere (e.g., Wray, 1999; Abouheif, 1999). In this section I will briefly outline an alternative view where most of these developmental homologies comprise elements of a developmental toolkit responsible for cell-type specification, rather than more complex morphogenetic pathways. This view has already been discussed at greater length in Erwin and Davidson (2002).

Erwin and Davidson (2002) suggested that when the developmental patterns are considered in detail the similarities between the developmental processes of various bilaterian clades often lie in differentiated cell types rather than developmental morphogenesis. In other words, in the formation of such specialized cell types as neurons and associated ganglions, the slow contractile cells of heart muscles, photoreceptors, digestive and secretory cells, etc., the production of these differentiated cell types is clearly conserved. Morphogenetic pathways, however, involve laying down a general pattern in part of a body and then progressively deploying the appropriate genetic programs to build a particular structure through coordinated gene expression, cell movement, and cell division. Today morphogenetic pathways establish general patterning prior to cell type differentiation, but as Gehring and Ikeo (1999) also noted, the development of metazoan morphogenesis is likely to have followed a pattern of intercalary evolution, with the spatial and temporal patterning of morphogenesis intercalated between the initial regulatory triggers and the final specification of cell types. We suggested, for example, that the ancestral role of *Pax6* was likely to be in initiating the genes coding for visual pigments. The prediction arising from this point of view is that the actual gene regulatory networks, once they are decoded, should be clade-specific rather than shared across bilaterians.

Some of the shared developmental toolkit and components that are re-used when needed for particular functions are deployed widely in development, others only within certain cell types. An example of the latter is the basic helix-loop-helix (bHLH) signalling family of transcriptional activators and related repressors which establishes proneural cellular differentiation (Rebeiz *et al.*, 2005). In *Drosophila* this family includes *achaete, scute* and *atonal*, and bHLH genes have been found in Cnidaria as well.

If this alternative view is correct, as some of the evidence reviewed above suggests, then the urbilaterian may have been less complex than often imagined. Consequently it would be less likely to leave traces in the rock record and the discrepancy between molecular clock dates and the first appearance of bilaterian fossil in the rock record is less troubling for paleontologists.

More recent comparative studies of the gene regulatory networks (GRNs) associated with the highly conserved developmental genes discussed in this paper suggest an additional perspective on the origin of metazoa. Eric Davidson's group at CalTech have dissected the gene regulatory network for endomesodern development in the echinoid *Strongylocentrotus purpuratus* (Davidson, 2001; Davidson *et al.*, 2002). This allowed them to more rapidly establish structure of the same network in starfish (Hinman *et al.* 2003). Comparison of these results suggests an interesting hierarchical structure to GRNs at the center of which are highly conserved networks of genes responsible for the spatial patterning of critical morphologic fields within a developing embryo (Davidson and Erwin, 2006). These highly conserved components are termed kernels. Other components of this hierarchical structure include plug-ins, for small segments that are repeatedly re-used for a diversity of developmental functions, switches that act as input/output controllers, and finally the differentiation gene batteries at the end of networks the activity of which produces specific cell types. A comparison of *Drosophila* and vertebrate heart regulatory networks suggests that they display a similar hierarchical structure (Davidson and Erwin 2006).

Metazoan kernels for key body patterning appear to have largely been established during the Ediacaran–Cambrian metazoan radiation. After this time regulatory changes appear to have been shunted to upstream and downstream components of the regulatory network. Formation of species and genera may largely reflect changes in differentiation gene batteries and I/O switches. As discussed by Davidson and Erwin (2006), one implication of this hypothesis is that the Linnean taxonomic hierarchy is an imperfect reflection of an underlying regulatory structure. Phyla and some classes may reflect the establishment of kernels, and the inflexibility of the kernels to evolutionary modification may be the primary reason for the apparent morphological stability of metazoan phyla.

5. CONCLUSIONS

The presence of eight genes in the *Hox* cluster and a *ParaHox* cluster of four genes in the Urbilateria seems indisputable and the preservation of such gene duplications strongly suggests A/P differentiation to a level somewhat greater than cnidarians. The Lowe *et al.* (2003) analysis of *Saccoglossus* suggests the urbilaterian at best possessed only a simple nerve net rather than the complex, tripartite CNS advocated by others. If the developmental toolkit approach is correct, it may have only been a suite of neuronal cell types. Both ciliary and rhapdomeric photoreceptors now seem to have been present in the urbilaterian, although their roles remain unclear and no more than a very simple photoreceptor, similar to that in cnidarians, seems required by the available data. The regulatory framework for endomesoderm formation is highly conserved across bilateria (Davidson *et al.* 2002; Levine and Davidson, 2005). Heart formation genes seem plausibly interpreted as specifying contractile muscle cells, and the genes now involved in segmentation may have been involved in establishing positional boundaries within the developing embryo.

Thus despite the suggestions of many developmental biologists that a relatively complex bilaterian ancestor is virtually required by the developmental data, alternative interpretations do exist for many of the highly conserved developmental control genes. These permit the elucidation of a less complex urbilaterian, with a variety of differentiated cell types and some degree of morphogenesis and positional pattern-formation (which, in any case, would be required by information from cnidarians). This level of developmental complexity also permits a level of morphologic complexity which is less likely to leave a trace in the fossil record as either a trace or body fossil. Consequently the fossil record would be more agnostic about the age of the urbilaterian ancestor.

In either case, the challenge for comparative developmental biologists is to unravel the clade-specific regulatory pathways leading to complex morphogenesis and establish which of these are truly shared across bilaterians. This will require considerable study of largely neglected clades of metazoans, particularly among the lophotrochozoa. But just as important will be the adoption of a more rigorous approach to tracing the patterns of regulatory circuits, as exemplified by Levine and Davidson (2005).

The extent of the highly conserved elements in the urbilateria, whatever its complexity, poses another challenge to evolutionary biologists. Although the concept of macroevolution began as a distinct view of the origin of morphological novelties, it has evolved into a hierarchical view of the sorting (and potentially selection) of distinct evolutionary entities (Erwin, 2000). A growing body of developmental information suggests that at least some of the novelty of the Cambrian metazoan radiation may be explained not just by differential sorting or higher rates of origination, but by differences in the means of origination of developmental and morphological novelties.

ACKNOWLEDGEMENTS

I thank Eric Davidson, Kevin Peterson and David Krakauer for discussions on these issues, and Sam Bowring for general discussions of the Metazoan radiation. The manuscript benefited from comments by two anonymous reviewers and the editors. This research was supported in part by the Santa Fe Institute.

REFERENCES

Abouheif, E., 1999, Establishing homology criteria for regulatory gene networks: prospects and challenges, in: *Homology* (G. R. Bock and G. Cardew, eds.), Novartis Foundation Symposium, Wiley, Chichester, pp. 207–225.

Adams, M. D. *et al.*, 2000, The genome sequence of *Drosophila melanogaster*, *Science* **287**: 2185–95.

Aguinaldo, A. M. A., Turbeville, J. M., Linford, L. S., Rivera, M. C., Garey, J. R., Raff, R. A., and Lake, J. A., 1997, Evidence for a clade of nematodes, arthropods and other moulting animals, *Nature* **387**: 489–493.

Arendt, D., 2003, Evolution of eyes and photoreceptors, *Int. J. Dev. Biol.* **47**: 563–571.

Arendt, D., and Nübler-Jung, K., 1999, Comparison of early nerve cord development in insects and vertebrates, *Development* **126**: 2309–2325.

Arendt, D., Technau, U., and Wittbrodt, J., 2001, Evolution of the bilateria larval foregut, *Nature* **409**: 81–85.

Arendt, D., Tessmar-Raible, K., Snyman, H., Dorresteijn, A., and Wittbrodt, J., 2004, Ciliary photoreceptors with a vertebrate-type opsin in an invertebrate brain, *Science* **306**: 869–871.

Arendt, D., and Wittbrodt, J., 2001, Reconstructing the eyes of Urbilateria, *Proc. Roy. Soc. London, Ser. B* **356**: 1545–1563.

Baguñà, J., and Riutort, M., 2004, The dawn of bilaterian animals: the case of acoelomorph flatworms, *BioEssays* **26**: 1046–1057.

Baguñà, J., Ruiz-Trillo, I., Paps, J., Loukota, M., Ribera, C., Jondelius, U., and Riutort, M., 2001, The first bilaterian organisms: simple or complex? New molecular evidence, *Int. J. Dev. Biol.* **45**: S133–S134.

Balavoine, G., and Adoutte, A., 2003, The segmented *Urbilateria*: A testable scenario, *Integr. Comp. Biol.* **43**: 137–147.

Balavoine, G., de Rosa, R., and Adoutte, A., 2002, Hox clusters and bilaterian phylogeny, *Mol. Phylogenet. Evol.* **24**: 366–373.

Ball, E. E., Hayward, D. C., Saint, R., and Miller, D. J., 2004, A simple plan—cnidarians and the origins of developmental mechanisms, *Nat. Rev. Genet.* **5**: 567–577.

Baonza, A., and Freeman, M., 2002, Control of *Drosophila* eye specification by Wingless signalling, *Development* **129**: 5313–5322.

Barolo, S., and Posakony, J. W., 2002, Three habits of highly effective signaling pathways: principles of transcriptional control by developmental cell signaling, *Genes & Development* **16**: 1167–1181.

Baylies, M. K., and Michelson, A. M., 2001, Invertebrate myogenesis: looking back to the future of muscle development, *Curr. Opinion Genet. Dev.* **11**: 431–439.

Berson, D. M., Dunn, F. A., and Takao, M., 2002, Phototransduction by retinal ganglion cells that set the circadian clock, *Science* **295**: 1070–1073.

Bodmer, R., and Venkatesh, T. V., 1998, Heart development in *Drosophila* and vertebrates: conservation of molecular mechanisms, *Dev. Genet.* **22**: 181–186.

Botting, J. P., and Butterfield, N. J., 2005, Reconstructing early sponge relationships by using the Burgess Shale Fossil *Eiffelia globosa*, *Proc. Nat. Acad. Sci. USA* **102**: 1554–1559.

Brooke, N. M., Garcia-Fernandez, J., and Holland, P. W. H., 1998, The ParaHox gene cluster is an evolutionary sister of the Hox gene cluster, *Nature* **392**: 920–922.

Budd, G. E., 2001, Why are arthropods segmented? *Evol. Dev.* **3**: 332–342.

Buss, L. W., 1987, *The Evolution of Individulality*, Princeton University Press, Princeton, NJ.

C. elegans Sequencing Consortium, 1998, Genome sequence of the nematode *C. elegans*: a platform for investigating biology, *Science* **282**: 2012–2018.

Callaerts, P., Halder, G., and Gehring, W. J., 1997, PAX-6 in development and evolution, *Annu. Rev. Neurosci.* **20**: 483–532.

Carroll, S. B., 1995, Homeotic genes and the evolution of arthropods and chordates, *Nature* **376**: 479–485.

Castro, L. F. C., and Holland, P. W. H., 2003, Chromosomal mapping of ANTP class homeobox genes in Amphioxus: piecing together ancestral genomes, *Evol. Dev.* **5**: 459–465.

Chisholm, A., and Tessier-Lavigne, M., 1999, Conservation and divergence of axon guidance mechanisms, *Curr. Opinion Neurobiol.* **9**: 603–615.

Cohen-Gihon, I., Lancet, D., and Yanai, I., 2005, Modular genes with metazoan-specific domains have increased tissue specificity, *Trends Genet.* **21**: 210–213.

Collins, A. G., 2002, Phylogeny of Medusozoa and the evolution of cnidarian life cycles, *J. Evol. Biol.* **15**: 418–432.

Conway Morris, S., 1994, Early metazoan evolution: first steps to an integration of molecular and morphological data, in: *Early Life on Earth* (S. Bengtson, ed.), Columbia University Press, New York, pp. 450–459.

Conway Morris, S., 1998, Early metazoan evolution: reconciling paleontology and molecular biology, *Am. Zool.* **38**: 867–877.

Cook, C. E., Jimenez, E., Akam, M., and Salo, E., 2004, The Hox gene complement of acoel flatworms, a basal bilaterian clade, *Evol. Dev.* **6**: 154–63.

Copf, T., Schroder, R., and Averof, M., 2004, Ancestral role of caudal genes in axis elongation and segmentation, *Proc. Nat. Acad. Sci. USA* **101**: 17711–17715.

Davidson, E. H., 2001, *Genomic Regulatory Systems*, Academic Press, San Diego.

Davidson, E. H., and Erwin, D. H, 2006. Gene regulatory networks and the origin of animal body plans. *Science* **311**: 796-800.

Davidson, E. H., Rast, J. P., Oliveri, P., Ransick, A., Calestani, C., Yuh, C. H., Minokawa, T., Amore, G., Hinman, V., Arenas-Mena, C., Otim, O., Brown, C. T., Livi, C. B., Lee, P. Y., Revilla, R., Rust, A. G., Pan, Z., Schilstra, M. J., Clarke, P. J., Arnone, M. I., Rowen, L., Cameron, R. A., McClay, D. R., Hood, L., and Bolouri, H., 2002, A genomic regulatory network for development, *Science* **295**: 1669–1678.

Davis, G. K., and Patel, N. H., 1999, The origin and evolution of segmentation, *Trends Cell Biol.* **9**: M68–72.

Deganan, B. M., Leys, S. P., and Larraoux, C., 2005, Sponge development and antiquity of animal pattern formation, *Integr. Comp. Biol.* **45**: 335–341.

de Robertis, E. M., and Sasai, Y., 1996, A common plan for dorsoventral patterning in bilateria, *Nature* **380**: 37–40.

de Rosa, R., Grenier, J. K., Andreeva, T., Cook, C. E., Adoutte, A., Akam, M., Carroll, S. B., and Balavoine, G., 1999, Hox genes in brachiopods and priapulids and protostome evolution, *Nature* **399**: 772–776.

Eernisse, D. J., and Peterson, K. J., 2004, The History of Animals, in: *Assembling the Tree of Life* (J. Cracraft, and M. J. Donoghue, eds.), Oxford University Press, Oxford & New York, pp. 197–208.

Erwin, D. H. 1993, The origin of metazoan development: a palaeobiological perspective, *Biol. J. Linn. Soc.* **50**: 255–274.

Erwin, D. H., 2000, Macroevolution is more than repeated rounds of microevolution, *Evol. Dev.* **2**: 78–84.

Erwin, D. H. 2005. The Origin of Animal Bodyplans. in: *Form and Function. Essays in Honor of Adolf Seilcher* (D. E. G. Briggs, e.d), Yale University Press, New Haven, pp. 67–80.

Erwin, D. H., and Davidson, E. H., 2002, The last common bilaterian ancestor, *Development* **129**: 3021–3032.

Erwin D. H., Valentine J.W., Jablonski D. 1997. The origin of animal bodyplans. *Am. Sci.* **85**: 126–37

Erwin, D. H., Valentine, J. W., and Sepkoski, J. J., Jr., 1987, A comparative study of diversification events: the early Paleozoic vs. the Mesozoic, *Evolution* **41**: 1177–1186.

Fernald, R. D., 2000, Evolution of eyes, *Curr. Opinion Neurobiol.* **10**: 444–450.

Ferrier, D. E., and Holland, P. W., 2002, *Ciona intestinalis* ParaHox genes: evolution of Hox/ParaHox cluster integrity, developmental mode, and temporal colinearity, *Mol. Phylogenet. Evol.* **24**: 412–417.

Ferrier, D. E., and Minguillon, C., 2003, Evolution of the *Hox/ParaHox* gene clusters, *Int. J. Dev. Biol.* **47**: 605–611.

Finkelstein, R., and Boncinelli, E., 1994, From fly head to mammalian forebrain: the story of *otd* and *Otx*, *Trends Genet.* **10**: 310–315.

Finnerty, J. R., 2003, The origins of axial patterning in the metazoa: how old is bilateral symmetry, *Int. J. Dev. Biol.* **47**: 523–529.

Finnerty, J. R., Pang, K., Burton, P., Paulson, D., and Martindale, M. Q., 2004, Origins of bilateral symmetry: Hox and *dpp* expression in a sea anemone, *Science* **304**: 1335–1337.

Galliot, B., 2000, Conserved and divergent genes in apex and axis development of cnidarians, *Curr. Opinion Genet. Dev.* **10**: 629–637.

Galliot, B., and Miller, D., 2000, Origin of anterior patterning: How old is our head, *Trends Genet.* **16**: 1–5.

Gehring, W. J., 1996, The master control gene for morphogenesis and evolution of the eye, *Genes to Cells* **1**: 11–15.

Gehring, W. J., 2004, Historical perspective on the development and evolution of eyes and photoreceptors, *Int. J. Dev. Biol.* **48**(8–9): 707–717.

Gehring, W. J., and Ikeo, K., 1999, Pax 6. mastering eye morphogenesis and eye evolution, *Trends Genet.* **15**: 371–376.

Gellon, G., and McGinnis, W., 1998, Shaping animal body plans in development and evolution by modulation of Hox expression patterns, *BioEssays* **20**: 116–125.

Gerhart, J., 1999, Signaling pathways in development, *Teratology* **60**: 226–239.

Ghysen, A., 2003, The origin and evolution of the nervous system, *Int. J. Dev. Biol.* **47**: 555–562.

Giribet, G., 2003, Molecules, development and fossils in the study of metazoan evolution; Articulata versus Ecdysozoa revisited, *Zoology* **106**: 303–326.

Groger, H., and Schmid, V., 2001, Larval development in Cnidaria: A connection to Bilateria, *Genesis* **29**: 110–114.

Halanych, K. M., 2004, The new view of animal phylogeny, *Annu. Rev. Ecol. Systemat.* **35**: 229–256.

Halder, G., Callaerts, P., and Gehring, W. J., 1995a, Induction of ectopic eyes by targeted expression of the eyeless gene in *Drosophila*, *Science* **267**: 1788–1792.

Halder, G., Callaerts, P., and Gehring, W. J., 1995b, New perspectives on eye evolution, *Curr. Opinion Genet. Dev.* **5**: 602–609.

Hanson, I. M., 2001, Mammalian homologues of the *Drosophila* eye specification genes, *Seminar in Cell. Dev. Biol.* **12**: 475–484.

Harvey, R. P., 1996, NK-2 homeobox genes and heart development, *Dev. Biol.* **178**: 203–216.

Haun, C., Alexander, J., Stainier, D. Y., and Okkema, P. G., 1998, Rescue of *Caenorhabditis elegans* pharyngeal development by a vertebrate heart specification gene, *Proc. Nat. Acad. Sci. USA* **95**: 5072–5075.

Hinman, V. F., Nguyen, A. T., Cameron, R. A., and Davidson, E. H. 2003, Developmental gene regulatory network architecture across 500 million years of echinoderm evolution. *Proc. Nat. Acad. Sci. U S A* **100**: 13356–13561.

Hirth, F., Kammermeier, L., Frei, E., Walldorf, U., Noll, M., and Reichert, H., 2003, An urbilaterian origin of the tripartite brain: developmental genetic insights from *Drosophila*, *Development* **130**: 2365–2373.

Hirth, F., and Reichert, H., 1999, Conserved genetic programs in insect and mammalian brain development, *BioEssays* **21**: 677–684.

Hoegg, S., and Meyer, A., 2005, Hox clusters as models for vertebrate genome evolution. *Trends Genet.* **21**, 421–424.

Hoffmann, J. A., 2003, The immune response of *Drosophila*, *Nature* **426**: 33–38.

Holland, L. Z., 2000, Body-plan evolution in the Bilateria: early antero-posterior patterning and the deuterostome-protostome dichotomy, *Curr. Opinion Genet. Dev.* **10**: 434–442.

Holland, L. Z., Kene, M., Williams, N. A., and Holland, N. D., 1997, Sequence and embryonic expression of the amphioxis engrailed gene (*AmphiEn*): the metameric pattern of transcription resembles that of its segment-polarity homolog in *Drosophila*, *Development* **124**: 1723–1732.

Holland, N. D., 2003, Early central nervous system evolution: an era of skin brains? *Nat. Rev. Neurosci.* **4**: 617–627.

Holland, N. D., Venkatesh, T. V., Holland, L. Z., Jacobs, D., and Bodmer, R., 2003, *AmphiNk1-tin*, an amphioxus homeobox gene expressed in myocardial progenitors: insights into the evolution of the vertebrate heart, *Dev. Biol.* **255**: 128–137.

Holland, P. W. H., Garcia-Fernandez, J., Williams, N. A., and Sidow, A., 1994, Gene duplications and the origins of vertebrate development, *Development* Suppl.: 125–133.

Holley, S. A., Jackson, P. D., Sasai, Y., Lu, B., De Robertis, E. M., Hoffmann, F. M., and Ferguson, E. L., 1995, A conserved system for dorsal-ventral patterning in insects and vertebrates involving *sog* and *chordin*, *Nature* **376**: 249–253.

Horowitz, N. H., 1945, On the evolution of biochemical syntheses. *Proc. Nat. Acad. Sci. USA* **31**:153–157.

Hughes, A. L., 2003, 2R or not 2R: testing hypotheses of genome duplication in early vertebrates, *J. Struct. Funct. Genomics* **3**: 85–93.

International Human Genome Sequencing Consortium, 2004, Finishing the euchromatic sequence of the human genome, *Nature* **431**:931–935

Kaiser, D., 2001, Building a multicellular organism, *Annu. Rev. Genet.* **35**: 103–23.

Kimmel, C. B., 1996, Was *Urbilateria* segmented? *Trends Genet.* **12**: 329–332.

King, N., 2004, The unicellular ancestry of animal development, *Dev. Cell* **7**: 313–325.

King, N., Hittinger, C. T., and Carroll, S. B., 2003, Evolution of key cell signaling and adhesion protein families predates animal origins, *Science* **301**: 361–363.

Knoll, A. H., and Carroll, S. B., 1999, Early animal evolution: emerging views from comparative biology and geology, *Science* **284**: 2129–2137.

Kourakis, M. J., and Martindale, M. Q., 2000, Combined-method phylogenetic analysis of Hox and ParaHox genes of the metazoa, *J. Exp. Zool.* **288**: 175–191.

Krieger, J., and Breer, H., 1999, Olfactory reception in invertebrates, *Science* **286**: 720–723.

Kumar, J. P., and Moses, K., 2001, Eye specification in *Drosophila*: perspectives and implications, *Seminar Cell. Dev. Biol.* **12**: 469–474.

Kusserow, A., Pang, K., Strum, C., Hrouda, M., Lentfer, J., Schmidt, H. A., Technau, U., von Haeseler, A., Hobmayer, B., Martindale, M. Q., and Holstein, T. W., 2005, Unexpected complexity of the *Wnt* gene family in a sea anemone, *Nature* **433**: 156–160.

Lacalli, T. C., 2001, New perspectives on the evolution of protochordate sensory and locomotory systems, and the origin of brains and heads, *Proc. Roy. Soc. London, Ser. B* **356**: 1565–1572.

Land, M. F., and Fernald, R. D., 1992, The evolution of eyes, *Annu. Rev. Neurosci.* **15**: 1–29.

Levine, M., and Davidson, E. H., 2005, Gene regulatory networks for development, *Proc. Nat. Acad. Sci. USA* **102**:4936–4942.

Levine, M., and Tjian, R., 2003, Transcription regulation and animal diversity, *Nature* **424**: 147–51.

Lichtneckert, R., and Reichert, H., 2005, Insights into the urbilaterian brain: conserved genetic patterning mechanisms in insect and vertebrate brain development, *Heredity*: 1–13.

Lowe, C. J., Wu, M., Salic, A., Evans, L., Lander, E., Stange-Thomann, N., Gruber, C. E., Gerhart, J., and Kirschner, M., 2003, Anteroposterior patterning in hemichordates and the origins of the chordate nervous system, *Cell* **113**: 853–865.

Lundin, L. G., 1999, Gene duplications in early metazoan evolution, *Seminars Cell Dev. Biol.* **10**: 523–530.

Martindale, M. Q., Finnerty, J. R., and Henry, J. Q., 2002, The Radiata and the evolutionary origins of the bilaterian body plan, *Mol. Phylogenet. Evol.* **24**: 358–365.

Martindale, M. Q., Pang, K., and Finnerty, J. R., 2004, Investigating the origins of triploblasty: 'mesodermal' gene expression in a diploblastic animal, the sea anemone *Nematostella vectensis* (phylum, Cnidaria; class, Anthozoa), *Development* **131**: 2463–2474.

McGinnis, W., and Krumlauf, R., 1992, Homeotic genes and axial patterning, *Cell* **68**: 283–302.

Metzger, R. J., and Krasnow, M. A., 1999, Genetic control of branching morphogenesis, *Science* **284**: 1635–1639.

Miller, D. J., Hayward, D. C., Reece-Hoyes, J. S., Scholten, I., Catmull, J., Gehring, W. J., Callaerts, P., Larsen, J. E., and Ball, E. E., 2000, *Pax* gene diversity in the basal cnidarian *Acropora millepora* (Cnidaria, Anthozoa): implications for the evolution of the *Pax* gene family, *Proc. Nat. Acad. Sci. USA* **97**: 4475–4480.

Minelli, A., 2003, The origin and evolution of appendages, *Int. J. Dev. Biol.* **47**: 573–581.

Minelli, A., and Fusco, F., 2004, Evo-devo perspectives on segmentation: model organisms and beyond, *Trends Ecol. Evol.* **19**: 423–429.

Müller, M., Weizsäker, E. V., and Campos-Ortega, J. A., 1996, Expression domains of a zebrafish homologue of the *Drosophila* pair-rule gene *hairy* corresponds to primordia of alternating somites. *Development* **122**: 2071–2078.

Müller, P., Yanze, N., Schmid, V., and Spring, J., 1999, The homeobox gene *Otx* of the jellyfish *Podocoryne carnea*: role of a head gene in striated muscle and evolution, *Dev. Biol.* **216**: 582–594.

Müller, W. E. G., 2001, How was the metazoan threshold crossed: the hypothetical Urmetazoan? *Comp. Biochem. Physiol. A* **129**: 433–460.

Müller, W. E., Wiens, M., Adell, T., Gamulin, V., Schroder, H. C., and Muller, I. M., 2004, Bauplan of urmetazoa: basis for genetic complexity of metazoa, *Int. Rev. Cytol.* **235**: 53–92.

Nielsen, C., 1999, Origin of the chordate central nervous system and the origin of chordates, *Dev. Genes Evol.* **209**: 198–205.

Nilsson, D. E., and Pelger, S., 1994, A pessimistic estimate of the time required for an eye to evolve, *Proc. Roy. Soc., Biol. Sci.* **256**: 53–58.

Niwa, N., Hiromi, Y., and Okabe, M., 2004, A conserved developmental program for sensory organ formation in *Drosophila melanogaster*, *Nat. Genet.* **36**: 293–297.

Oelgeschlager, M., Larrain, J., Geissert, D., and De Robertis, E. M., 2000, The evolutionarily conserved BMP-binding protein twisted gastrulation promotes BMP signalling, *Nature* **405**: 757–763.

Panda, S., Hogenesch, J. B., and Kay, S. A., 2002, Circadian rhythms from flies to human, *Nature* **417**: 329–335.

Panda, S. et al. 2005, Illumination of the melanopsin signaling pathway. *Science* **307**: 600–604.

Panganiban, G., Sebring, A., Nagy, L., and Carroll, S. 1995, The development of crustacean limbs and the evolution of arthropods. *Science* **270**:1363–1366.

Panganiban, G. E. F., Irvine, S. M., Lowe, C., Roehl, H., Corley, L. S., Sherbon, B., Grenier, J. K., Fallon, J. F., Kimble, J., Walker, M., Wray, G. A., Swalla, B. J., Martindale, M. Q., and Carroll, S. B., 1997, The origin and evolution of animal appendages, *Proc. Nat. Acad. Sci. USA* **94**: 5162–5166.

Panganiban, G., and Rubenstein, J. L. R., 2002, Developmental functions of the Distal-less/Dlx homeobox genes, *Development* **129**: 4371–4386.

Patient, R. K., and McGhee, J. D., 2002, The GATA family (vertebrates and invertebrates), *Curr. Opinion Genet. Dev.* **12**: 416–422.

Peterson KJ, Butterfield NJ. 2005. Origin of the Eumetazoa: testing ecological predictions of molecular clocks against the Proterozoic fossil record. *Proc. Nat. Acad. Sci. USA* **102**: 9547–52

Peterson, K. J., Lyons, J. B., Nowak, K. S., Takacs, C. M., Wargo, M. J., and McPeek, M. A. 2004. Estimating metazoan divergence times with a molecular clock. *Proc. Nat. Acad. Sci. USA* **101**:6536–6541.

Philip, G. K., Creevey, C. J., and Mcinerney, J. O. 2005. The Opisthokonta and Ecdysozoa may not be clades: stronger support for the grouping of plant and animal than for animal and fungi and stronger support for the coelomata than Ecdysozoa. *Mol. Biol. Evol.* **22**: 1175–1184.

Philippe, H., Lartillot, N., and Brinkmann, H. 2005. Multigene Analyses of Bilaterian Animals Corroborate the Monophyly of Ecdysozoa, Lophotrochozoa, and Protostomia. *Mol. Biol. Evol.* **22**:1246–1253.

Piatigorsky, J., and Kozmik, Z., 2004, Cubozoan jellyfish: an Evo/Devo model for eyes and other sensory systems, *Int. J. Dev. Biol.* **48**: 719–729.

Plachetzki, D. C., Serb, J. M. & Oakley, T. H. (2005). New insights into the evolutionary history of photoreceptor cells. *Trends Ecol. Evol.* **20**, 465–467.

Popadic, A., Rusch, D., Peterson, M. D., Rogers, B. T., and Kaufman, T. C., 1996, Origin of the arthropod mandible, *Nature* **380**:395.

Prasad, B. C., and Reed, R. R., 1999, Chemosensation: molecular mechanisms in worms and mammals, *Trends Genet.* **15**: 150–153.

Prince, V., 2002, The Hox Paradox: More complex(es) than imagined, *Dev. Biol.* **249**: 1–15.

Quiring, R., Walldorf, U., Kloter, U., and Gehring, W. J., 1994, Homology of the eyeless gene of *Drosophila* to the Small eye gene in mice and Aniridia in humans, *Science* **265**: 785–789.

Ranganayakulu, G., Elliott, D. A., Harvey, R. P., and Olson, E. N., 1998, Divergent roles for NK-2 class homeobox genes in cardiogenesis in flies and mice, *Development* **125**: 3037–3048.

Rebeiz, M., Stone, T., and Posakony, J. W., 2005, An ancient transcriptional regulatory linkage, *Dev. Biol.* **281**: 299–308.

Reichert, H., and Simeone, A., 2001, Developmental genetic evidence for a monophyletic origin of the bilaterian brain, *Proc. Roy. Soc. London, Ser. B* **356**: 1533–1544.

Ruiz-Trillo, I., Riutort, M., Littlewood, D. T. J., Herniou, E. A., and Baguna, J., 1999, Acoel flatworms: earliest extant bilaterian metazoans, not members of Platyhelminthes, *Science* **283**: 1919–1923.

Salwini-Plawen, L. V., and Mayr, E., 1977, On the evolution of photoreceptors and eyes, *Evol. Biol.* **10**: 207–263.

Schierwater, B., and Kuhn, K., 1998, Homology of Hox genes and the zootype concept in early Metazoan evolution, *Mol. Phylogenet. Evol.* **9**: 375–381.

Scott, M. P., 1994, Intimations of a creature, *Cell* **79**: 1121–1124.

Seaver, E. C., 2003, Segmentation: mono- or polyphyletic? *Int. J. Dev. Biol.* **47**: 583–595.

Seo, H. C., Edvardsen, R. B., Maeland, A. D., Bjordal, M., Jensen, M. F., Hansen, A., Flaat, M., Weissenbach, J., Lehrach, H., Wincker, P., Reinhardt, R., and Chourrout, D., 2004, Hox cluster disintegration with persistent anteroposterior order of expression in Oikopleura dioica, *Nature* **431**: 67–71.

Shankland, M., and Seaver, E. C., 2000, Evolution of the bilaterian body plan: what have we learned from annelids? *Proc. Nat. Acad. Sci. USA* **97**: 4434–4437.

Shenk, M. A., and Steele, M. A., 1993, A molecular shapshot of the metazoan 'Eve', *Trends Biochem. Sci.* **18**: 459–463.

Shoichet, S. A., Malik, T. H., Rothman, J. H., and Shivdasani, R. A., 2000, Action of the Caenorhabditis elegans GATA factor END-1 in Xenopus suggests that similar mechanisms initiate endoderm development in ecdysozoa and vertebrates, *Proc. Nat. Acad. Sci. USA* **97**: 4076–4081.

Shubin, N. H., Tabin, C., and Carroll, S., 1997, Fossils, genes and the evolution of animal limbs, *Nature* **388**: 639–648.

Slack, J. M. W., Holland, P. W. H., and Graham, C. F., 1993, The zootype and the phylotypic stage, *Nature* **361**: 490–492.

Smith, K. M., Gee, L., Blitz, I. L., and Bode, H. R., 1999, CnOtx, a member of the Otx gene family, has a role in cell movement in hydra, *Dev. Biol.* **212**: 392–404.

Spring, J., Yanze, N., Josch, C., Middel, A. M., Winninger, B., and Schmid, V., 2002, Conservation of *Brachyury, Mef2* and *Snail* in the myogenic lineage of jellyfish: a connection to the mesoderm of Bilateria, *Dev. Biol.* **244**: 372–384.

Steele, R. E., 2002, Developmental signaling in Hydra: what does it take to build a "simple" animal? *Dev. Biol.* **248**: 199–219.

Stollenwerk, A., Schoppmeier, M., and Damen, W. G. M., 2003, Involvement of Notch and Delta genes in spider segmentation, *Nature* **423**: 863–865.

Strausfeld, N. J., and Hildebrand, J. G., 1999, Olfactory systems: common design, uncommon origins, *Curr. Opinion Neurobiol.* **9**: 634–639.

Suga, H., Katoh, K., and Miyata, T., 2001, Sponge homologs of vertebrate protein tyrosine kinases and frequent domain shufflings in the early evolution of animals before the parazoan-eumetazoan split. *Gene* **280**: 195–201.

Tanaka, M., Kasahara, H., Bartunkova, S., Schinke, M., Komuro, I., Inagaki, H., Lee, Y., Lyons, G. E., and Izumo, S., 1998, Vertebrate homologs of *tinman* and *bagpipe*: roles of the homeobox genes in cardiovascular development, *Dev. Genet.* **22**: 239–249.

Tautz, D., 2004, Segmentation, *Dev. Cell* **7**: 301–312.

Technau, U., and Scholz, C. A., 2003, Origin and evolution of endoderm and mesoderm, *Int. J. Dev. Biol.* **47**: 531–539.

Technau, U., Rudd, S., Maxwell, P., Gordon, P. M. K., Saina, M., Grasso, L. C., Hayward, D. C., Sensen, C. W., Saint, R., Holstein, T. W., Ball, E. E. & Miller, D. J. 2005, Maintenance of ancestral complexity and non-metazoan genes in two basal cnidarians. *Trends Genet.* **21**: 633–639.

Treisman, J. E., 2004, How to make an eye, *Development* **121**: 3823–3827.

True, J. R., and Carroll, S. B., 2002, Gene co-option in physiological and morphological evolution, *Annu. Rev. Cell Dev. Biol.* **18**: 53–80.
Valentine, J. W., 2004, *On the Origin of Phyla*, University of Chicago Press, Chicago.
Valentine, J. W., and Erwin, D. H., 1987, Interpreting great developmental experiments: the fossil record, in: *Development as an Evolutionary Process* (R. A. Raff, ed.), A. R. Liss, Inc., New York, pp. 71–107.
Valentine, J. W., and Jablonski, D., 2003, Morphological and developmental macroevolution: a paleontological perspective, *Int. J. Dev. Biol.* **47**: 517–522.
Valentine, J. W., Jablonski, D., and Erwin, D. H., 1999, Fossils, molecules and embryos: new perspectives on the Cambrian explosion, *Development* **126**: 851–859.
Wikramanayake, A. H., Hong, M., Lee, P. N., Pang, K., Byrum, C. A., Bince, J. M., Xu, R., and Martindale, M. Q., 2003, An ancient role for nuclear beta-catenin in the evolution of axial polarity and germ layer segregation, *Nature* **426**: 446–450.
Wilmer, P., 1990, *Invertebrate Relationships: Patterns in Animal Evolution*, Cambridge University Press, Cambridge.
Wolpert, L., 1990, The evolution of development, *Biol. J. Linn. Soc.* **39**: 109–124.
Wray, G. A., 1999, Evolutionary dissociations between homologous genes and homologous structures, in: *Homology* (G. R. Bock and G. Cardew, eds.), Wiley, Chichester, pp. 189–203.
Yanze, N., Spring, J., Schmidli, C., and Schmid, V., 2001, Conservation of *Hox/ParaHox*-related genes in the early development of a cnidarian, *Dev. Biol.* **236**: 89–98.
Zaret, K., 1999, Developmental competence of the gut ectoderm: genetic potentiation by GATA and HNF3/Forkhead proteins, *Dev. Biol.* **209**: 1–10.
Zhang, J. M., Chen, L., Krause, M., Fire, A., and Paterson, B. M., 1999, Evolutionary conservation of MyoD function and differential utilization of E proteins, *Dev. Biol.* **208**: 465–472.

Chapter 7

Molecular Timescale of Evolution in the Proterozoic

S. BLAIR HEDGES, FABIA U. BATTISTUZZI AND JAIME E. BLAIR

Department of Biology and NASA Astrobiology Institute, Pennsylvania State University, University Park, PA 16802, USA.

1. Introduction	199
2. Molecular Clock Methods	201
3. Molecular Timescales	203
3.1 Prokaryotes	203
3.2 Eukaryotes	205
3.3 Land Plants	212
3.4 Fungi	213
3.5 Animals	215
4. Astrobiological Implications	217
4.1 Complexity	217
4.2 Global glaciations	219
4.3 Oxygen and the Cambrian explosion	221
5. Conclusions	221
Acknowledgements	222
References	222

1. INTRODUCTION

The late Precambrian (Neoproterozoic; 1000–543 Million years ago, Ma) was a transitional time in Earth history and the evolution of eukaryotes. Although atmospheric oxygen levels initially rose in the early Proterozoic

(~2300 Ma), perhaps to as much as 10% of present levels, a second major increase occurred at some point in the late Neoproterozoic (Canfield, 2005). By the early Phanerozoic, the atmospheric oxygen level was close to that of the present (Berner et al., 2003). There has been speculation for years that the sudden appearance of many animal phyla in the early Phanerozoic (Cambrian Explosion) was causally tied to an increase in atmospheric oxygen, and that small and soft-bodied animals may have existed for a lengthy period before the Phanerozoic (e.g., Nursall, 1959). Nonetheless, the leading explanation for the Cambrian Explosion is that it represents a rapid evolutionary radiation of animals in the latest Precambrian or earliest Phanerozoic (Gould, 1989; Conway Morris, 2000).

Molecular clocks, which measure times of divergence between species from sequence data, have focused even greater attention on the Cambrian Explosion. Such studies in the last three decades have frequently found divergences among animal phyla to be hundreds of millions of years earlier than predicted by the fossil record (Brown et al., 1972; Runnegar, 1982b; Wray et al., 1996; Wang et al., 1999; Hedges et al., 2004; Pisani et al., 2004; Blair et al., 2005). Moreover, plants and fungi, two groups with a fossil record that, at least until recently, has been firmly rooted in the Phanerozoic, have also been suggested to have deep roots in the Neoproterozoic using molecular clocks (Berbee and Taylor, 2001; Heckman et al., 2001; Hedges et al., 2004; Padovan et al., 2005). Not all molecular clock studies agree, and several recent studies (Aris-Brosou and Yang, 2002, 2003; Douzery et al., 2004; Peterson et al., 2004; Peterson and Butterfield, 2005) have obtained results more closely in line with the majority of fossil evidence (i.e., late Neoproterozoic).

Differing results among molecular clock studies are mirrored by differing opinions among palaeontologists as to the validity of the earliest fossils of multicellular life. This exemplifies the point that debates in this area are not necessarily between different fields (e.g., molecular evolution versus palaeontology) but are often within fields. For example, the earliest eukaryote fossils, assigned to the genus *Grypania* (Han and Runnegar, 1992) and now dated to ~1900 Ma are considered by some paleontologists to be prokaryotes (Samuelsson and Butterfield, 2001; Sergeev et al., 2002), and none of the many trace fossils of animals from deep in the Precambrian (e.g., Seilacher et al., 1998) are widely accepted (Jensen et al., 2005). Nonetheless, the fossil record of fungi and land plants has been pushed back tens to hundreds of millions of years in recent studies (Yang et al., 2004; Butterfield, 2005; Yuan et al., 2005), and a greater diversity of animal fossils has come to light from Neoproterozoic Lagerstätten, especially phosphorites (Xiao et al., 1998; Chen et al., 2004). It is only to be expected that claims of the earliest fossil of any major group of organism will be controversial and subject to continued scrutiny.

Concurrent with developments in molecular dating and palaeontology have been discoveries in Earth history that have placed additional attention on the Neoproterozoic as a time period of great interest for the evolution of complex life. Chief among these has been the elucidation of a series of global glaciations (Snowball Earth events) which would have greatly restricted the habitable area for life on Earth and may have influenced patterns of speciation and macroevolution (Hoffman et al., 1998; Hedges, 2003). However, as with Neoproterozoic molecular clocks and fossils, there has been disagreement among geologists as to mechanisms and extent of the glaciations (Hyde et al., 2000; Young, 2002; Poulsen, 2003; Poulsen and Jacob, 2004).

Despite the uncertainties and debates within different fields, the interdisciplinary nature of this research in Neoproterozoic geobiology and astrobiology has been appealing to many and has invigorated the field. Here, we will review the current state of knowledge regarding evolutionary relationships and times of origin of organisms in the Proterozoic as they bear on these questions of how complex life evolved in the face of a changing planetary environment.

2. MOLECULAR CLOCK METHODS

Methods for estimating time from molecular sequence data have evolved over the four decades since the original proposal of a molecular clock (Zuckerkandl and Pauling, 1962). At the basic level, molecular clocks provide a means of estimating the divergence time of species based on rates of sequence change in genes and genomes. These rates are usually established first by calibration with the fossil record.

The same mechanism driving radiometric clocks, stochasticity, is believed to be the engine of molecular clocks (Kimura, 1983). Of course, different genes and regions of the genome evolve at different rates, as a result of constraints imposed from natural selection, just as different isotopes decay at different rates. In both cases, it is this diversity of rates that allows use of these methods at different timescales. Nonetheless, molecular clocks are more variable than radiometric clocks, and this variation has been responsible for much of the debate over the use of clocks in recent years.

Fortunately, data sets have become larger and methods have become more sophisticated in parallel with developments in the field of molecular evolution (reviewed in Hedges and Kumar, 2003; Kumar, 2005). The importance of the size of a data set (number of sites and genes) cannot be stressed enough, because a large sample is needed for deriving any mean estimate from a stochastic mechanism. Time estimates based on a single gene or small number of genes may have large associated errors or undetected biases, just as time estimates derived from a small number of radioisotope decays (in practice, the number of decays is not a limiting factor). Hence, it is preferable to estimate time from large numbers of genes.

While it is possible to restrict analysis to genes evolving at a constant rate among lineages or branches, a larger data set can be used by including those genes having rate variation among branches. This difference has led to the distinction of global clocks (rate constancy throughout tree) from local clocks (rate variation in localized parts of the tree) (Hedges and Kumar, 2003). The latter contains a great diversity of methods (Hasegawa et al., 1989; Takezaki et al., 1995; Sanderson, 1997; Schubart et al., 1998; Thorne et al., 1998), some of which are referred to as "relaxed clock" methods. Notwithstanding claims that time can be estimated "without a clock" (Sanderson, 2003; Bell et al., 2005), the term "clock"—defined in most dictionaries as "a device used for measuring and indicating time"—applies appropriately to any method of estimating time.

The introduction of minimum and maximum constraints for calibrations in some local clock methods (Sanderson, 1997; Kishino et al., 2001) has helped to focus greater attention on confidence intervals. However at the same time this has exposed weaknesses in these sources of data. In particular, the maximum time of divergence is almost never known with any certainty, although it has been widely used in molecular clock studies in recent years. As discussed elsewhere (Hedges and Kumar, 2004), this fact and the lack of knowledge of the probability distribution for most calibrations have resulted (mostly likely) in many underestimates and overestimates of time. Earlier molecular clock studies, in contrast, often used minimum calibration points, which are less subject to error, and interpreted their results with respect to that constraint (Hedges and Kumar, 2004). Nonetheless, it is preferable to obtain a mean estimate for the time of divergence rather than the minimum estimate, and therefore further improvements in these local clock methods seem likely.

Finally, a new method of incorporating statistical error of time estimates has been developed based on bootstrapping (Kumar et al., 2005). This error estimation method can be used with any molecular clock method, and includes all types of variance, such as that resulting from differences in gene sampling, site sampling, error contributed by distance estimation procedures, variance in rates among lineages, and any error associated with calibrations. No previous molecular clock studies have incorporated all of these errors. With such methods, time estimates in future studies will likely have larger confidence intervals but should be more realistic.

3. MOLECULAR TIMESCALES

Some remarkable fossil discoveries have been made of Neoproterozoic organisms in recent years (Xiao *et al.*, 1998; Chen *et al.*, 2004; Yuan *et al.*, 2005) but the fossil record of this time period, and of earlier time periods, is much poorer than that of the Phanerozoic. For this reason, it is more difficult to estimate times of divergence between lineages from the fossil record alone, placing more importance on obtaining such estimates using molecular clocks. Below, we review those studies that have estimated divergence times of organisms that lived during the Precambrian, with emphasis on studies using relatively large data sets. Such information is critical to understanding how features of the planetary environment, including global glaciations, plate tectonics, and changes in atmospheric gases, influenced or were impacted by the evolution of life at that time.

3.1 Prokaryotes

There have been several molecular clock studies of prokaryotes that have estimated divergence times in the Precambrian. Without exception, all have found deep divergences, prior to the Neoproterozoic, of the major groups of archaebacteria and eubacteria (Doolittle *et al.*, 1996; Feng *et al.*, 1997; Hedges *et al.*, 2001; Sheridan *et al.*, 2003; Battistuzzi *et al.*, 2004). One study (Sheridan *et al.*, 2003) used nucleotide variation in the gene for the small subunit ribosomal RNA whereas the other studies used amino acid sequences of multiple proteins. Taxon sampling was relatively limited in the earlier protein studies (Doolittle *et al.*, 1996; Feng *et al.*, 1997; Hedges *et al.*, 2001), with the recent study (Battistuzzi *et al.*, 2004) having the largest number of taxa. Also, it is now known that there are major rate differences among groups of prokaryotes and between prokaryotes and eukaryotes (Kollman and Doolittle, 2000; Hedges *et al.*, 2001). Some time estimates (Doolittle *et al.*, 1996; Feng *et al.*, 1997; Sheridan *et al.*, 2003) were made without accounting for those rate differences, and another (Hedges *et al.*, 2001) was made using a two rate model. The recent analysis (Battistuzzi *et al.*, 2004) was performed using a Bayesian local clock method that permitted rate variation among branches and the results of that study are reproduced here (Fig. 1). Genes that showed obvious evidence of lateral gene transfer were avoided. As can be seen, and not surprisingly, all of the major groups of prokaryotes were extant by the onset of the Neoproterozoic and therefore their major metabolic activities (e.g., anoxygenic photosynthesis, oxygenic photosynthesis, methanogenesis, aerobic methanotrophy, etc.) were present. In most cases, this only reinforces what has already been revealed with geologic and other evidence (Knoll, 2004).

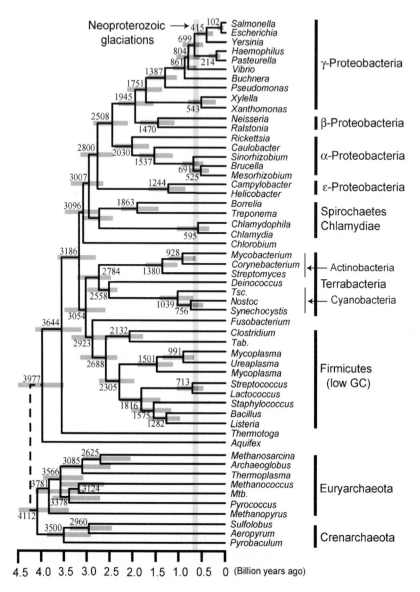

Figure 1. A timetree of prokaryotes constructed by a Bayesian analysis of proteins from complete genomes (7600 amino acids, total) (Battistuzzi et al., 2004). Gray horizontal bars are Bayesian credibility intervals.

Of potential interest to astrobiology is the clustering of three major groups (Cyanobacteria, Deinococcales, and Actinobacteria) which presumably had a common ancestor that was terrestrial approximately 3000 Ma (Fig. 1). Besides sharing photoprotective compounds, all three share a high resistance to dehydration and have species that are currently terrestrial. These three groups were named, collectively, Terrabacteria (Battistuzzi *et al.*, 2004). This supports the paleontological and geological evidence that prokaryotes colonized the land surface in the Precambrian (Horodyski and Knauth, 1994; Watanabe *et al.*, 2000) and therefore their metabolic and erosional activities should be considered, as well as potential interactions with terrestrial eukaryotes.

3.2 Eukaryotes

The relationships of the major lineages of eukaryotes have become much better known during the last decade as more sequence data have been gathered and analyzed (reviewed in Baldauf *et al.*, 2000; Hedges, 2002; Keeling *et al.*, 2005). Because the relationships of single-celled eukaryotes (protists) are intimately tied to the relationships of multicellular eukaryotes (algae, plants, fungi, animals), it is usually more convenient to discuss this subject in terms of overall (higher-level) eukaryote phylogeny, as will be done here. The land plants, fungi, and animals will be discussed in separate sections.

The ease of sequencing ribosomal RNA (rRNA), and especially the small subunit, meant that an initial molecular view and framework of eukaryote phylogeny, from molecules, was based on that gene. For eukaryotes, those trees defined a crown consisting of plants, animals, fungi, and related protists, and a series of lineages along the stem or base of the tree, with the diplomonad *Giardia* as the earliest branch (Sogin *et al.*, 1989; Schlegel, 1994). Later analyses using complex models and different genes showed that some—but not all—basal branching lineages (e.g., microsporidia, *Dictyostelium*) actually belong higher in the tree, and their misplacement was the result of long-branch attraction or other biases (e.g., Philippe and Germot, 2000). Subsequently, much recent attention has been placed on building trees with as many genes and taxa as possible, and using different types of analyses, including complex substitution models. This has brought welcomed stability to some aspects of the tree, and remarkable volatility to others.

Some major questions that were once controversial have now been answered to the satisfaction of many in the field. For example, animals and fungi appear to be closest relatives (opisthokonts) to the exclusion of plants (although see Philip *et al.*, 2005), and red algae are on the "plant lineage" and not basal to the divergence of plants and opisthokonts as previously thought. There is growing support that amoebozoans are the closest relatives of opisthokonts (e.g., Amaral Zettler *et al.*, 2001; Baldauf, 2003), and that microsporidia are the closest relatives of fungi.

Recently, there has been an effort to summarize these and other aspects of the eukaryote tree in the form of a five-group arrangement: plants, unikonts, chromalveolates, rhizarians, and excavates (Keeling, 2004; Keeling et al., 2005). Under this scheme, Plantae is defined by the presence of plastids acquired by primary endosymbiosis and includes the land plants, charophytes, chlorophytes, rhodophytes, and glaucophytes. The unikonts are defined by the presence of a single cilium-bearing centriole and include the opisthokonts (animals, fungi, choanoflagellates, ichthyosporeans, and nuclearids) and the amoebozoans. Rhizaria includes the cercozoans, foraminiferans, polycistines, and acanthareans. Chromalveolates include the alveolates (e.g., ciliates, apicomplexans, and dinoflagellates) and stramenopiles (e.g., brown algae, diatoms, haptophytes, and cryptomonads). The excavates include the discicristates (e.g., euglenids and kinetoplastids), oxymonads, and metamonads (e.g., diplomonads, parabasalids, and *Carpediemonas*); the content of the excavates and relationships among the included taxa are particularly controversial (see below). The above arrangement also agrees with a division of eukaryotes into unikonts and bikonts (Richards and Cavalier-Smith, 2005). The amount of evidence supporting inclusion of different taxa varies considerably, from hundreds of genes in some cases (e.g., animals joining with fungi) to relatively small amounts of morphological or molecular data in other cases (e.g., excavates).

As a point of discussion, this five-group arrangement serves a useful purpose. However, a major problem is that it avoids the question of the root. Technically, without a root there can be no evolutionary polarity or claim of monophyly (i.e., no "five groups"). In fact, analyses of the largest sequence data sets, using complex models of evolution, show with statistical significance that the root lies within one of these five groups, the excavates (Hedges et al., 2001; Bapteste et al., 2002; Hedges et al., 2004), which breaks up the monophyly of the bikonts. This supports the earlier proposal of a basal position for *Giardia* (a diplomonad), based on rRNA sequences and cytological arguments (Sogin et al., 1989). Because complex models of evolution have been used in these recent studies, there is no clear evidence yet that long-branch attraction or other substitutional biases are responsible for this root position.

One analysis demonstrated sensitivity of the topology to removal of fast-evolving sites, but those results were inconclusive because of a lack of statistical support for most nodes (Arisue et al., 2005). With the great age of these lineages and the fact that long-branch attraction and other substitutional biases may lead to incorrect groupings (Philippe et al., 2000), it is worth being cautious in interpreting any results, even if statistically significant. Future analyses of large numbers of taxa and genes, and testing of hypotheses concerning substitutional biases, should help better resolve the tree of eukaryotes. However, the weight of the current sequence evidence argues, significantly, against the monophyly of at least one of the five groups, the "excavates," and in favour of a root between the metamonads (or at least the diplomonads) and other eukaryotes (Fig. 2A). Thus, a six-group classification would divide the excavates into the discicristates and metamonads.

Another possible location of the eukaryote root, between opisthokonts and all other eukaryotes, has been proposed based on a gene fusion event, joining dihydrofolate reductase and thymidylate synthase (Philippe et al., 2000; Stechmann and Cavalier-Smith, 2002) in many bikont eukaryotes. Evidence of a fusion of three genes in the pyrimidine biosynthetic pathway in unikonts led those same authors to revise their rooting scheme to include amoebozoans with opisthokonts (i.e., all unikonts) in the root (Stechmann and Cavalier-Smith, 2003). However, the subsequent finding of that triple gene fusion in a red alga (Matsuzaki et al., 2004), which is clearly not related to opisthokonts or amoebozoans, undermined the usefulness of that gene fusion character.

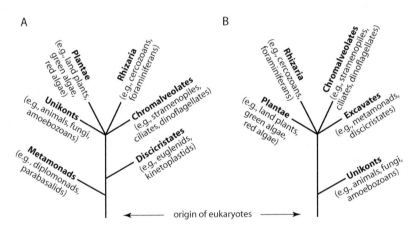

Figure 2. Two alternative hypotheses for the phylogenetic tree of eukaryotes. (A). The metamonad root, reflecting a six-group classification. This tree is favoured by phylogenetic analyses of DNA sequence data. (B). The unikont root, reflecting a five-group classification (metamonads and discicristates are combined into "excavates"). Under unikont rooting, the non-unikont eukaryotes (bikonts) are monophyletic.

Even more recently, evidence from the gene structure of myosin genes has been marshalled to further support a root between unikonts and bikonts (Richards and Cavalier-Smith, 2005) (Fig. 2B). In this case, gene and domain evolution is complex and there is homoplasy among the data. Moreover, most of the characters proposed as support for the root actually support the largely uncontroversial grouping of animals, fungi, and amoebozoans (unikonts), which does not define the root position. Both of the two characters indicated as supporting the alternative branch ("bikonts"), which is critical for the claim of unikont rooting, are problematic. One character involves the two-gene fusion, but this turns out to be absent in species that are critical to defining the root (e.g., metamonads), and the clear case of homoplasy involving the triple-gene fusion shows that gene fusions in general are not necessarily reliable characters. The second character is an insertion of 60 amino acids in bikonts. However, only two of the 13 bikonts examined (*Trypanosoma* and *Phytophthora*) had this insertion and neither was a metamonad. Furthermore, a recent study (Hampl *et al.*, 2005) claimed to recover excavate monophyly but close scrutiny shows that those authors fixed the root to unikonts and therefore they did not actually test excavate monophyly with a prokaryote outgroup.

Time estimation of protist evolution has lagged behind that of other groups largely because of the complexity of relationships and slower accumulation of sequence data. Recently, a sequence analysis of the phylogeny and divergence times of eukaryotes, including the major groups of protists, was made using 22–188 proteins per node (Hedges *et al.*, 2004), (Fig. 3). Divergence times were estimated using both global and local (including Bayesian) clock methods, and the genes were analysed separately and as a single "supergene." The diplomonad *Giardia* was found to be basal to the plant-animal-fungi clade, with significant bootstrap support, in Bayesian, likelihood, and distance analyses of 39 proteins (Hedges *et al.*, 2004). Two other protists lineages, the euglenozoans (105 proteins; 38,492 amino acids) and alveolates (73 proteins; 27,497 amino acids) were also found to be basal to the plant-animal-fungi clade with significant support.

In a separate phylogenetic analysis (Bapteste *et al.*, 2002) of a similar amount of sequence data (123 proteins, 25,023 amino acids; albeit with some missing sequences) and with a greater number of taxa, the same higher-level structure of the "protist tree" was found. In both cases, the results contradict the "five-group" classification of protists (Keeling, 2004; Keeling *et al.*, 2005) and opisthokont or unikont rooting of eukaryotes (Stechmann and Cavalier-Smith, 2002, 2003; Richards and Cavalier-Smith, 2005). A solution to this problem with the five-group classification is to separate the discicristates from a restricted excavate group (metamonads, and possibly oxymonads and malawimonads if future studies show them to be related). Therefore, if either the unikonts or the metamonads form the root of eukaryotes (Fig. 2), it does not contradict this six-group classification scheme. It is also possible that the root of eukaryotes is at yet another position, such as (for example) between the discicristates (e.g., euglenids) and all other eukaryotes.

The timetree (phylogeny scaled to evolutionary time) of eukaryotes shows that plants diverged from the animal-fungi clade approximately 1600 Ma and that animals diverged from fungi approximately 1500 Ma (Fig. 3), reflecting a relative consistency in these time estimates found in studies using large numbers of genes (Wang et al., 1999; Hedges et al., 2004; Blair et al., 2005). In this timetree, the divergence of red algae (Rhodophyta) from the land plant lineage was approximately 1400 Ma, which is consistent with the date (1200 Ma) for the first fossils of red algae (Butterfield, 2000). Although plastids were obtained by some clades of protists through secondary endosymbiotic events, they arose initially on the plant lineage through primary endosymbiosis between a protist and a cyanobacterium. The date of that event is constrained to approximately 1500–1600 Ma (Fig. 3). The alveolates and euglenozoans branch more basally (~1900 Ma) in the timetree of eukaryotes, while the most basal branch (diplomonads) is dated to ~2300 Ma. A separate analysis of genes involved in the mitochondrial symbiotic event dated that event as 1840 ± 200 Ma (Hedges et al., 2001), and together with these data (Hedges et al., 2004) suggest a date of ~1800–2300 Ma for the origin of mitochondria.

The time estimates in this timetree (Fig. 3) compare closely with those in an analysis that focused on divergences among algae (Yoon et al., 2004). In that study, DNA and protein sequences of several plastid genes were analyzed with a local clock method (rate smoothing) and the primary plastid endosymbiotic event was found to be "before 1558 Ma." The split of red algae from green algae was found to be 1474 Ma, also comparing closely with that found in the other study, 1428 Ma (Hedges et al., 2004), despite different genes and methods. In some earlier timing studies (Feng et al., 1997; Nei et al., 2001), sequences from different clades of protists were combined and therefore the results are not comparable, although time estimates for their hybrid protist lineages, ~1500–1700 Ma, are similar in general to those here (Fig. 3).

However, a recent time estimation study using a relatively large data set (129 proteins and 36 taxa) obtained younger dates, with the most basal branches among eukaryotes (in this case, between opisthokonts and all other eukaryotes) splitting only 950–1259 Ma (Douzery et al., 2004). In particular, the split of red from green algae was dated as 928 (825–1061) Ma, which is only about 60% as old as the date obtained in those two other studies (Hedges et al., 2004; Yoon et al., 2004) and which directly conflicts with the oldest fossil of red algae at 1200 Ma (Butterfield, 2000). Their dates for the splits between green algae and land plants (729 Ma) and stramenopiles and alveolates (872 Ma) were also younger than the earliest fossils of those groups (e.g., green algae and stramenopiles), 1000 Ma (Woods et al., 1998; Kumar, 2001). Douzery et al. (2004) explained the conflict by attributing uncertainty (723–1267 Ma) to the geologic dating of the red algal fossil, and citing that same reference (Butterfield, 2000). However, this is incorrect because the paleontological reference (Butterfield, 2000) instead lists a date of 1198 ± 24 Ma for the fossil and claims that it is a refinement of an earlier interval spanning 723–1267 Ma.

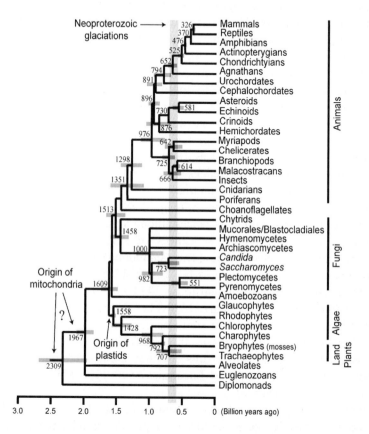

Figure 3. A timetree of eukaryotes based on several molecular studies. Divergence times of deuterostome animals are from Blair and Hedges (2005b), those of arthropods are from Pisani *et al.* (2004), the divergence time of chytrid fungi from higher fungi is from Heckman *et al.* (2001), that of glaucophyte algae from rhodophytes + chlorophytes (and their terrestrial descendants) is from Yoon *et al.* (2004), and the position of choanoflagellates and amoebozoans is constrained by phylogeny. Other divergence times, including those of algae, fungi, plants, other animals, and other protists, are from Hedges *et al.* (2004). The time of origin of the mitochondrion, and its debated position, is discussed elsewhere (Hedges *et al.*, 2001; Hedges *et al.*, 2004). The time of origin of the plastid is constrained at the base of the plastid-bearing clade (Hedges *et al.*, 2004). Gray horizontal bars are 95% confidence intervals.

One possible reason as to why the red algae time estimate of Douzery et al. (2004) conflicts with the fossil date is because they rooted their tree to a unikont (amoebozoan). Although they considered such a rooting to be correct, and a kinetoplastid rooting to be a "reconstruction artefact," they nonetheless calculated divergence times with the latter rooting for comparison. In doing so, they obtained an older date (899–1191 Ma, 95% credibility interval) for the chlorophyte-rhodophyte split. Nonetheless, even that estimate nearly conflicts with the fossil record (1200 Ma) and is 30–40% younger than the dates obtained by others (Hedges et al., 2004; Yoon et al., 2004) for this split.

An additional explanation for the young dates in that analysis (Douzery et al., 2004) is that the calibrations used were applied incorrectly. For calibrations, they used minimum and maximum constraints based on the upper and lower time boundaries of the geologic periods containing the earliest fossils of a lineage. There are at least two problems with that approach. First, the geologic periods used were more inclusive than documented for the fossils. For example, they used the Devonian period (354–417 Ma) for the split of mammals and actinopterygian fishes. However, the fossil data are much better constrained than that, with the earliest fossils defining that split occurring in the very earliest Devonian, or more likely, late Silurian, 425 Ma (Donoghue et al., 2003). 425 Ma is 20% older than 354 Ma, the minimum date used in the study (Douzery et al., 2004). Even if the earliest fossils were in the Devonian, their date can usually be ascertained to a much finer level (e.g., age, stage, epoch) than major geologic period, and therefore this general approach is flawed and will result in an underestimate of divergence time.

The second problem is the assignment of a maximum date (constraint) for the calibration to the maximum age of the geologic period. For the time estimation analyses, assignment of a maximum calibration constraint means that the true divergence did not happen earlier than that time. But evolutionary biologists, including palaeontologists, usually never interpret the fossil record as a literal history of life, and most would agree that the true divergences occurred earlier (in many cases, even in earlier periods) than the first fossil occurrences. This approach of assigning a maximum close to the time of the first fossil occurrence would only be valid if the conclusions of the study were that the resulting times of divergence represented minimum estimates rather than mean estimates (for more discussion of this topic, see Hedges and Kumar, 2004). However, Douzery et al. (2004) interpreted their resulting time estimates as mean (true) times of divergence and drew attention to the conflict (difference) between their time estimates and other published dates that are older. If interpreted as minimum time estimates, their estimates would not be in conflict with older time estimates. It is likely that these two problems with calibration methodology, combined with a forced unikont rooting, explain why those time estimates for protists and other eukaryotes (Douzery *et al.*, 2004) are much younger than other published analyses (e.g., Hedges *et al.*, 2004; Yoon *et al.*, 2004).

3.3 Land Plants

Normally land plants would not even be mentioned in a volume concerning the Precambrian, because evolutionary biologists have long considered that these organisms arose in the Phanerozoic. However, a molecular clock analysis of 54 proteins (5526 amino acids) found that the divergence of mosses (bryophytes) and vascular plants occurred 703 ± 45 Ma (Heckman *et al.*, 2001). Presumably the common ancestor of those two groups was a land plant, providing a minimum estimate for the colonization of land by plants. Subsequently, those data were analyzed further with local clock methods, including Bayesian and likelihood rate smoothing, and a similar date was obtained (707 ± 98 Ma; 95% confidence interval, 515–899 Ma) (Fig. 3) (Hedges *et al.*, 2004).

In contrast, a separate analysis of 27 plastid genes (Sanderson, 2003) resulted in a younger time estimate for the origin of land plants. The time estimate was presented as a range (425–490 Ma) although it was in reality two point estimates: 425 Ma using a global clock approach and 490 Ma using a local clock (likelihood rate smoothing). The errors or confidence intervals on those point estimates were not presented, but logically they must have extended older than 490 Ma and younger than 425 Ma. Nonetheless, it was concluded that "the nearness of these molecular age estimates to the first fossil evidence for land plants contrasts sharply with the results of Heckman *et al.* (2001)." However, a better agreement with the fossil record does not necessarily mean that those younger dates are closer to the true time of divergence.

One major difference in these two studies, not mentioned in the second (Sanderson, 2003), was that different lineages of basal land plants were compared: mosses in one case (Heckman *et al.*, 2001) and liverworts in the other (Sanderson, 2003). Because the relationships of the major clades of land plants (mosses, liverworts, hornworts, and tracheophytes) are not yet resolved (Hedges, 2002), it is not known whether the two studies addressed the same evolutionary divergence, except that in both cases the common ancestor was presumably a land plant. However, an even more significant difference in these two studies was in the calibrations used. The first study used an external calibration (1576 Ma for the plant-animal-fungi divergence) whereas the second study (Sanderson, 2003) used internal calibrations (330

Ma for the angiosperm-gymnosperm divergence and 125 Ma for "crown group eudicot angiosperms").

The external calibration of the first study was from an earlier molecular clock estimate (Wang *et al.*, 1999), derived in turn from an unusually robust vertebrate fossil calibration (Hedges *et al.*, 1996; Kumar and Hedges, 1998; Hedges and Kumar, 2004; van Tuinen and Hadly, 2004). As detailed in the previous section, a 1500–1600 Ma split of plants, animals, and fungi has been a relatively consistent result of several recent molecular clock studies using large numbers of genes, and is constrained by a 1200 Ma fossil (Butterfield, 2000). In contrast, the internal calibration used in the second study (Sanderson, 2003) is less robust for several reasons: (1) it is based primarily on the fossil record of only one of the two daughter lineages (stem group conifers), (2) the early evolution of the angiosperm clade (the other lineage) has been controversial from a phylogenetic standpoint and may have tens of millions of years of missing (or unidentified) fossil record (Stewart and Rothwell, 1993; Doyle, 1998; Crane *et al.*, 2004), and (3) new discoveries are extending the early fossil record of stem seed plants by tens of millions of years (Gerrienne *et al.*, 2004) indicating caution in interpreting the current fossil record of seed plants as being complete. For these reasons, molecular clock studies have often chosen the angiosperm-gymnosperm divergence as a time to estimate rather than calibrate (e.g., Savard *et al.*, 1994; Goremykin *et al.*, 1997; Soltis *et al.*, 2002). Although the title of the Sanderson (2003) study was "molecular data from 27 proteins do not support a Precambrian origin of land plants," those data and analyses would result in a Precambrian time estimate if the calibration date were only 11% older (366 Ma instead of 330 Ma).

Finally, new fossil discoveries of the earliest land plants are bringing the group closer (within ~30 Ma) to a Precambrian origin without help from molecular clocks. Recently land plant megafossils have been discovered from the Ordovician (Wellman *et al.*, 2003) and early Middle Cambrian (Yang *et al.*, 2004), which predate the previously oldest megafossils of land plants by about 50–80 Ma. Fossil spores suggested to be of land plants have been known from the Cambrian (Strother and Beck, 2000) and Ordovician (Wellman *et al.*, 2003), but were controversial, and therefore these new megafossils provided support to the interpretation that these fossil spores are of land plants.

3.4 Fungi

Like plants, fungi are not often mentioned in discussions of life in the Precambrian. However, fungi have generally fewer morphological characters that are taxonomically useful and a poorer fossil record than animals or plants, and therefore there is less of a tendency to interpret the fungal fossil record in a literal sense. Although an early molecular clock analysis of fungi using the small subunit rRNA gene (Berbee and Taylor, 1993) obtained relatively young times of divergence, close to fossil record times, an updated analysis using a refined calibration (Berbee and Taylor, 2001) showed deeper divergences, in the Proterozoic. Likewise, a time estimation analysis of up to 88 protein-coding genes (Heckman et al., 2001; Hedges et al., 2004) and another analysis of the rRNA gene (Padovan et al., 2005) have also found Precambrian divergences among fungi (Fig. 3).

As is typical in comparing the results of molecular clock studies, variation in time estimates usually can be ascribed to the use of different calibrations. For example, differences between the early rRNA studies (Berbee and Taylor, 1993, 2001) and the study of multiple proteins (Heckman et al., 2001) might at first be thought to relate to the different data sets, but the calibrations used were quite different. When rRNA data (Padovan et al., 2005) were analyzed using the same calibration as in the protein study, 1576 Ma for the split of animals and fungi, the time estimates of the two studies were in relatively close agreement.

One particularly useful fossil constraint for fungi is a Devonian (~400 Ma) sordariomycete ("pyrenomycete"), an ascomycotan (Taylor et al., 1999). Because the next youngest sordariomycete is considerably younger, and because the Rhynie Chert is an example of exceptional fossil preservation, there is no reason to assume that the sordariomycete lineage diverged from other fungi immediately prior to 400 Ma. Instead, it is best to interpret that calibration point (or constraint) as a minimum time of separation. However, in either case (minimum or as a fixed calibration), this fossil calibration still results in time estimates of fungal divergences deep in the Proterozoic (Heckman et al., 2001; Padovan et al., 2005).

The higher-level relationships of fungi have not yet been fully resolved (Hedges, 2002; Lutzoni et al., 2004), and therefore it is not surprising that the various time estimation studies also differ somewhat in their estimates of relationships. Nonetheless, chytrids (Chytridiales) appear to be the most basal group of living fungi, diverging from other fungi ~1400 Ma (Fig. 3). The other major lineages of fungi, including the two largest groups—Ascomycota and Basidiomycota—are estimated to have split ~900–1000 Ma (Hedges et al., 2004; Blair et al., 2005). However, the relationships of those lineages, and of several that may have arisen even earlier (Blastocladiales, Glomeromycota, and "Zygomycota") remain unresolved (Heckman et al., 2001; Hedges, 2002; Lutzoni et al., 2004; Padovan et al., 2005).

With so many major lineages of fungi appearing hundreds of millions of years prior to the Phanerozoic, the virtual absence of Precambrian fungal fossils has been surprising. It has been suggested that such fossils do exist in collections, but have been misidentified. For example, much of the late Precambrian Ediacara "fauna" has been interpreted alternatively as representing marine lichens or fungi (Retallack, 1994; Peterson et al., 2003). More recently, Precambrian fungal fossils have been described by two other groups. In one case, fossils dated to ~850 Ma and 1450 Ma have been identified as being "probable fungi" (Butterfield, 2005), and in the other, fossils from phosphorite and dated to 551–635 Ma were interpreted as lichenized fungi (Yuan et al., 2005). Such fossils fall short of documenting a diversity of Precambrian fungi implied by the time estimation studies, but they support the contention that fungi existed in the Precambrian and lay the ground for future studies of Precambrian fungi.

3.5 Animals

Estimating divergence times among animal phyla has been confounded by an incomplete understanding of phylogenetic relationships within the kingdom. In recent years, a new phylogeny of animals has been proposed, based predominantly on small subunit ribosomal RNA sequence analyses, that divides the bilaterally-symmetric animals (bilaterians) into three main groups: deuterostomes, edysozoans, and lophotrochozoans (Aguinaldo et al., 1997; de Rosa et al., 1999; Mallatt et al., 2004). Despite a lack of strong statistical support for Ecdysozoa, this new animal phylogeny has had a large influence on studies of metazoan evolution and development, and has led some researchers to suggest that the last common ancestor of the bilaterians may have been a complex organism (Balavoine and Adoutte, 2003). Other studies using larger numbers of genes have supported and refuted certain aspects of this new phylogeny (Blair et al., 2002; Wolf et al., 2004; Philip et al., 2005; Philippe et al., 2005). Perhaps most controversial has been the position of nematodes (round worms) and platyhelminths (flatworms). These two phyla lack true body cavities and are traditionally placed basal to most other bilaterian phyla. Molecular evidence has mostly supported the elevation of platyhelminths (excluding Acoela) into protostomes (specifically within Lophotrochozoa), but there is currently no consensus as to the position of nematodes.

A number of studies over the past four decades have used molecular clocks to time divergences among animal phyla (e.g., Brown et al., 1972; Runnegar, 1982a; Wray et al., 1996; Hedges et al., 2004; Blair et al., 2005). Such analyses have consistently indicated deep origins for animal phyla (~800–1200 Ma), much earlier than predicted by the fossil record (i.e., Cambrian Explosion, ~520 Ma). Recently, some studies have proposed substantially younger molecular time estimates (Aris-Brosou and Yang, 2002, 2003; Douzery et al., 2004; Peterson et al., 2004; Peterson and

Butterfield, 2005). These studies claimed that through careful consideration of potential biases in both rate modelling and calibration, they produced molecular divergence times that were consistent (or more so) with the fossil record. However, upon closer inspection, these studies suffer from methodological biases that cast doubt on their results.

Most recent molecular clock analyses (e.g., Douzery *et al.*, 2004; Hedges *et al.*, 2004; Blair and Hedges, 2005a; Blair *et al.*, 2005) have used sequences concatenated from multiple genes, thus avoiding potential statistical biases from averaging multiple single-gene estimates (the "mean of the ratios" problem). However, the criticism (Rodriguez-Trelles *et al.*, 2002) that previous studies (e.g., Wang *et al.*, 1999) were biased in that manner is incorrect, because those studies addressed asymmetry in distributions of time estimates by using medians and modes, and eliminating outliers. Even without such corrections, the simulations of Rodriguez-Trelles *et al.* (2002) showed that there was relatively little bias under most normal conditions (parameters). Also, the results from concatenated-gene studies (e.g., Hedges *et al.*, 2004) corroborated the results of those earlier studies (e.g., Wang *et al.*, 1999), indicating that such statistical biases are not responsible for old (~1 Ga) divergence time estimates among animal phyla.

Differences in how fossil calibrations are applied probably explain most of the variation in time estimates among studies. As discussed above (section 3.2), one study estimating animal divergence times (Douzery *et al.*, 2004) used fossil time constraints that were substantially younger than the fossils themselves, producing artificially younger time estimates. In some cases (Peterson *et al.*, 2004; Peterson and Butterfield, 2005), younger divergence times among animals were attributed to the use of calibrations from the invertebrate fossil record, rather than from vertebrates. However, other studies have also used invertebrate fossil calibrations and did not recover such young divergence times among animal phyla (Hedges *et al.*, 2004; Pisani *et al.*, 2004; Blair and Hedges, 2005a). Also, the young times found by Peterson and Butterfield (2005) are likely the result—in large part—of their decision to use molecular differences for timing without any statistical correction for hidden substitutions (multiple hits).

A related methodological issue involves the estimation of evolutionary rates among lineages. Two recent studies claimed that evolutionary rates were higher during the time of the Cambrian Explosion, which when accounted for in rate models allowed for younger divergence times to be recovered (Aris-Brosou and Yang, 2002, 2003). However, simulations (Ho *et al.*, 2005) have suggested that this higher rate was an artefact of the particular method used in those two studies. Other problems associated with rate modelling in those studies have been discussed elsewhere (Blair and Hedges, 2005a).

Finally, although we have noted (above) possible explanations for why these recent molecular studies (Aris-Brosou and Yang, 2002, 2003; Douzery *et al.*, 2004; Peterson *et al.*, 2004; Peterson and Butterfield, 2005) have erred in their analyses, we wish to draw attention to a simple criticism that applies generally. They fail a basic test of consistency because they yield time estimates that are contradicted by the fossil record. For example, the Douzery *et al.* (2004) study estimated that the origin of various groups of algae (e.g., red, green) was hundreds of millions of years after their first fossil occurrences (see Section 3.2). When additional taxa were added to the data set of Aris-Brosou and Yang (2003), the divergence of animals and plants was found to be 671 Ma, nearly a half-billion years younger than the fossil constraint for that divergence (1200 Ma). In the other studies (Peterson *et al.*, 2004; Peterson and Butterfield, 2005), the relevant data were not assembled by those authors to conduct such a consistency test. Considering this, the relatively small size of that data set, and especially the lack of statistical corrections for multiple substitutions, these results likewise are placed in question.

The consistency test demonstrates that those studies are biased to produce young dates and therefore those time estimates are unreliable. Thus if any of these young time estimates for animal evolution are to be seriously considered, it is incumbent upon those authors to explain why their results are not consistent with other aspects of the fossil record. Not considering these aberrant results, molecular clocks continue to support a long history of animal evolution in the Proterozoic (Fig. 3).

4. ASTROBIOLOGICAL IMPLICATIONS

4.1 Complexity

It is logical to assume and expect that life begins in a simple state of organization and, through natural selection, develops greater complexity. For several reasons it is of interest to astrobiologists to know if there is any general and predictable pattern to this rise in complexity, because it would bear on our expectations of the existence of complex life (e.g., animal life) elsewhere in the Universe. For example, if the rise in complexity occurs quickly and easily, the probability that complex life occurs elsewhere is much higher than if it takes billions of years to develop complex life. Ward and Brownlee (2000), using this logic (in part), concluded that complex life is rare in the universe even though simple (prokaryote-like) life may be common.

The conclusion that complex life takes a long time to develop was based on a literal reading of the fossil record, which shows that most animal phyla first appeared in the earliest Phanerozoic, the Cambrian Explosion, nearly four billion years after the Earth was formed. However, most molecular phylogenies and timescales in the last four decades have indicated a deeper (Proterozoic) origin for the major groups of animals, as discussed above (Section 3.5). Also, an earlier origin of plants, fungi, and the major lineages of protists has been estimated (Fig. 3). But how does this new information bear on the rise in complexity? Biological complexity can be defined in many ways, including shape, size, number of cells, and number of genes, among many possibilities. However, the most common measure used to compare complexity across all of life is the number of cell types (Bonner, 1988; Bell and Mooers, 1997; McShea, 2001). Using this measure, and by estimating the number of cell types of common ancestors in the timetree of life (Figs. 1–3), it is possible to construct a contour for the rise in complexity of life on Earth (Hedges et al., 2004) (Fig. 4). This shows that complexity began to rise much earlier in time, roughly 2–2.5 billion years after the Earth was formed. The animal grade of complexity then rose more rapidly (>10 cell types) between 1000–1500 Ma, not 500–600 Ma as predicted by a literal reading of the fossil record.

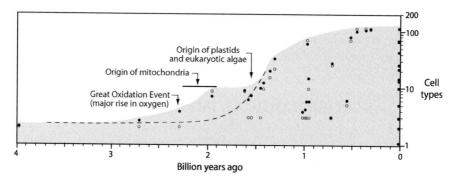

Figure 4. Increase in the maximum number of cell types throughout the history of life (after Hedges *et al.*, 2004). Data points are from living taxa (time zero) and common ancestors (earlier points) estimated with squared-change parsimony (solid circles) and linear parsimony (hollow circles) using a molecular timetree (Hedges *et al.*, 2004). The dashed line shows an alternative interpretation based on uncertainty as to the level of complexity of ancestors of early branching eukaryotes.

Perhaps a key factor in this rise of complexity was the major rise in oxygen in the early Proterozoic (~2300 Ma) (Holland, 2002). The mitochondrion appeared soon thereafter (Hedges *et al.*, 2001; Hedges *et al.*, 2004), allowing eukaryotes to gain much more energy in cellular respiration compared with glycolysis. This symbiotic event (Great Respiration Event) is estimated to have occurred 2300–1800 Ma (Fig. 4).

That additional energy source may have provided the fuel for the rise in complexity, feeding the associated energy requirements (e.g., cell-signalling, mobility, etc.). The addition of the plastid at 1500–1600 Ma, through a symbiotic event with a cyanobacterium, then gave eukaryotes the ability to produce oxygen. This was the beginning of eukaryotic algae and almost certainly led to an increase in eukaryotic diversity and biomass. The parallel diversification of animals and fungi after 1500 Ma are likely related to this Great Algification Event.

Returning to the original question, these new insights from molecular clocks increase the probability that complex life exists elsewhere in the universe because the time required for complexity—in our single example on Earth—is less than previously thought. If the rise in complexity is tied to an energy source such as oxygen, as suggested, then a further consideration must be the time required to evolve the biological machinery for producing that energy (oxygenic photosynthesis or some other process).

4.2 Global glaciations

The defining aspect of the Neoproterozoic, in terms of the planetary environment, appears to have been the multiple episodes of global glaciations (Hoffman *et al.*, 1998). Certainly, the freezing of the entire Earth, or most of it, is an event that must have had a major impact on life at that time. As this review has shown, all of the major groups and subgroups of prokaryotes living today must have experienced these glaciations (Fig. 1), as well as did a great diversity of eukaryotes (Fig. 3). At least three global glaciations have been identified, at ~713 Ma, ~636 Ma, and ~580 Ma (Hoffmann *et al.*, 2004), with each presumably lasting 10^5–10^7 years and triggered by either a geological or biological mechanism.

The geological trigger (Hoffman *et al.*, 1998) involves the long term carbon cycle and an unusual configuration of the continents. The long term carbon cycle normally provides a buffer for major shifts in temperature. Erosion of land exposes silicate minerals (e.g., $CaSiO_3$) and releases calcium, which combines with carbon dioxide in rainwater to form limestone. Return of the carbon to the atmosphere is delayed for millions of years until released by volcanism associated with subduction. If erosion is accelerated, then the carbon dioxide levels in the atmosphere are lowered, hence lowering the temperature. If erosion is slowed, as when land areas are covered with ice, atmospheric carbon dioxide increases and surface temperature increases. The presence of continents at high latitudes improves the buffering mechanism because they freeze over more quickly and provide an early brake on any tendency toward global cooling.

This geological trigger model proposes that it was the unusual configuration of the continents clustered at low latitudes (Rodinia) that led to the global glaciations (Hoffman et al., 1998; Schrag et al., 2002). An absence of high latitude continents removed the "early brakes" for global cooling and permitted ice sheets to extend further towards the equator than usual. Models show that if ice sheets (reflecting energy away from Earth) extend below ~30° north and south latitude they will quickly reach the equator (Snowball Earth) from a runaway albedo effect. Millions of years of volcanism, releasing carbon dioxide, may have been required to raise the temperature sufficiently to escape from a Snowball Earth (Hoffman et al., 1998).

The biological trigger model (Heckman et al., 2001; Hedges, 2003) also involves the long term carbon cycle but proposes that the critical excursions in rates of erosion came from biological sources, not the configuration of continents. Those biological sources probably were the fungi (including lichens) and land plants that evolved and colonized the land in the Proterozoic, although prokaryotes (e.g., cyanobacteria) also may have occupied land areas and may have contributed. The enhancement of weathering by organisms is well-established, and lichens can increase rates 10–100 times by themselves (Schwartzman and Volk, 1989; Schwartzman, 1999). Even prior to the elaboration of the Snowball Earth model (Hoffman et al., 1998) and to the molecular evidence for an early origin of fungi and land plants (Heckman et al., 2001), biological weathering has been considered in discussions of temperature changes in the Proterozoic (Carver and Vardavas, 1994; Retallack, 1994).

The exact configuration of the continents in the Neoproterozoic is far from certain (Meert and Powell, 2001) but the cyclic nature of the glaciations, over a long period (713–580 Ma) is not consistent with an unusual configuration of continents. On the other hand, a biological trigger may explain such cycles (Heckman et al., 2001; Hedges, 2003). During each glaciation, most life on land would have been eliminated, followed by a post-Snowball recovery period that included expansion and diversification of life, and increased weathering leading to the next Snowball. If the carbon isotope excursion at the Precambrian–Cambrian boundary is included as a fourth event, even though it did not appear to be a full Snowball Earth, the time between each event appears to have decreased: 78 my (713–635 Ma), 55 my (635–580 Ma), and 37 my (580–543 Ma). The significance of such a decrease is unknown, but could be interpreted as an increasing development of the land biota, reducing the recovery time between successive glaciations.

4.3 Oxygen and the Cambrian explosion

A simple explanation for the Cambrian Explosion is that it records, in a literal sense, the evolutionary diversification (phylogenetic branching) of animal phyla in the latest Proterozoic and early Cambrian (Gould, 1989; Conway Morris, 2000). Alternatively, if the molecular time estimates showing a deep origin of animal phyla in the Proterozoic (e.g., Fig. 3) are correct, then an explanation is required as to why we see an explosion of fossils in the Cambrian, and almost nothing before that time. A leading theory is that the Cambrian Explosion reflects a major rise in oxygen levels (e.g., Nursall, 1959; Cloud, 1976; Runnegar, 1982a; Canfield and Teske, 1996). Oxygen would have been a limiting factor for body size and the production of hard parts (Rhoads and Morse, 1971; Bengtson and Farmer, 1992; Bengtson, 1994), although Ca^{2+} levels may also have been important (Brennan *et al.*, 2004). A corollary of this theory is that the early representatives of animals in the Proterozoic were small and soft-bodied (although still complex, with many cell types; see above), explaining why they have been difficult to identify in the fossil record. Although the exact timing of the increase in oxygen is not yet established, most agree that it happened in the Neoproterozoic, and probably in the latter half of that time period (Knoll, 2003; Canfield, 2005).

Recently, it was proposed that this Neoproterozoic rise in oxygen, as a trigger of the Cambrian Explosion, was caused by the colonization of land by organisms, principally fungi and plants (Heckman *et al.*, 2001; Hedges, 2003). Specifically, enhancement of weathering and burial of terrestrial carbon were mentioned as potential mechanisms. This model has been elaborated further to show that selective weathering of phosphorus also can accomplish a rise in oxygen (Lenton and Watson, 2004). Further research is needed to constrain the timing of the rise in oxygen and to search for fossil evidence of a Neoproterozoic land flora predicted by molecular analyses.

5. CONCLUSIONS

Knowledge of the phylogeny and times of origin of major groups of organisms in the Neoproterozoic would help answer many questions about the rise in complex life and its interaction with the planetary environment. Although the fossil record will improve with time, and it is the only source of information for extinct groups, it is unlikely to ever provide that knowledge by itself because of the vagaries of preservation. In contrast, such information is currently being obtained from the genomes of organisms, and molecular timetrees will only improve in the future as more genomes are sequenced and analytical methods are refined. Nonetheless, it remains to be seen whether the current large and controversial gaps in the Proterozoic fossil record, implied by molecular timescales, will be reduced. The best evidence at present indicates that those gaps are real. Therefore, while it is true that the fossil record gives us some brilliant windows into the past, we should also expect a few dark hallways.

ACKNOWLEDGEMENTS

This work was supported by funding from the NASA Astrobiology Institute.

REFERENCES

Aguinaldo, A. M., Turbeville, J. M., Linford, L. S., Rivera, M. C., Garey, J. R., Raff, R. A., and Lake, J. A., 1997, Evidence for a clade of nematodes, arthropods and other moulting animals, *Nature* **387**: 489–493.

Amaral Zettler, L. A., Nerad, T. A., O'Kelly, C. J., and Sogin, M. L., 2001, The nucleariid amoebae: more protists at the animal-fungal boundary, *J. Eukaryot. Microbiol.* **48**: 293–297.

Aris-Brosou, S., and Yang, Z., 2002, Effects of models of rate evolution on estimation of divergence dates with special reference to the metazoan 18S ribosomal RNA phylogeny, *Syst. Biol.* **51**: 703–714.

Aris-Brosou, S., and Yang, Z., 2003, Bayesian models of episodic evolution support a late Precambrian explosive diversification of Metazoa, *Mol. Biol. Evol.* **20**: 1947–1954.

Arisue, N., Hasegawa, M., and Hashimoto, T., 2005, Root of the eukaryota tree as inferred from combined maximum likelihood analyses of multiple molecular sequence data, *Mol. Biol. Evol.* **22**: 409–420.

Balavoine, G., and Adoutte, A., 2003, The segmented Urbilateria: A testable scenario, *Integr Comp Biol* **43**: 137–147.

Baldauf, S. L., 2003, The deep roots of eukaryotes, *Science* **300**: 1703–1706.

Baldauf, S. L., Roger, A. J., Wenk-Siefert, I., and Doolittle, W. F., 2000, A kingdom-level phylogeny of eukaryotes based on combined protein data, *Science* **290**: 972–977.

Bapteste, E., Brinkmann, H., Lee, J. A., Moore, D. V., Sensen, C. W., Gordon, P., Durufle, L., Gaasterland, T., Lopez, P., Muller, M., and Philippe, H., 2002, The analysis of 100 genes supports the grouping of three highly divergent amoebae: *Dictyostelium, Entamoeba*, and *Mastigamoeba, Proc. Nat. Acad. Sci. USA* **99**: 1414–1419.

Battistuzzi, F. U., Feijao, A., and Hedges, S. B., 2004, A genomic timescale of prokaryote evolution: insights into the origin of methanogenesis, phototrophy, and the colonization of land, *BMC Evol Biol* **4**(1): 44 (doi:10.1186/1471-2148-4-44).

Bell, C. D., Soltis, D. E., and Soltis, P. S., 2005, The age of the angiosperms: a molecular timescale without a clock, *Evolution* **59**: 1245–58.

Bell, G., and Mooers, A. O., 1997, Size and complexity among multicellular organisms, *Biol. J. Linn. Soc.* **60**: 345–363.

Bengtson, S., 1994, The advent of animal skeletons, in: *Early life on Earth* (S. Bengston, ed.), Columbia University Press, New York, pp. 412–425.

Bengtson, S., and Farmer, J. D., 1992, The evolution of metazoan body plans, in: *The Proterozoic biosphere* (J. W. Schopf and C. Klein, eds.), Cambridge University Press, Cambridge, pp. 443–446.

Berbee, M. L., and Taylor, J. W., 1993, Dating the evolutionary radiations of the true fungi, *Can. J. Bot.* **71**: 1114–1127.

Berbee, M. L., and Taylor, J. W., 2001, Fungal molecular evolution: gene trees and geologic time, in: *The Mycota Vol. VIIB, Systematics and Evolution (D. J.* McLaughlin and E. McLaughlin, eds.), Springer-Verlag, New York, pp. 229–246.

Berner, R. A., Beerling, D. J., Dudley, R., Robinson, J. M., and Wildman, R. A., 2003, Phanerozoic atmospheric oxygen, *Annu. Rev. Earth and Planet. Sci.* **31**: 105–134.

Blair, J. E., and Hedges, S. B., 2005a, Molecular clocks do not support the Cambrian explosion, *Mol. Biol. Evol.* **22**: 387–90.

Blair, J. E., and Hedges, S. B., 2005b, Molecular phylogeny and divergence times of deuterostome animals, *Mol. Biol. Evol.* **22**: 2275–2284.

Blair, J. E., Ikeo, K., Gojobori, T., and Hedges, S. B., 2002, The evolutionary position of nematodes, *BMC Evol. Biol.* **2**: 7 (doi:10.1186/1471-2148-2-7).

Blair, J. E., Shah, P., and Hedges, S. B., 2005, Evolutionary sequence analysis of complete eukaryote genomes, *BMC Bioinformatics* **6**(1): 53 (doi:10.1186/1471-2105-6-53).

Bonner, J. T., 1988, *The evolution of complexity by means of natural selection*, Princeton University Press, Princeton, New Jersey.

Brennan, S. T., Lowenstein, T. K., and Horita, J., 2004, Seawater chemistry and the advent of biocalcification, *Geology* **32**: 473–476.

Brown, R. H., Richardson, M., Boulter, D., Ramshaw, J. A. M., and Jeffries, R. P. S., 1972, The amino acid sequence of cytochrome c from *Helix aspera* Müeller (Garden Snail), *Biochem. J.* **128**: 971–974.

Butterfield, N. J., 2000, *Bangiomorpha pubescens* n. gen., n. sp.: implications for the evolution of sex, multicellularity, and the Mesoproterozoic/Neoproterozoic radiation of eukaryotes, *Paleobiology* **26**: 386–404.

Butterfield, N. J., 2005, Probable proterozoic fungi, *Paleobiology* **31**(1): 165–182.

Canfield, D., 2005, The early history of atmospheric oxygen: Homage to Robert A. Garrels, *Annu. Rev. Earth and Planet. Sci.* **33**: 1–36.

Canfield, D. E., and Teske, A., 1996, Late Proterozoic rise in atmospheric oxygen concentration inferred from phylogenetic and sulphur-isotope studies, *Nature* **382**: 127–132.

Carver, J. H., and Vardavas, I. M., 1994, Precambrian glaciations and the evolution of the atmosphere, *Ann. Geophys.* **12**: 674–682.

Chen, J. Y., Bottjer, D. J., Oliveri, P., Dornbos, S. Q., Gao, F., Ruffins, S., Chi, H. M., Li, C. W., and Davidson, E. H., 2004, Small bilaterian fossils from 40 to 55 million years before the Cambrian, *Science* **305**: 218–222.

Cloud, P., 1976, Beginnings of biospheric evolution and their biogeochemical consequences, *Paleobiology* **2**: 351–387.

Conway Morris, S., 2000, The Cambrian "explosion": slow-fuse or megatonnage? *Proc. Nat. Acad. Sci. USA* **97**: 4426–4429.

Crane, P. R., Herendeen, P., and Friis, E. M., 2004, Fossils and plant phylogeny, *Am. J. Bot.* **91**: 1683–1699.

de Rosa, R., Grenier, J. K., Andreeva, T., Cook, C. E., Adoute, A., Akam, M., Carroll, S. B., and Balavoine, G., 1999, Hox genes in brachiopods and priapulids and protostome evolution, *Nature* **399**: 772–776.

Donoghue, P. C. J., Smith, M. P., and Sansom, I. J., 2003, The origin and early evolution of chordates: molecular clocks and the fossil record, in: *Telling the Evolutionary Time: Molecular Clocks and the Fossil Record* (P. C. J. Donoghue and M. P. Smith, eds.), CRC Press, Boca Raton, Florida, pp. 190–223.

Doolittle, R. F., Feng, D.-F., Tsang, S., Cho, G., and Little, E., 1996, Determining divergence times of the major kingdoms of living organisms with a protein clock, *Science* **271**: 470–477.

Douzery, E. J. P., Snell, E. A., Bapteste, E., Delsuc, F., and Philippe, H., 2004, The timing of eukaryotic evolution: Does a relaxed molecular clock reconcile proteins and fossils? *Proc. Nat. Acad. Sci. USA* **101**: 15386–15391.

Doyle, J. A., 1998, Molecules, morphology, fossils, and the relationship of angiosperms and Gnetales, *Mol. Phyl. Evol.* **9**: 448–462.

Feng, D.-F., Cho, G., and Doolittle, R. F., 1997, Determining divergence times with a protein clock: update and reevaluation, *Proc. Nat. Acad. Sci. USA* **94**: 13028–13033.

Gerrienne, P., Meyer-Berthaud, B., Fairon-Demaret, M., Streel, M., and Steemans, P., 2004, Runcaria, a middle Devonian seed plant precursor, *Science* **306**: 856–858.

Goremykin, V. V., Hansmann, S., and Martin, W. F., 1997, Evolutionary analysis of 58 proteins encoded in six completely sequenced chloroplast genomes: revised molecular estiamtes of two seed plant divergence times, *Plant Systemat. Evol.* **206**: 337–351.

Gould, S. J., 1989, *Wonderful life*, W. W. Norton, New York.

Hampl, V., Horner, D. S., Dyal, P., Kulda, J., Flegr, J., Foster, P., and Embley, T. M., 2005, Inference of the phylogenetic position of oxymonads based on nine genes: Support for Metamonada and Excavata, *Mol. Biol. Evol.* **22**: 2508–2518.

Han, T.-M., and Runnegar, B., 1992, Megascopic eukaryotic algae from the 2.1 billion-year-old Negaunee iron-formation, Michigan, *Science* **257**: 232–235.

Hasegawa, M., Kishino, H., and Yano, T., 1989, Estimation of branching dates among primates by molecular clocks of nuclear DNA which slowed down in Hominoidea, *J. Human Evol.* **18**: 461–476.

Heckman, D. S., Geiser, D. M., Eidell, B. R., Stauffer, R. L., Kardos, N. L., and Hedges, S. B., 2001, Molecular evidence for the early colonization of land by fungi and plants, *Science* **293**: 1129–1133.

Hedges, S. B., 2002, The origin and evolution of model organisms, *Nat. Rev. Genet.* **3**: 838–849.

Hedges, S. B., 2003, Molecular clocks and a biological trigger for the Neoproterozoic snowball Earths and Cambrian explosion., in: *Telling Evolutionary Time: Molecular Clocks and the Fossil Record* (P. C. J. Donoghue and M. P. Smith, eds.), Taylor and Francis, London, pp. 27–40.

Hedges, S. B., Blair, J. E., Venturi, M. L., and Shoe, J. L., 2004, A molecular timescale of eukaryote evolution and the rise of complex multicellular life, *BMC Evol Biol* **4**: 2 (doi:10.1186/1471-2148-4-2).
Hedges, S. B., Chen, H., Kumar, S., Wang, D. Y., Thompson, A. S., and Watanabe, H., 2001, A genomic timescale for the origin of eukaryotes, *BMC Evol. Biol.* **1**(1): 4 (doi:10.1186/1471-2148-1-4).
Hedges, S. B., and Kumar, S., 2003, Genomic clocks and evolutionary timescales, *Trends Genet.* **19**: 200–206.
Hedges, S. B., and Kumar, S., 2004, Precision of molecular time estimates, *Trends Genet.* **20**: 242–247.
Hedges, S. B., Parker, P. H., Sibley, C. G., and Kumar, S., 1996, Continental breakup and the ordinal diversification of birds and mammals, *Nature* **381**: 226–229.
Ho, S. Y. W., Phillips, M. J., Drummond, A. J., and Cooper, A., 2005, Accuracy of rate estimation using relaxed-clock models with a critical focus on the early metazoan radiation, *Mol. Biol. Evol.* **22**: 1355–1363.
Hoffman, P. F., Kaufman, A. J., Halverson, G. P., and Schrag, D. P., 1998, A Neoproterozoic snowball Earth, *Science* **281**: 1342–1346.
Hoffmann, K. H., Condon, D. J., Bowring, S. A., and Crowley, J. L., 2004, U–Pb zircon date from the Neoproterozoic Ghaub Formation, Namibia: Constraints on Marinoan glaciation, *Geology* **32**: 817–820.
Holland, H. D., 2002, Volcanic gases, black smokers, and the Great Oxidation Event, *Geochim. Cosmochim. Acta* **21**: 3811–3826.
Horodyski, R. J., and Knauth, L. P., 1994, Life on land in the Precambrian, *Science* **263**: 494–498.
Hyde, W. T., Crowley, T. J., Baum, S. K., and Peltier, W. R., 2000, Neoproterozoic "snowball Earth" simulations with a coupled climate/ice-sheet model, *Nature* **405**: 425–429.
Jensen, S., Droser, M. L., and Gehling, J. G., 2005, Trace fossil preservation and the early evolution of animals, *Palaeogeogr. Palaeoclimat. Palaeoecol.* **220**: 19–29.
Keeling, P. J., 2004, Diversity and evolutionary history of plastids and their hosts, *Am. J. Bot.* **91**: 1481–1493.
Keeling, P. J., Burger, G., Durnford, D. G., Lang, B. F., Lee, R. W., Pearlman, R. E., Roger, A. J., and Gray, M. W., 2005, The tree of eukaryotes, *Trends Ecol. Evol.* **20**: 670–676.
Kimura, M., 1983, *The neutral theory of molecular evolution*, Cambridge University Press, Cambridge, United Kingdom.
Kishino, H., Thorne, J. L., and Bruno, W. J., 2001, Performance of a divergence time estimation method under a probabilistic model of rate evolution, *Mol. Biol. Evol.* **18**: 352–361.
Knoll, A. H., 2003, The geobiological consequences of evolution, *Geobiology* **1**: 3–14.
Knoll, A. H., 2004, *Life on a Young Planet*, Princeton University Press, Princeton, NJ.
Kollman, J. M., and Doolittle, R. F., 2000, Determining the relative rates of change for prokaryotic and eukaryotic proteins with anciently duplicated paralogs, *J. Mol. Evol.* **51**: 173–181.
Kumar, S., 2001, Mesoproterozoic megafossil *Chuaria-Tawuia* association may represent parts of a multicellular plant, Vindhyan Supergroup, Central India., *Precambrian Res.* **106**: 187–211.

Kumar, S., 2005, Molecular clocks: four decades of evolution, *Nat. Rev. Genet.* **6**: 654–662.

Kumar, S., Filipski, A., Swarma, V., Walker, A., and Hedges, S. B., 2005, Placing confidence limits on the molecular age of the human-chimpanzee divergence, *Proc. Nat. Acad. Sci. USA* **102**: 18842–18847.

Kumar, S., and Hedges, S. B., 1998, A molecular timescale for vertebrate evolution, *Nature* **392**: 917–920.

Lenton, T. M., and Watson, A. J., 2004, Biotic enhancement of weathering, atmospheric oxygen and carbon dioxide in the Neoproterozoic, *Geophysical Research Letters* **31**: L05202 (doi: 10.1029/2003GL018802).

Lutzoni, F., Kauff, F., Cox, C. J., McLaughlin, D., Celio, G., Dentinger, B., Padamsee, M., Hibbett, D., James, T. Y., Baloch, E., Grube, M., Reeb, V., Hofstetter, V., Schoch, C., Arnold, A. E., Miadlikowska, J., Spatafora, J., Johnson, D., Hambleton, S., Crockett, M., Shoemaker, R., Hambleton, S., Crockett, M., Shoemaker, R., Sung, G. H., Lucking, R., Lumbsch, T., O'Donnell, K., Binder, M., Diederich, P., Ertz, D., Gueidan, C., Hansen, K., Harris, R. C., Hosaka, K., Lim, Y. W., Matheny, B., Nishida, H., Pfister, D., Rogers, J., Rossman, A., Schmitt, I., Sipman, H., Stone, J., Sugiyama, J., Yahr, R., and Vilgalys, R., 2004, Assembling the fungal tree of life: Progress, classification and evolution of subcellular traits, *Am. J. Bot.* **91**: 1446–1480.

Mallatt, J. M., Garey, J. R., and Shultz, J. W., 2004, Ecdysozoan phylogeny and Bayesian inference: first use of nearly complete 28S and 18S rRNA gene sequences to classify the arthropods and their kin, *Mol. Phyl. Evol.* **31**: 178–191.

Matsuzaki, M., Misumi, O., Shin-I, T., Maruyama, S., Takahara, M., Miyagishima, S. Y., Mori, T., Nishida, K., Yagisawa, F., Nishida, K., Yoshida, Y., Nishimura, Y., Nakao, S., Kobayashi, T., Momoyama, Y., Higashiyama, T., Minoda, A., Sano, M., Nomoto, H., Oishi, K., Hayashi, H., Ohta, F., Nishizaka, S., Haga, S., Miura, S., Morishita, T., Kabeya, Y., Terasawa, K., Suzuki, Y., Ishii, Y., Asakawa, S., Takano, H., Ohta, N., Kuroiwa, H., Tanaka, K., Shimizu, N., Sugano, S., Sato, N., Nozaki, H., Ogasawara, N., Kohara, Y., and Kuroiwa, T., 2004, Genome sequence of the ultrasmall unicellular red alga Cyanidioschyzon merolae 10D, *Nature* **428**: 653–657.

McShea, D. W., 2001, The hierarchical structure of organisms: a scale and documentation of a trend in the maximum, *Paleobiology* **27**: 405–423.

Meert, J. G., and Powell, C. M., 2001, Assembly and break-up of Rodinia: introduction to the special volume, *Precambrian Res.* **110**: 1–8.

Nei, M., Xu, P., and Glazko, G., 2001, Estimation of divergence times from multiprotein sequences for a few mammalian species and several distantly related organisms, *Proc. Nat. Acad. Sci. USA* **98**: 2497–2502.

Nursall, J. R., 1959, Oxygen as a prerequisite to the origin of the Metazoa, *Nature* **183**: 1170–1172.

Padovan, A. C. B., Sanson, G. F. O., Brunstein, A., and Briones, M. R. S., 2005, Fungi evolution revisited: Application of the penalized likelihood method to a Bayesian fungal phylogeny provides a new perspective on phylogenetic relationships and divergence dates of ascomycota groups, *J. Mol. Evol.* **60**: 726–735.

Peterson, K. J., and Butterfield, N. J., 2005, Origin of the Eumetazoa: Testing ecological predictions of molecular clocks against the Proterozoic fossil record, *Proc. Nat. Acad. Sci. USA* **102**: 9547–9552.

Peterson, K. J., Lyons, J. B., Nowak, K. S., Takacs, C. M., Wargo, M. J., and McPeek, M. A., 2004, Estimating metazoan divergence times with a molecular clock, *Proc. Nat. Acad. Sci. USA* **101**: 6536–6541.

Peterson, K. J., Waggoner, B., and Hagadorn, J. W., 2003, A fungal analog for Newfoundland Ediacaran fossils?, *Integr. Comp. Biol.* **43**: 127–136.

Philip, G. K., Creevey, C. J., and McInerney, J. O., 2005, The Opisthokonta and the Ecdysozoa may not be clades: Stronger support for the grouping of plant and animal than for animal and fungi and stronger support for the Coelomata than Ecdysozoa, *Mol. Biol. Evol.* **22**: 1175–1184.

Philippe, H., and Germot, A., 2000, Phylogeny of eukaryotes based on ribosomal RNA: long-branch attraction and models of sequence evolution, *Mol. Biol. Evol.* **17**: 830–834.

Philippe, H., Lartillot, N., and Brinkmann, H., 2005, Multigene analyses of bilaterian animals corroborate the monophyly of Ecdysozoa, Lophotrochozoa, and Protostomia, *Mol. Biol. Evol.* **22**: 1246–1253.

Philippe, H., Lopez, P., Brinkmann, H., Budin, K., Germot, A., Laurent, J., Moreira, D., Muller, M., and Le Guyader, H., 2000, Early-branching or fast-evolving eukaryotes? An answer based on slowly evolving positions, *Proc. Roy. Soc. London B Biol.* **267**: 1213–1221.

Pisani, D., Poling, L. L., Lyons-Weiler, M., and Hedges, S. B., 2004, The colonization of land by animals: molecular phylogeny and divergence times among arthropods, *BMC Biol* **2**(1): 1 (doi:10.1186/1741-7007-2-1).

Poulsen, C. J., 2003, Absence of a runaway ice-albedo feedback in the Neoproterozoic, *Geology* **31**: 473–476.

Poulsen, C. J., and Jacob, R. L., 2004, Factors that inhibit snowball Earth simulation, *Paleoceanography* **19**: PA4021 (doi:10.1029/2004PA001056).

Retallack, G. J., 1994, Were the Ediacaran fossils lichens, *Paleobiology* **20**: 523–544.

Rhoads, D. C., and Morse, J. W., 1971, Evolutionary and ecological significance of oxygen-deficient marine faunas, *Lethaia* **4**: 413–428.

Richards, T. A., and Cavalier-Smith, T., 2005, Myosin domain evolution and the primary divergence of eukaryotes, *Nature* **436**: 1113–1118.

Rodriguez-Trelles, F., Tarrio, R., and Ayala, F. J., 2002, A methodological bias toward overestimation of molecular evolutionary time scales, *Proc. Nat. Acad. Sci. USA* **99**: 8112–8115.

Runnegar, B., 1982a, The Cambrian explosion: animals or fossils? *J. Geol. Soc. Australia* **29**: 395–411.

Runnegar, B., 1982b, A molecular-clock date for the origin of the animal phyla, *Lethaia* **15**: 199–205.

Samuelsson, J., and Butterfield, N. J., 2001, Neoproterozoic fossils from the Franklin Mountains, northwestern Canada: stratigraphic and paleobiological implications, *Precambrian Res.* **107**: 235–251.

Sanderson, M. J., 1997, A nonparametric approach to estimating divergence times in the absence of rate constancy, *Mol. Biol. Evol.* **14**: 1218–1231.

Sanderson, M. J., 2003, Molecular data from 27 proteins do not support a Precambrian origin of land plants, *Am. J. Bot.* **90**: 954–956.

Savard, L., Li, P., Strauss, S. H., Chase, M. W., Michaud, M., and Bousquet, J., 1994, Chloroplast and nuclear gene-sequences indicate late Pennsylvanian time for the last common ancestor of extant seed plants, *Proc. Nat. Acad. Sci. USA* **91**: 5163–5167.

Schlegel, M., 1994, Molecular phylogeny of eukaryotes, *Trends Ecol. Evol.* **9**: 330–335.

Schrag, D. P., Berner, R. A., Hoffman, P. F., and Halverson, G. P., 2002, On the initiation of a snowball Earth, *Geochem. Geophys. Geosys.* **3**: 4 (doi: 10.1029/2001GC000219).

Schubart, C. D., Diesel, R., and Hedges, S. B., 1998, Rapid evolution to terrestrial life in Jamaican crabs, *Nature* **393**: 363–365.

Schwartzman, D., and Volk, T., 1989, Biotic enhancement of weathering and the habitability of Earth, *Nature* **340**: 457–460.

Schwartzman, D. W., 1999, *Life, Temperature, and the Earth*, Columbia University Press, New York.

Seilacher, A., Bose, P. K., and Pfluger, F., 1998, Triploblastic animals more than 1 billion years ago: trace fossil evidence from India, *Science* **282**: 80–83.

Sergeev, V. N., Gerasimenko, L. M., and Zavarzin, G. A., 2002, The Proterozoic history and present state of cyanobacteria, *Microbiology* **71**: 623–637.

Sheridan, P. P., Freeman, K. H., and Brenchley, J. E., 2003, Estimated minimal divergence times of the major bacterial and archaeal phyla, *Geomicrobiol. J.* **20**(1): 1–14.

Sogin, M. L., Gunderson, J. H., Elwood, H. J., Alonso, R. A., and Peattie, D. A., 1989, Phylogenetic meaning of the kingdom concept: an unusual ribosomal RNA from *Giardia lamblia*, *Science* **243**: 75–77.

Soltis, P. S., Soltis, D. E., Savolainen, V., Crane, P. R., and Barraclough, T. G., 2002, Rate heterogeneity among lineages of tracheophytes: Integration of molecular and fossil data and evidence for molecular living fossils, *Proc. Nat. Acad. Sci. USA* **99**: 4430–4435.

Stechmann, A., and Cavalier-Smith, T., 2002, Rooting the eukaryote tree by using a derived gene fusion, *Science* **297**: 89–91.

Stechmann, A., and Cavalier-Smith, T., 2003, The root of the eukaryote tree pinpointed, *Curr. Biol.* **13**: R665–R666.

Stewart, W. N., and Rothwell, G. W., 1993, *Paleobotany and the Evolution of Plants*, Cambridge University Press, Cambridge, UK.

Strother, P. K., and Beck, J. H., 2000, Spore-like microfossils from Middle Cambrian strata: expanding the meaning of the term cryptospore, in: *Pollen and Spores: Morphology and Biology* (M. M. Harley, C. M. Morton, and S. Blackmore, eds.), Royal Botanic Gardens, Kew, England, pp. 413–424.

Takezaki, N., Rzhetsky, A., and Nei, M., 1995, Phylogenetic test of the molecular clock and linearized trees, *Mol. Biol. Evol.* **12**: 823–833.

Taylor, T. N., Hass, H., and Kerp, H., 1999, The oldest fossil ascomycetes, *Nature* **399**: 648.

Thorne, J. L., Kishino, H., and Painter, I. S., 1998, Estimating the rate of evolution of the rate of molecular evolution, *Mol. Biol. Evol.* **15**: 1647–1657.

van Tuinen, M., and Hadly, E. A., 2004, Error in estimation of rate and time inferred from the early amniote fossil record and avian molecular clocks, *J. Mol. Evol.* **59**: 267–276.

Wang, D. Y., Kumar, S., and Hedges, S. B., 1999, Divergence time estimates for the early history of animal phyla and the origin of plants, animals and fungi, *Proc. Roy. Soc. London. B Biol.* **266**: 163–171.

Ward, P. D., and Brownlee, D., 2000, *Rare Earth*, Copernicus, New York.

Watanabe, Y., Martini, J. E., and Ohmoto, H., 2000, Geochemical evidence for terrestrial ecosystems 2.6 billion years ago, *Nature* **408**: 574–8.

Wellman, C. H., Osterloff, P. L., and Mohluddin, U., 2003, Fragments of the earliest land plants, *Nature* **425**: 282–285.

Wolf, Y. I., Rogozin, I. B., and Koonin, E. V., 2004, Coelomata and not ecdysozoa: Evidence from genome-wide phylogenetic analysis, *Genome Res.* **14**: 29–36.

Woods, K. N., Knoll, A. H., and German, T. N., 1998, Xanthophyte Algae from the Mesoproterozoic/Neoproterozoic Transition: Confirmation and Evolutionary Implications., *GSA Annu. Meeting Abstr. with Progr.* **30**: A232.

Wray, G. A., Levinton, J. S., and Shapiro, L. H., 1996, Molecular evidence for deep Precambrian divergences among metazoan phyla, *Science* **274**: 568–573.

Xiao, S. H., Zhang, Y., and Knoll, A. H., 1998, Three-dimensional preservation of algae and animal embryos in a Neoproterozoic phosphorite, *Nature* **391**: 553–558.

Yang, R.-D., Mao, J.-R., Zhang, W.-H., Jiang, L.-J., and Gao, H., 2004, Bryophyte-like fossil (*Parafunaria sinensis*) from Early-Middle Cambrian Kaili formation in Guizhou Province, China, *Acta Bot. Sinica* **46**: 180–185.

Yoon, H. S., Hackett, J. D., Ciniglia, C., Pinto, G., and Bhattacharya, D., 2004, A molecular timeline for the origin of photosynthetic eukaryotes, *Mol. Biol. Evol.* **21**: 809–18.

Young, G. M., 2002, Stratigraphic and tectonic settings of Proterozoic glaciogenic rocks and banded iron-formations: Relevance to the snowball Earth debate, *J. Afr. Earth Sci.* **35**: 451–466.

Yuan, X. L., Xiao, S. H., and Taylor, T. N., 2005, Lichen-like symbiosis 600 million years ago, *Science* **308**: 1017–1020.

Zuckerkandl, E., and Pauling, L., 1962, Molecular disease, evolution, and genetic heterogeneity., in: *Horizons in Biochemistry* (M. Marsha and B. Pullman, eds.), Academic Press, New York, pp. 189–225.

Chapter 8

A Neoproterozoic Chronology

GALEN P. HALVERSON

Laboratoire des Mécanismes et Transferts en Géologie, Université Paul Sabatier, 31400 Toulouse, France.

Present Address: Geology and Geophysics, School of Earth and Environmental Sciences, The University of Adelaide, Adelaide 5005, South Australia, Australia.

1. Introduction	232
2. Constructing the Record	233
2.1 The Neoproterozoic Sedimentary Record	233
2.2 The $\delta^{13}C$ Record	236
2.3 Bases for Correlation	238
3. Review of the Neoproterozoic	242
3.1 The Tonian (1000–720? Ma)	242
3.2. The Cryogenian (720?–635 Ma)	245
.2.1. The Sturtian Glaciation	245
3.2.2. The Interglacial	248
3.2.3. The Marinoan Glaciation	250
3.3. The Ediacaran Period (635–542 Ma)	253
3.3.1. The Post-Marinoan Cap Carbonate Sequence	253
3.3.2. The Early Ediacaran	254
3.3.3. The Gaskiers Glaciation	258
3.3.4. The Terminal Proterozoic	260
4. Conclusions	261
Acknowledgements	262
References	262

S. Xiao and A.J. Kaufman (eds.), Neoproterozoic Geobiology and Paleobiology, 231–271.
© 2006 Springer.

1. INTRODUCTION

The Neoproterozoic Era, spanning from 1000 to 542 Ma, has emerged as one of the epic chapters in Earth's history. From the breakup of a supercontinent to the radiation of macroscopic faunas, the magnitude of the tectonic, climatic, and biological upheavals during this time period were so great and their consequences so profound that they challenge the canon of uniformitarianism that has long held sway in the interpretation of past geological processes. In a field where the observations are extraordinary and the unanswered questions compelling, controversy is inevitable. With nearly 460 million years of Earth history to explore, it is doubtful the disagreements will abate soon, and those controversies that do subside will yield to others as we pry deeper and deeper into the geological record.

The subject of the Neoproterozoic time scale has been reviewed by Knoll (2000), who outlined the principles behind the calibration and correlation of late Proterozoic events and the inherent difficulties in parsing Precambrian time. Despite these challenges, the IUGS has recently ratified the Ediacaran Period (Knoll *et al.*, 2004, Knoll *et al.*, 2006), which spans from the end of the Marinoan glaciation (635 Ma) (Condon *et al.*, 2005) to the end of the Precambrian (542 Ma) (Amthor *et al.*, 2003). This is the first period yet formally defined for the Precambrian and reflects decades of research and discussion among a great number of geologists. It serves as a reminder of how much work is entailed in dividing Precambrian time in a meaningful way (Knoll *et al.*, 2004).

The rest of the Neoproterozoic remains to be subdivided in this manner, although the definition of the base of the Cryogenian is now being actively discussed. In the meantime, a chronology for the Neoproterozoic, even an imperfect one, is important to frame the discussion and map out directions for future research. In this contribution a temporal framework is proposed for the major events, thus far recognized, that shaped the surface of the Neoproterozoic earth and set the environmental stage for the proliferation of animals. The template for this chronology is a composite record of the secular variation in the $\delta^{13}C$ composition of seawater through this interval, as obtained from the isotopic analyses of marine carbonates spanning from >900 to 542 Ma, and modified from Halverson *et al.* (2005) in light of new isotopic and radiometric data. The timing of the major events remains murky and some correlations are not yet certain. Nonetheless, the available data is sufficient to warrant a review of the Neoproterozoic. The aim of this work is not to supplant the geological time scale, establish a precise chronology, or to resolve any of the major controversies surrounding the Neoproterozoic glaciations, but rather to highlight the most interesting events and least understood intervals of this time period within a temporal context.

2. CONSTRUCTING THE RECORD

2.1 The Neoproterozoic Sedimentary Record

The ultimate repository of clues about the history of the earth is the stratigraphic record. Beyond the complication that all sedimentary rocks of Precambrian age have been, to varying degrees, folded, faulted, metamorphosed and otherwise altered from their initial depositional state, the stratigraphic record is inherently incomplete and no single succession spans the entire Neoproterozoic. This shortfall is exacerbated by the lack of a detailed biostratigraphy in the Neoproterozic (Knoll and Walter, 1992) and the relative rarity of volcanic beds that are datable by U–Pb geochronology—the single source of precise and absolute dates on sedimentary successions. Therefore, reconstructing the chronology of the Neoproterozoic unavoidably requires making correlations, both within single sedimentary basins and between widely separated successions, in order to fill in the gaps in the record. Fortunately, Neoproterozoic sedimentary rocks are widespread, rimming the former cratonic fragments dispersed during the break-up of the Rodinian supercontinent (Hoffman, 1991). The geological record exists, and the challenge is to fit all the pieces together.

As interest in the Neoproterozoic began to grow, Knoll and Walter (1992) predicted that carbon isotope chemostratigraphy would prove to be an invaluable means for making correlations, particularly when coupled to other tools, such as sequence stratigraphy and other marine proxy records. These authors pointed out that Neoproterozoic stratigraphic record is amenable to using $\delta^{13}C$ chemostratigraphy to make correlations because of the preponderance of large amplitude changes in the $\delta^{13}C$ composition of seawater, at least during the latter half of the era, compared to intrinsic variability (due to diagenetic and hydrologic causes) in the composition of coeval carbonates. Judging from the voluminous outpouring of Neoproterozoic carbon isotope data over the past 15 years (Shields and Veizer, 2002), Knoll and Walter's (1992) prediction has been borne out.

Due to the relatively low cost and ease of the analytical measurements, carbon isotopic analyses are now a routine component of any stratigraphic project that includes carbonate rocks. However, their utility in the Neoproterozoic has not been fully realized in large part because at least some of the major anomalies are closely tied to episodes of glaciation, which themselves are not well constrained chronostratigraphically and are in many cases difficult to correlate. That is to say, the windfall of isotopic data has confirmed that the glaciations are associated with negative $\delta^{13}C$ anomalies (but not necessarily *vice versa*) and that these anomalies are reproducible, but has not resolved the debate over the number of glaciations (e.g. Kennedy *et al.*, 1998). Other proxy records, namely $^{87}Sr/^{86}Sr$ and $\delta^{34}S$, are also important tools in Neoproterozoic carbonate stratigraphy (Walter *et al.*, 2000) and are considered in the correlations presented here, but detailed discussion of these records is beyond the scope of this work.

By virtue of many new, useful radiometric dates (Table 1) and an ever growing database of detailed carbon isotopic and stratigraphic data from well studied successions, such as the Huqf Supergroup in Oman, the Adelaide Rift Complex in South Australia, and the Otavi Group in Namibia, the chronological picture of the Neoproterozoic is coming into focus. The combination of stratigraphic, biostratigraphic, and chemostratigraphic data tipped the debate over the number of glaciations in favor of three (Knoll, 2000, Hoffman and Schrag, 2002, Xiao *et al.*, 2004, Halverson *et al.*, 2005), and a suite of new, crucially situated U–Pb ages (Thompson and Bowring, 2000; Bowring *et al.*, 2003; Hoffmann *et al.*, 2004; Zhou *et al.*, 2004; Condon *et al.* 2005)(Table 1) now confirm that there were at least three glaciations. The older two of these glaciations are conventionally known as the Sturtian and Marinoan events, and the youngest is referred to here as the Gaskiers glaciation, after the 580 Ma glacigenic Gaskiers Formation (Bowring *et al.*, 2003) in Newfoundland. However, even as some recent radiometric ages have confirmed the synchroneity of the end of at least one glaciation (Condon *et al.*, 2005), others, namely from Australia (Schaefer and Burgess, 2003; Kendall *et al.*, 2005), Tasmania (Calver *et al.*, 2004), and Idaho (Lund *et al.*, 2003; Fanning and Link, 2004), have challenged the popular notion that all the Neoproterozoic glaciations can be easily binned into discrete and distinct events and have called into question the use of this terminology. The names for the three glacial epochs, though likely to be modified in the future, are left unchanged here so as not to confuse the reader with new and unagreed upon nomenclature.

Table 1. (on Page 235) Summary of radiometric ages pertinent to the construction of the Neoproterozoic $\delta^{13}C$ record and chronology. For an exhaustive but somewhat dated review of Neoproterozoic radiometric ages, see Evans (2000). See Condon *et al.* (2005) and Knoll *et al.* (2005) for more specific and recent discussions of late Neoproterozoic (Ediacaran) ages. *These two ages are from the same unit and differ as a result of different analytical methods employed (Kendall and Creaser, 2004) and underscore the need to refine and test the Re–Os method. †A refined age on the Ghubrah of ca. 712 Ma was quoted in Allen et al. (2002), and another of 711.8 ± 1.6 was cited in Kilner et al. (2005), but the isotopic measurements have not been published. (z) = zircon, (a) = apatite. MC = MC–ICP–MS.

A Neoproterozoic Chronology

Table 1. Compilation of Neoproterozoic radiometric ages.

Age	Reference	Succession/Location	Method	Dated rock	Significance
542±0.6	Amthor et al. 2003	Huqf Spgp. (Oman)	U-Pb z TIMS	Ash in Ara Fm.	Precambrian-Cambrian boundary excursion.
543±0.3	Amthor et al. 2003	Huqf Spgp. (Oman)	U-Pb z TIMS	Ash in Ara Fm.	Precambrian-Cambrian boundary excursion.
543±1	Grotzinger et al. 1995	Nama Gp. (southern Namibia)	U-Pb z TIMS	Ash in Spitskop Mb.	Max. age Precambrian-Cambrian boundary.
545±1	Grotzinger et al. 1995	Nama Gp. (southern Namibia)	U-Pb z TIMS	Ash in Spitskop Mb.	Age on "+2‰ plateau."
549±1	Grotzinger et al. 1995	Nama Gp. (southern Namibia)	U-Pb z TIMS	Ash in Zaris Fm.	Max. age of "+2‰ plateau."
550±0.7	Condon et al. 2005	Doushantou Fm. (south China)	U-Pb z TIMS	Ash bed	Min. age of Shuram/Wonoka anomaly.
555±0.3	Martin et al. 2000	Siberia	U-Pb z TIMS	Ash bed	Age constraint on *Kimberella* Sp.
575	Bowring et al. 2003	Conception Gp. (Newfoundland)	U-Pb z TIMS	Ash bed	Max. age of Ediacaran biota
575±3	Calver et al. 2004	Grassy Group (King Island)	U-Pb SHRIMP	Mafic intrusion	Min. age of Marinoan? Glacials.
579±14	Chen et al. 2004	S. China (Doushantou Fm.)	Pb-Pb a MC	Phosphorite beds	Doushantuo biota.
580	Bowring et al. 2003	Conception Gp. (Newfoundland)	U-Pb z TIMS	Ash beds	Age of Gaskiers glaciation.
582±4	Calver et al. 2004	Togari Group (Tasmania).	U-Pb SHRIMP	Rhyodacite flow	Max. age of Marinoan (?) glaciation.
*592±14	Schaefer & Burgess 2003	Amadeus Basin (central Australia)	Re-Os TIMS	Organic-rich shale	Max. age of Marinoan glaciation?
596±2	Thompson & Bowring 2000	Boston Bay Gp. (Squantum Mb.)	U-Pb z TIMS	Detrital zircons	Max. age on Squantum (Gaskiers) glacials.
599±4.2	Barford et al. 2002	S. China (Doushantou Fm.)	Pb-Pb a MC	Phosphorite beds	Doushantuo biota.
601±4	Dempster et al. 2002	Dalradian (Tayvallich Volcanics)	U-Pb z TIMS	Mafic volcanics	Max. age of Loch na cille Boulder Bed (Gaskiers?)
608±4.7	Kendall et al. 2004	Windermere Spg. (W. Canada)	Re-Os TIMS	Organic-rich shale	Age of Marinoan(?) glaciation?
620±14	Bingen et al. 2005	Hedmark Gp. (S. Norway)	U-Pb z LA-ICP	Detrital zircons	Max. age of Moelv Tillite
621±7	Zhang et al. 2005	Doushantou Fm. (south China)	U-Pb SHRIMP	Ash bed	Age above Marinoan cap dolostone.
633±0.5	Condon et al. 2005	Doushantou Fm. (south China)	U-Pb z TIMS	Ash bed	Age above Marinoan cap dolostone.
635±0.6	Condon et al. 2005	Doushantou Fm. (south China)	U-Pb z TIMS	Ash bed	Age within Marinoan cap dolostone.
636±0.6	Hoffmann et al. 2004	Swakop Gp. (central Namibia)	U-Pb z TIMS	Ash bed	Only direct age on the Marinoan glaciation
643±2.4	Kendall et al. 2005	Tapley Hill Fm. (Australia)	Re-Os TIMS	Organic-rich shale	Min. age of Sturtian glaciation?.
*658±5.5	Kendall & Creaser 2004	Amadeus Basin (central Australia)	Re-Os TIMS	Organic-rich shale	Max. age of Marinoan glacials?
663±4	Zhou et al. 2004	Datango Fm. (south China)	U-Pb SHRIMP	Ash bed	Max. age of Nantuo glacials?
667±5	Fanning and Link 2004	Pocatello Fm. (Idaho)	U-Pb SHRIMP	Ash bed	Min. age Sturtian/Max. age Marinoan?
684±4	Lund et al. 2003	Gospel Peaks Sq. (Idaho)	U-Pb z TIMS	Rhyodacite flow	Age on Sturtian glaciation?
685±7	Lund et al. 2003	Gospel Peaks Sq. (Idaho)	U-Pb z TIMS	Rhyodacite flow	Age on Sturtian glaciation?
709±4	Fanning and Link 2004	Pocatello Fm. (Idaho)	U-Pb SHRIMP	Tuff breccia	Age on Sturtian glaciation?
†723±16/10	Brasier et al. 2000	Ghubrah Fm. (Oman)	U-Pb z TIMS	Reworked tuff	Age within Sturtian glacials?
735±5	Key et al. 2001	Katanga Supergroup (Zambia)	U-Pb SHRIMP	Mafic volcanics	Age of end of Sturtian glaciation?
741±6	Frimmel et al. 1996	Gariep Belt (S. Namibia)	Pb-Pb z TIMS	Rhyolite flow	Age of end of Sturtian glaciation?
746±2	Hoffman et al. 1996	Naauwpoort Fm. (N. Namibia)	U-Pb z TIMS	Rhyolite flow	Max. age constraint on Sturtian glaciation
755±3	Windgate et al. 2000	Mundine Dykes (Pilbara craton)	U-Pb SHRIMP	Mafic dikes	Min. age separation between Laurentia and Australia
760±1	Halverson et al. 2005	Otavi Group (northern Namibia)	U-Pb z TIMS	Ash bed	Age on pre-Sturtian δ^{13}C record
777±7	Preiss 2000	Boucat Volcanics (S. Australia)	U-Pb SHRIMP	Rhyolite	Min. age constraint on Bitter Springs anomaly?
780±2.3	Harlan et al. 2003	Mackenzie Mts. Spg. (NW Caada)	U-Pb z TIMS	Mafic dikes	Inferred age of Little Dal Basalt.
802±10	Fanning et al. 1986	Callana Gp. (S. Australia)	U-Pb SHRIMP	Rook Tuff	Age constraint on Bitter Springs anomaly?
827±6	Wingate et al. 1998	Gairdner Dykes (S. Australia)	U-Pb SHRIMP	Mafic dikes	Correlative with volcanics in upper Bitter Springs Fm.?

2.2 The $\delta^{13}C$ Record

Fig. 1A presents an up-to-date version of the Neoproterozoic composite $\delta^{13}C$ record, modified from Halverson et al. (2005). This record includes new data from the Little Dal Group (Mackenzie Mountain Supergroup), northwest Canada, and incorporates new radiometric ages, most importantly, two U–Pb dates from the basal Doushantuo Formation—the cap carbonate sequence to the Marinoan-aged Nantuo glacials in south China (Condon et al., 2005) and ages on the Sturtian glaciation that suggest an age closer to 710–700 Ma (Fanning and Link, 2004). The principal difference between this compilation and that presented by Halverson et al. (2005) is the assumption that the Petrovbreen Member diamictite—the older of two glacigenic units in Svalbard—may be Sturtian in age rather than Marinoan, as argued in Halverson et al. (2004). However, this difference does not change significantly the overall structure of the $\delta^{13}C$ curve. More problematic to this compilation are persistent uncertainties regarding the timing, nature, and global correlation of the Sturtian glaciation. The Sturtian glaciation is here assumed to span from ca. 715 to 700 Ma (Fig. 1A) based on radiometric constraints from pre-Marinoan glacial deposits in Oman (Brasier et al., 2000; Allen et al., 2002) and Idaho (Fanning and Link, 2004) (Table 1). Although the glaciations are treated as discrete events for the sake of constructing the compilation (i.e., no data are included within the Sturtian and Marinoan glacial intervals), it is becoming inceasingly apparent that the pre-Marinoan record is much more complex.

Figure 1. (on Page 237) (A) Composite $\delta^{13}C$ record based on correlations shown in (B) and modified from Halverson et al. (2005) with new data included from the Little Dal and Coates Lake groups in NW Canada. This compilation is based on the correlation (B) of the Petrovbreen Member diamictite in Svalbard with the Sturtian glacials in Namibia (Chuos) and northwest Canada (Rapitan). The implication of this correlation is that negative $\delta^{13}C$ anomalies precede both the Sturtian and Marinoan glaciations. Symbols on the top line in (A) indicate prescribed ages used in constructing the timescale: star = direct age constraint; triangle = age constraint correlated from other succession with high degree of confidence; X = age constraint correlated from other succession with a moderate degree of confidence; diamond = arbitrary age constraint. The time scale is interpolated linearly between all imposed ages. Solid horizontal lines indicate duration of the contribution of carbon isotope data each from each of the four successions used in this compilation (NW Canada: Little Dal and Coates Lake Group; Svalbard: Akademikerbreen Group; N Namibia: Abenab and Tsumeb Subgroups; Oman: Huqf Supergroup). Solid + dashed lines show inferred time span of the Neoproterozoic sedimentary succession at each location (note that although the Oman sequence extends below the Sturtian, the interglacial record is almost completely absent; Le Guerroué et al., 2005). (B) Simplified stratigraphic sections of successions from which the carbon isotope data in (A) are derived, showing the correlations used as a basis for the compilation. U–Pb age constraints (in Ma) are shown in boxes. CLG = Coates Lake Group; RG = Rapitan Group; Om = Ombombo Subgroup; Ug = Ugab Subgroup.

A Neoproterozoic Chronology

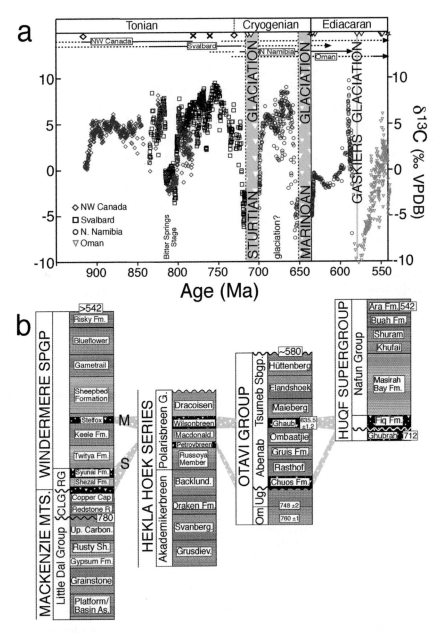

Notwithstanding the ambiguity remaining in some correlations, the advantage of these compilations over previously published $\delta^{13}C$ records for the Neoproterozoic is that they are constructed from a limited number of thick, carbonate-rich successions for which high-resolution isotopic data are available. For all carbon isotope data, ages were assigned *a posteriori*

through linear interpolation of fixed ages from successions from which the data is derived and assumed ages for the beginning and end of the Sturtian glaciation and the beginning of the Marinoan glaciation. Unfortunately, firm radiometric ages from these successions are few, and most of the calibration dates are correlated into the composite record from other successions, which unavoidably entails the risk of miscorrelation.

This method is not ideal and the resulting time scale is surely inaccurate in places, but the relative position of the data should be correct (apart from some mismatch across the intervals where correlations are made). Additional radiometric ages from other successions can then be applied to the record with varying degrees of confidence, based on correlation with the carbon isotope record and other considerations (such as other isotopic data).

Clearly, the composite record is far from a finished project, and just as the version here differs from alternatives presented in Halverson *et al.* (2005), so too will this version give way to improved compilations as new data become available and correlations are tested. In order to facilitate construction of improved records and integrations this record with other data sets, all $\delta^{13}C$ data from NW Canada, Svalbard, and Namibia included in the record are available at *http://www.igcp512.com* as composite sections.

2.3 Bases for Correlation

Due to the recognition of glacial deposits of clearly Sturtian and Marinoan affinity (Hoffman and Prave, 1996; Kennedy *et al.* 1998, Hoffman *et al.* 1998b) and the abundance of carbonate section spanning the two glacial horizons in the Otavi Group, the Neoproterozoic succession of northern Namibia serves as the backbone of the composite carbon isotope record (Fig. 1). The correlations between Cryogenian sequences used here fundamentally rest upon the assumption that the Chuos and Ghuab diamictites in Namibia are equivalent to the Sturtian and Marinoan glacials in Australia and the Rapitan and Stelfox glacials in NW Canada (Kennedy *et al.*, 1998; Hoffman and Schrag, 2002; Halverson *et al.*, 2005), although, as discussed below, new radiometric ages (including a 607.8 ± 4.7 Ma Re–Os age on shales from the purported equivalent of the upper diamictite in the Mackenzie Mountains; Kendall *et al.*, 2004) have challenged this model. Since most of the data shown in the compilation are indubitably pre- and post-Cryogenian, the uncertainties in correlation do not profoundly affect the overall structure of the $\delta^{13}C$ record.

A U–Pb zircon age of 635.5 ± 1.2 Ma on the Ghaub glacials in central Namibia (Hoffmann *et al.*, 2004) provides a key time constraint on the Marinoan glaciation. The thick (< 2 km) Tsumeb Subgroup, overlying the Ghaub glacials, presents an unrivaled post-Marinoan carbonate record. The age of the top of this passive margin sequence is poorly constrained, but is presumed to approximate (Halverson *et al.*, 2005) the ca. 580 Ma onset of continental collision on the western margin of the Congo craton (Goscombe *et al.*, 2003). Two pre-Sturtian U–Pb ages from the Naauwpoort Volcanics (746 ± 2 Ma; Hoffman *et al.*, 1996) and the Ombombo Subgroup (760 ± 1 Ma; Halverson *et al.*, 2005) are useful time markers within the Otavi Group but are not applied to the $\delta^{13}C$ compilation due to difficulty in correlating the fragmentary pre-Sturtian record from Namibia with the much more complete but virtually undated records in Svalbard and northwest Canada.

Whereas Halverson *et al* (2004, 2005) suggested that the Polarisbreen diamictites (Petrovbreen Member and Wilsonbreen Formation) collectively correlated with the Marinoan glaciation, more recent data suggest instead that the lower of these diamictites predates the Marinoan glaciation (Halverson *et al.*, in review). If the Petrovbreen Member represents the Sturtian glaciation in Svalbard (e.g. Kennedy *et al.*, 1998), then it follows that both the Marinoan and Sturtian glaciations were preceded by negative $\delta^{13}C$ anomalies of similar magnitude, thus minimizing the use of a pre-glacial anomaly as a correlation tool. Furthermore, purported glendonites between the two glacial intervals (Halverson *et al.*, 2004) could be roughly coeval with recently discovered glendonites in strata between the Rapitan and Stelfox glacials in NW Canada (James *et al.*, 2005), and perhaps account for the growing body of evidence for glaciation at ca. 680 Ma (e.g. Lund *et al.*, 2003; Zhou *et al.*, 2004; Fanning and Link, 2004; Kendall *et al.*, 2005).

Although this correlation does not dramatically alter the shape of the $\delta^{13}C$ record, it does have important implications for the ages of other North Atlantic glacial deposits and the duration and completeness of the pre-Sturtian records in Svalbard and northwest Canada, as discussed below. Irrespective of whether the Petrovbreen Member is Sturtian, Marinoan, or something in between, the Akademikerbreen Group in Svalbard is entirely pre-Sturtian in (Halverson *et al.*, 2005), meaning that the Hekla Hoek Series preserves a very complete (2 km) carbonate record (Knoll and Swett, 1990) for a period within the Neoproterozoic that is not well understood (Figs. 1–2).

Although the Neoproterozoic succession in northwest Canada is not well dated, close similarities between the Sturtian and Marinoan cap carbonate sequences, the interglacial $\delta^{13}C$ record, and strontium isotope data support the correlation between the Rapitan and Ice Brook (Stelfox) glacials in northwest Canada and the Chuos and Ghaub glacial in Namibia (Kennedy *et al.*, 1998; Hoffman and Schrag, 2002). It follows from this correlation that the Coates Lake Group in northwestern Canada is pre-Sturtian in age (Figs. 1–2). The Rapitan and Coates Lake groups are separated by an unconformity (Jefferson and Ruelle, 1986), which means that the latter likely does not preserve a complete record leading into the Sturtian glaciation. The contact between the Coates Lake and Little Dal groups is also unconformable (Fig. 2), and given that the former was deposited during a phase of regional extension (Jefferson and Ruelle, 1986), the time span between the top of the Little Dal carbonates and the base of the Coates Lake carbonates could be significant. Locally, the Little Dal Basalt, which is inferred to be ~780 Ma based on geochemical similarity to mafic dikes and sills that intrude the Mackenzie Mountain Supergroup (Jefferson and Parrish, 1989, Harlan et al., 2003), occurs at this contact and appears to be conformable with the top of the Little Dal carbonates (Aitken, 1981). The Little Dal Basalt thus provides a potentially useful calibration point in the $\delta^{13}C$ record.

The Huqf Supergroup in Oman is one of the best documented and most complete stratigraphic sections spanning the Ediacaran Period (Gorin et al., 1982), and the carbonate-rich, latest Neoproterozoic section is superbly preserved in outcrop and drill core (Amthor et al., 2003, Le Guerroué et al., 2006). Radiometric ages from the Precambrian–Cambrian boundary interval pin the age of the boundary at 542 Ma and constrain the duration of the negative $\delta^{13}C$ anomaly associated with the boundary to < 1 m.y. (Amthor et al., 2003). Oman was also one of the first places (along with South Australia) where the large, post-Marinoan Shuram (or Wonoka) negative $\delta^{13}C$ anomaly (Halverson et al., 2005) was first documented; the $\delta^{13}C$ record from the Huqf Supergroup (Burns and Matter, 1993; Amthor et al., 2003; Cozzi et al., 2004; Le Guerroué et al., 2006) is among the most complete spanning this anomaly.

The Fiq glacials and overlying Masirah Bay Formation cap carbonate sequence are equivalent to the Ghaub-Maieberg in Namibia (Leather et al., 2002, Hoffman and Schrag, 2002, Allen et al., 2005) and constitute one tie point between these two successions. Unfortunately, since the Masirah Bay Formation (cap carbonate sequence) is predominantly siliciclastic above the Haddash cap dolostone (Allen and Leather, 2006) and the Tsumeb Subgroup in Namibia appears to be truncated beneath the Shuram/Wonoka anomaly

---

Figure 2. (on Page 241) Pre-Sturtian composite stratigraphic and $\delta^{13}C$ records from Northeast Svalbard (Halverson et al., 2005), the Mackenzie Mountains (this paper), and central Australia (Hill et al., 2000). The correlation shown implies that the succession in the Mackenzie Mountains preserves a significantly older record of $\delta^{13}C$ than found in Svalbard and Australia. G1 and S1 designate the isotopic shifts and associated sequence boundaries (in Svalbard), that define the so-called Bitter Springs Stage (Halverson et al., 2005). COATES L = Coates Lake Group; RR = Redstone River Formation. Note the change in scale between the Coates Lake and Little Dal Groups.

A Neoproterozoic Chronology

241

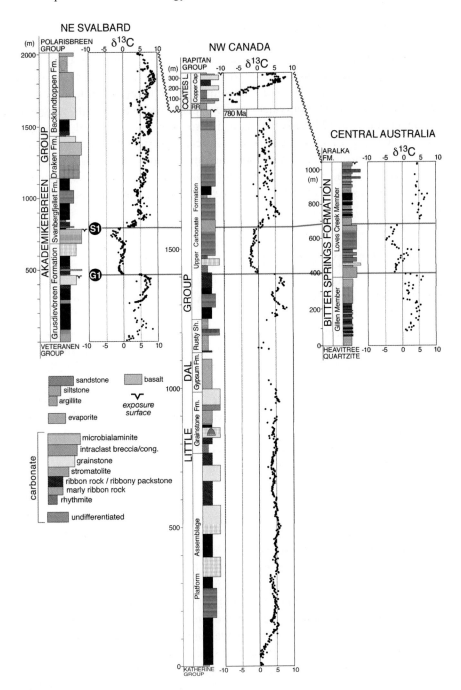

(Halverson et al., 2005), it is impossible to tie the complementary Nafun and Tsumeb $\delta^{13}C$ records precisely. However, the compilation of $\delta^{13}C$ data from the Nafun Group supports the argument that there was only one major $\delta^{13}C$ anomaly in the middle Ediacaran period (Le Guerroué et al., 2006). Thus, the correlation between a sharp downturn in $\delta^{13}C$ in the upper Kuiseb Formation (basin facies equivalent of the upper Tsumeb Subgroup) proposed by Halverson et al. (2005) is maintained here. It should be noted, however, that Condon et al. (2005) proposed a significantly different time scale for the Wonoka/Shuram anomaly, based on radiometric and carbon isotopic data from south China, indicating instead that the nadir of this anomaly significantly post-dates the Gaskiers glaciation and is perhaps as young ca. 555 Ma.

3. REVIEW OF THE NEOPROTEROZOIC

3.1 The Tonian (1000–720? Ma)

The chronometrically defined base of the Neoproterozoic (1000 Ma) approximately coincides with the boundary between the Middle and Upper Riphean (Knoll, 2000). The Meso-Neoproterozoic boundary interval is not well studied, but carbonate successions spanning it do occur in northwestern and southeastern Siberia. Carbon isotope data from these successions show a first order shift in $\delta^{13}C$ late in the Mesoproterozoic towards more ^{13}C-enriched values (Bartley et al., 2001), following a prolonged interval of stable values near 0‰ (Buick et al., 1895; Brasier and Lindsay, 1995). This shift in steady state carbon isotopic composition and increase in variability broadly coincides with the amalgamation of the Rodinia supercontinent and a decrease in marine $^{87}Sr/^{86}Sr$ (Kuznetsov et al., 1997; Bartley et al., 2001; Semikhatov et al., 2002).

Early Neoproterozoic (Tonian) sediments, including thick carbonate successions, occur across northwestern Canada and in northeastern Alaska in epicratonic basins of indeterminate origin (Aikten, 1981; Rainbird et al., 1996). Carbon and strontium isotopic data from the Shaler Supergroup on Victoria Island (Asmerom et al., 1991) established that the Tonian ocean was generally ^{13}C-enriched and unradiogenic. A new data set from the equivalent but better exposed Little Dal and Coates Lake groups in the Mackenzie Mountain fold belt provides a more detailed and continuous record through much of the Tonian (Fig. 2). Although age constraints on these successions are limited, the Little Dal Group is cross-cut by 780 Ma mafic dikes and sills and capped by a basalt of presumably equivalent age (Jefferson and Parrish, 1989, Harlan et al., 2003), giving a key minimum age on this stack of carbonates. The overlying Coates Lake Group, deposited during a phase of regional extension, is highly variable in thickness, and in places absent, beneath the overlying Rapitan Group (Aitken, 1981; Jefferson and Ruelle, 1986). It is constrained to be younger than 780 Ma and older than the Sturtian glaciation.

Other important pre-Sturtian successions are found in northeastern Svalbard and central Australia. The Mackenzie Mountains, Svalbard, and Australia successions all exhibit ^{13}C-enriched values typical of the Tonian, but also include a prominent interval of low δ^{13}C (Fig. 2), informally referred to as the *Bitter Springs Stage* (Halverson *et al.*, 2005), since it was first documented in full in the Bitter Springs Formation in central Australia (Hill *et al.*, 2000). The most detailed stratigraphic and isotopic record across this isotopic stage is found in Svalbard, where it spans 325 m of section and both the sharp decline and rise in δ^{13}C that define it coincide with sequence boundaries (G1 and S1, respectively; Fig. 2) that have been recognized across the entire belt of otherwise conformable carbonate strata (Halverson *et al.*, 2005). The nadir in δ^{13}C values above the G1 boundary coincides with a 1-m interval of formerly aragonitic seafloor cements, reminiscent of the Marinoan cap carbonates. The negative δ^{13}C shift is reproduced precisely in both the Little Dal Group and the Bitter Springs Formation, and in the former it coincides with a major flooding surface, reinforcing evidence from Svalbard that the perturbation to the carbon cycle is related to a global scale event that also drove large changes in sea level (Maloof *et al.*, 2006). The subsequent positive shift (S1) is a virtual mirror image of the G1 shift (Fig. 2). Whereas the transition to positive values is partially base-truncated in Svalbard, presumably due to exposure and non-deposition (Halverson *et al.*, 2005), it is recorded in full in the Mackenzie Mountains (Fig. 2).

Due to the association between δ^{13}C anomalies and glaciation in the Neoproterozoic (Knoll *et al.*, 1986, Kaufman *et al.*, 1997, Hoffman *et al.*, 1998a, b), it is tempting to speculate that the Bitter Springs isotope stage is related to a glaciation. Indeed, this possibility cannot be ruled out, even though no sedimentological evidence for glaciation has been found. However, paleomagnetic data from Svalbard suggest instead that the Bitter Springs isotope stage and associated fluctuations in sea level may be related to a pair of large-scale true polar wander (TPW) events (Maloof *et al.*, 2006). Detailed and reprudicible paleodeclination data spanning the G1 boundary in Svalbard indicate a large rotation (>50°) of Laurentia relative to the earth's spin axis at this time. Less well-placed data suggest a return to pre-G1 paleodeclinations above the S1 boundary (Maloof *et al.*, 2006). While it is conceivable that these rotations were generated by continental drift, this process would have required 10s of millions of years, which is inconsistent with the smooth δ^{13}C profile across the G1 boundary and lack of evidence for extensive erosional truncation (Halverson et al., 2005). Furthermore, whereas a plate-tectonic explanation for the paleomagnetic data would not account for the global changes in sea level, the TPW hypothesis does predict such a correlation, since variable sea level changes attend true polar wander as a function of position relative to the TPW rotation axis (Mound et al., 1999). Li et al. (2004) have independently proposed an episode of large scale TPW at ca. 800 Ma, initiated by the growth of a mantle superplume at high latitudes and as a possible trigger for the Cryogenian glaciations. Whereas it is not hard to imagine that a TPW event of this magnitude would drastically impact global climate and carbon cycling, any explanation for the connection between the perturbations between the $\delta^{13}C$ record and the TPW event remains speculative (Maloof et al., 2006).

Based on the suggested correlation between the pre-Sturtian successions in Svalbard, the Mackenzie Mountains and the Bitter Springs Formation (Fig. 2), and therefore the implied relative depositional rates in each succession, the base of the Little Dal Group is significantly older than that of either the Akademikerbreen Group or the Bitter Springs Formation. This interpretation is consistent with the fact the the trend of gently fluctuating $\delta^{13}C$ near 5‰ in the lower Little Dal Group is not seen beneath the G1 boundary in either Svalbard or Australia (Fig. 2). The base of the Little Dal Group is only constrained to be > 1003 ± 4 Ma from detrital zircons in the underlying Katherine Group (Rainbird et al., 1996). Barring extremely rapid depositional rates, this correlation implies that the lower Little Dal Group samples the hitherto poorly documented early Tonian, perhaps extending back beyond 900 Ma.

The $\delta^{13}C$ record from the lower Little Dal Group shows a prominent rise from 0‰ at the base to a mean of 5‰. Thus, it appears that average $\delta^{13}C$ values may have again dropped to near 0‰ in the early Neoproterozoic, following the rise to 3‰ in the late Mesoproterozoic (Bartley et al., 2001). Therefore, the lower Little Dal Group may record an important transition to the higher and more variable values characteristic of the Neoproterozoic (Kaufman and Knoll, 1995). In any case, the data from the Little Dal Group suggest that relatively high rates of organic carbon burial were the norm for most of the first half of the Neoproterozoic, well in advance of the earliest direct evidence for glaciation. If this 5‰ rise in $\delta^{13}C$ records a first order shift in the average isotopic composition of the ocean, then the limited stratigraphic range over which it occurs suggests that it was driven by a stepwise change in the mode of organic matter production and/or burial rather than being accomplished by gradual changes in the reservoir size of dissolved inorganic carbon in the Proterozoic ocean (e.g. Bartley and Kah, 2004).

Another important feature of the isotopic record from the Mackenzie Mountains is the profile through the Coates Lake Group, which gives a glimpse of the poorly constrained record prior the Sturtian glaciation. However, since the Coates Lake Group was deposited during active extension (Jefferson and Ruelle, 1986) and both the upper and lower contacts are unconformable, it is difficult to appraise the exact temporal relationship between this record and the onset of glaciation. $\delta^{13}C$ values in the lower part of the Coates Lake Group are extremely negative (Fig. 2), broadly coincide with a major deepening event, and are largely preserved in redeposited carbonates, raising the concern that these values do not record the evolution of marine $\delta^{13}C$. On the other hand, the very smooth rise in $\delta^{13}C$ in the Coates Lake Group would be difficult to produce by secondary processes.

If the $\delta^{13}C$ profile in the Coates Lake Group approximates the evolution of marine $\delta^{13}C$, one would expect the trend to appear within the highly conformable Svalbard stratigraphy. One clue as to how to draw this correlation comes from the isotopic profile through Bed-groups 19 and 20 in the Eleanor Bay Group of East Greenland, in which a negative anomaly of similar magnitude also coincides with a major deepening event (Fairchild et al., 2000) prior to the oldest glacial deposits in the succession. Bed-groups 19 and 20 are presumably equivalent to the Russøya Member in Svalbard (Fairchild and Hambrey, 1995), which displays a similar stratigraphic and isotopic profile, but without the highly negative values at the base. An alternative correlation is with the negative $\delta^{13}C$ anomaly that occurs directly beneath the Petrovbreen Member diamicites in Svalbard (Halverson et al., 2004), although this correlation would imply very deep erosion at this level, for which there is no evidence. Resolving this correlation will be key to reconstructing the evolution of $\delta^{13}C$ leading into the Cryogenian glaciations.

3.2   The Cryogenian (720?–635 Ma)

3.2.1   The Sturtian Glaciation

The age of the Sturtian glaciation is problematic. The best maximum age constraint has long been provided by a 746 ± 2 Ma U–Pb date on the Naauwpoort Volcanics, which occur well below the Chuos Formation on the southern margin of the Congo craton in northern Namibia (Hoffman et al., 1996). This age statistically overlaps with a pair of other ages from southern Africa that are purported to constrain the minimum age of the Sturtian glaciation: a Pb–Pb evaporation age of 741 ± 6 on the Rosh Pinah Volcanics in the Gariep belt in southern Namibia (Frimmel et al., 1996) and a 735 ± 5

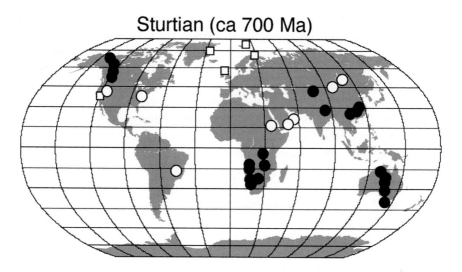

Figure 3. Distribution of glacial deposits of Sturtian (or at least, pre-Marinoan) age. Filled black circles represent glacial deposits of virtually certain Sturtian age, open circles of possible Sturtian age, and open squares of Sturtian age based on the correlation of the Petrovbreen Member diamictite in Svalbard (and the Surprise diamictite in Death Valley) with the Sturtian glacials, as shown in Fig. 1B.

Ma age from volcanic breccias at the top of the Grand Conglomerat in Zambia (Key *et al.*, 2001). Taken together, these three ages suggest that the Sturtian glaciation ended 740–735 Ma and was short-lived. However, this constraint grossly conflicts with radiometric age constraints on purported Sturtian glacial deposits elsewhere. The glacigenic Ghubrah Formation in Oman is separated by an angular unconformity from the Fiq glacials above (Le Guerroué *et al.*, 2005), which are confidently ascribed a Marinoan age (Allen *et al.*, 2005). A reworked tuff within the Ghubrah Formation has been dated quite imprecisely at 723 +16/–10 Ma (Brasier *et al.*, 2000), but new analyses suggest a more precise age of ca 712 Ma (Allen *et al.*, 2002; Kilner *et al.*, 2005). Comparable ages are found in Idaho, where a tuff within the Scout Mountain Member of the Pocatello Formation is dated at 709 ± 4 Ma and a clast within the upper diamictite is dated at 717 ± 3 (Fanning and Link, 2004). If the conventional interpretation that the Pocatello Formation is equivalent to the Rapitan Group glacials in the northern Cordillera (Crittenden *et al.*, 1971) is correct, then these ages suggest that the Sturtian glaciation was much younger, and definitely ongoing at 710 Ma. Thus, either the minimum age constraints on the Sturtian glaciation from southern Africa are incorrect, perhaps due to poor stratigraphic control (Master *et al.*,

2005), or they document an earlier phase of diamictite deposition that may or may not have been glacial.

Lund *et al.* (2003) obtained even younger age of ca 685 Ma from volcanics in the Edwardsburg Formation in central Idaho, that is lithologically similar to the Scout Member Member further south and that they correlate with the Rapitan Group and equivalent rocks along the length of the Cordillera. These authors thus argue that either the Sturtian glaciation was even younger—still ongoing at ca. 685 Ma—or diachronous. However, the fact that the Rapitan Group, Chuos, and Sturtian (*sensu strictu*) glacials are overlain by cap carbonates that preserve a record of post-glacial transgression and a negative carbon isotope anomaly implies that mere diachroneity of glaciation is not the answer to the conundrum posed by the conflicting ages. If the interpetation by Jiang *et al.* (2003) that the Tiesiao diamictite in south China is part of the Sturtian glacial sequence there, then the U–Pb zircon date of 663 ± 4 Ma from an ash bed at the base of the overlying Datangpo Fm. (Zhou *et al.,* 2004) supports the model that the Sturtian glaciation continued well beyond 700 Ma. This age is also statistically identical to an age of 667 ± 5 Ma (Fanning and Link, 2004) from an ash bed that occurs above the Scout Mountain Member diamictite in Idaho, and beneath a distinctive pink dolostone unit with inorganic sea floor cements and associated negative $\delta^{13}C$ anomaly (Lorentz *et al.,* 2004). Taken together, these data support the interpretation that the Sturtian glaciation continued to ca. 670 Ma.

On the other hand, the discovery of purported glendonites in the interglacial strata of northwest Canada (James *et al.*, 2005) supports the likelihood that cold conditions resumed sometime after the Sturtian glaciation ended and prior to the onset of the Marinoan glaciation. Thus, it is equally conceivable that a less severe glaciation punctuated the interval between the Sturtian and Marinoan glaciations. The Edwardsburg Formation diamictites, lacking an obvious cap carbonate sequence, may thus represent a post-Sturtian, pre-Marinoan glaciation.

The mystery of the timing and distribution of Sturtian glacial deposits (Fig. 4) will only be solved with more well-placed and precise radiometric ages, ideally from Namibia, Australia, or northwest Canada, where the stratigraphic context of both the Sturtian and Marinoan glaciations is well constrained. Such dates will also be vital to establishing the global extent of the Sturtian glacial deposits, which is now highly uncertain. Halverson *et al.* (2004) proposed that Sturtian glacial deposits are largely if not entirely absent from the North Atlantic region, and perhaps Death Valley. However, this conclusion was based on the argument that the Petrovbreen Member diamictite in Svalbard represents the early stage of the Marinoan glaciation (Halverson *et al.*, 2004), stemming from the observation that this unit was preceded by a negative $\delta^{13}C$ anomaly, much like pre-Marinoan rocks elsewhere. However, more recent isotopic and radiometric data from other successions do not favor this correlation (Halverson *et al.*, in press), and it seems that the Petrovbreen Member more likely represents a pre-Marinoan glaciation.

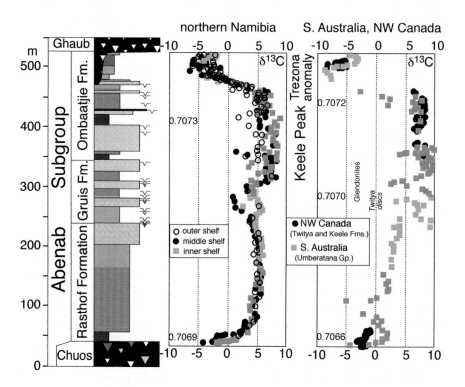

Figure 4. Composite stratigraphic column of the Abenab Subgroup (Otavi Group, Namibia), with carbon isotopic data from the shelf margin, middle shelf, and inner shelf, illustrating the lateral variability in absolute $\delta^{13}C$ values, but reproducibility of trends. Composite $\delta^{13}C$ records for the equivalent interval in South Australia (McKirdy *et al.*, 2001) and NW Canada (modified from Hoffman and Schrag, 2002) show remarkably similar trends. Summary of least radiogenic $^{87}Sr/^{86}Sr$ values (Kaufman *et al.*, 1997; McKirdy *et al.*, 2001; Yoshioka *et al.*, 2003; Halverson *et al.*, in review) available from these interglacial successions support the correlation. Glendonite occurrences in the transition between the Twitya and Keele Formation in Canada were reported by James *et al.* (2005).

3.2.2   The Interglacial

The ambiguity in the age of the Sturtian glaciation and its correlation in the North Atlantic region clouds the picture of the interval between these two glaciations. However, since the interglacial picture is clearer in Namibia, Australia and northern Canada, the record from these successions is emphasized here.

In Namibia, the Chuos glacials are overlain by the 200–400 m thick Rasthof Formation, a single shoaling upward sequence distinguished by its graphite color and microbial facies, which include chaotic stromatolites and sublittoral microbialites (Hoffman and Schrag, 2002; Hoffman and Halverson, in press). A sharp rise in $\delta^{13}C$ from values as low as –4‰ is typically preserved in a basal limestone rhythmite member (Yoshioka et al., 2003), and an abrupt positive shift in $\delta^{13}C$ commonly, but not always, coincides with the change in facies to stromatolites (Hoffman and Halverson, in press) (Fig. 4). $\delta^{13}C$ values continue to increase to the top of the Rasthof Formation, and following a few small excursions to ~1‰ in the overlying Gruis Formation and lower Ombaatjie Formation, they plateau at 7–9‰ (the *Keele Peak* of Kaufman et al., 1997). In the uppermost Ombaatjie Formation, $\delta^{13}C$ declines by >10‰ over 20–50 m of section, just below the Ghaub Formation or equivalent glacial unconformity (Halverson et al., 2002).

Detailed data sets from a transect from the margin zone on the southwestern edge of the Congo craton to the northern Otavi platform confirm the reproducibility of these trends (Fig. 4), but also indicate variation in the magnitude of the signals (Halverson et al., 2005), which are interpreted to represent increasing restriction northward, away from the shelf margin (Halverson, et al., 2002). Fortunately, the lateral variability is significantly smaller than the secular variation in $\delta^{13}C$, but it still serves as a useful reminder of the intrinsic noise in the carbon isotopic signal preserved in shallow water carbonates.

The major features of the interglacial carbon isotopic record are reproduced in the Twitya and Keele formations of northwest Canada and the Umberatana Group of South Australia (Fig. 4), confirming that these are global seawater signals. The fact that the post-Sturtian cap carbonates and associated negative $\delta^{13}C$ anomalies occur in all three, widely separated successions adds an important constraint in the discussion of the timing of the Sturtian glaciation. Insofar as these record global events, at least one major phase of pre-Marinoan glaciation ended synchronously. If there was a post-Sturtian, pre-Marinoan glaciation, the question is whether there is any stratigraphic expression of this event (apart from glendonite occurrences) in any of these successions.

The biostratigraphic record of the interglacial interval is quite sparse. It is established that acritarch diversity is low following the first of the Cryogenian glaciations (Knoll, 1994; Vidal and Moczydlowska-Vidal, 1997). Interestingly, the oldest discoid fossils (the Twitya discs), linked by some to the Ediacaran biota (Gehling et al., 2000) occur in the interglacial interval in Canada (Hofmann et al., 1990), but have not been found in any other indisputably pre-Marinoan strata (Xiao, 2004a). The >60 m.y. difference in age between these occurrences and the oldest Ediacaran fossils in Newfoundland is reason to question their Ediacaran affinity.

3.2.3 The Marinoan Glaciation

The Marinoan glaciation is presaged by a ~10‰ decline in $\delta^{13}C$ (Fig. 4). In Namibia, where this negative isotope shift (the *Trezona* anomaly) has been documented in over 20 stratigraphic sections, it spans 30–50 m of predominantly shallow, platform-facies carbonates and is estimated to have lasted ~0.5 to 0.75 m.y. (Halverson et al., 2002, Hoffman and Halverson, in press). In places, the anomaly is truncated at the Ghaub glacial surface. Together, these data and other field observations suggest that the full drop in $\delta^{13}C$ predated any major fluctuations in sea level related to the growth of Marinoan ice sheets (Halverson et al., 2002). Therefore, it appears that the negative carbon isotope excursions and glaciation are not related in a simple cause and effect manner, but rather that both may have been consequences of a separate forcing mechanism. The relative timing of the onset of glaciation and the Trezona anomaly in Namibia rules out the hypothesis that the glaciation was triggered simply by biologically-driven reduction of atmospheric pCO_2 levels (Halverson et al., 2002). Schrag et al. (2002) proposed instead that a prolonged interval (>100 kyr) of elevated methane flux to the atmosphere and subsequent collapse of the methane source could have accounted for both the carbon isotope anomaly and climate destabilization, leading to the onset of the Marinoan glaciation. However, as pointed out by Pavlov et al. (2003), a potentially fatal flaw in this hypothesis is that it relies on an unrealistic residence time for methane in the Neoproterozoic atmosphere. No other viable model has yet linked the *Trezona* anomaly to the onset of the Marinoan glaciation.

Paleomagnetic data on Marinoan glacials consistently indicate deposition in low latitudes (Sohl et al., 1998, Evans, 2000, Trindade et al., 2003, Kilner et al., 2005). The widespread geographic distribution (Fig. 5) of Marinoan glacial deposits confirms that this was a global event. In northern Namibia, the Marinoan (Ghaub Formation) glacial deposits are typically absent or scant on the continental platform, but thicker and continuous in foreslope settings (Hoffman and Halverson, in press). A similar pattern is seen in the Stelfox Member in NW Canada (Aitken, 1991a,b), but the equivalent Vreeland diamictites to the south in British Columbia comprise up to 800 m of foreslope glaciomarine sediments, including abundant extrabasinal clasts (McMechan, 2000). Such evidence of intense erosion of the continents during the Neoproterozoic glaciations, coupled with the initial vision of a snowball Earth as a static, dry, snow-starved world (Hoffman et al., 1998) has led to obvious criticism of the snowball model (Christie-Blick et al., 1999; McMechan, 2000).

While it now seems that dynamic, wet-based glaciers capable of eroding and depositing large volumes of sediment would have been active several hundred thousand years after an initial snowball freezeover (Hoffman, 2000; Donnadieu et al., 2003), evidence for open water conditions from the Fiq Member glacials in the Oman Mountains on the Arabian Peninsula is more difficult to reconcile with a completely ice-covered ocean (Leather et al., 2002). The Fiq Member consists of a range of non-glacial to glaciomarine facies preserved in six transgressive-regressive cycles, several of which contain wave-generated ripples, demonstrating that open water conditions prevailed intermittently during this glaciation (Leather et al., 2002, Allen et al., 2005).

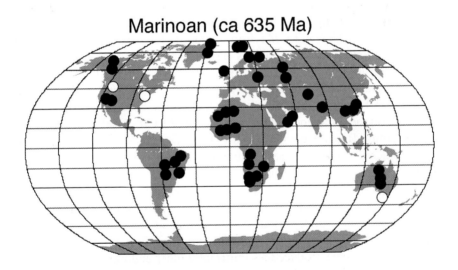

Figure 5. Distribution of glacial deposits of Marinoan age. Filled black circles represent glacial deposits of virtually certain Marinoan age and open circles of possible Marinoan age.

On the other hand, the evidence for ice lines at tidewater in the low latitudes for hundreds of thousands of years during the Marinoan glaciation (Sohl et al., 1998) attests to the fact that this glaciation was exceptional. Considering the climatic stabilizing effect (with the stable mode being a snowball) imposed by extensive ice cover at low latitudes (Caldeira and Kasting, 1992; Ikeda and Tajika, 1999), the question is not whether or not the Marinoan glaciation resembled the familiar Pleistocene glaciations (e.g. Leather et al., 2002), but how would a snowball, or at least global, glaciation, have evolved. That is, could open water conditions in the tropics have existed intermittently or at some phase during a snowball glacial cycle?

Aside from the controversy over the snowball Earth hypothesis, another subject of debate is the age of the Marinoan glacials in Australia. The onset of the Marinoan glaciation is constrained to post-date 663 ± 4 Ma based on an age from below the Nantuo glacials in south China (Zhou et al., 2004). A Re–Os age of 643.0 ± 2.4 Ma (Kendall et al., 2005) from the basal Tapley Hill Formation (the cap carbonate to the Sturtian glacials in South Australia) implies a significantly younger maximum age constraint. This age is difficult to reconcile with firm age constraints of 635 Ma for the end of the Marinoan glaciation in Namibia (Hoffmann et al., 2004) and south China (Condon et al., 2005), unless the Elatina diamictite is much younger, as argued by some authors (e.g. Calver et al., 2004). Similarly, a suite of other ages from Australia suggest that the Elatina and other glacials across Australia that have long been considered correlative (Dunn et al., 1971) may in fact be much younger, and possibly equivalent to the Gaskiers glaciation: a Re–Os age of 592 ± 14 Ma from black shales beneath the glacigenic Olympic Formation in the Amadeus Basin (Schaefer and Burgess, 2003), a U–Pb zircon age of 575 ± 3 Ma from intrusives into the Cotton Breccia (Grassy Group) on King Island (Calver et al., 2004), and another U–Pb Zircon age of 582 ± 4 Ma from a rhyodacite flow beneath the Croles Hill diamictite on the north coast of Tasmania (Calver et al., 2004).

The Cotton Breccia and Croles Hill Diamictite off the southern mainland are considered equivalent to the Elatina glacials by many (Calver and Walter, 2000), but if these ages and correlations are all correct, they imply that the true Marinoan glaciation is 55 m.y. younger than what is regarded as the Marinoan glaciation in Namibia (Ghaub Fm.) and south China (Nantuo Fm.). On the other hand, various considerations argue against assigning an age of 580 Ma to the Marinoan glacials in Australia. First, Kendall and Creasier (2004) have produced a significantly older and more precise Re–Os date of 658 ± 5.5 Ma on the same shales beneath the Olympic Formation dated by Schaefer and Burgess (2003). Whereas it is impossible to choose which Re–Os dates are more accurate, the contrasting ages suggests that the 592 ± 14 Ma age is not a firm maximum age constraint on Marinoan glaciation and highlight the need for both additional U–Pb ages and independent verification of the Re–Os method when applied to dating Proterozoic black shales.

Another problem for the proposed 580 Ma age for the Marinoan glaciation is that Tasmania's paleogeographic position relative to the Australian craton is undetermined (Calver and Walter, 2000). Given also

A Neoproterozoic Chronology   253

that the firm maximum age constraint of 582 ± 4 Ma in Tasmania is on a diamictite lacking a cap carbonate (which is very unusual for a Marinoan glacial deposit), extrapolating the Tasmania ages to the classic Marinoan diamictites on mainland Australia is tenuous at best. This correlation also has unsavory implications for the stratigraphy of South Australia. For example, it suggests that the equivalent of the ca. 635 Ma Ghaub glaciation, Maieberg cap carbonate, and associated $\delta^{13}C$ anomalies should occur somewhere between the Marinoan and Sturtian diamictites, even though no trace of any of these features is apparent in any Australian succession. Furthermore, a 580 Ma age on the Elatina diamictite in South Australia would imply that the entire Ediacaran section there was deposited in less than 40 m.y. and raises the question of why Ediacaran fossils occur so far above the Marinoan glacials in South Australia when they appear within 5 m.y. after the end of the Gaskiers glaciation in Newfoundland (Bowring *et al.*, 2003, Narbonne and Gehling, 2003) and 200 m above the top of the Mortensnes Formation in northern Norway (Farmer *et al.*, 1992).

Since it is established that there was a glaciation at 580 Ma (Bowring *et al.*, 2003), there is little reason to doubt that the Tasmanian diamictites are glacial in origin. However, the lack of geochronological support for their correlation to mainland Australia and the unlikely stratigraphic scenarios that these correlations imply are sufficient to cast serious doubt on the hypothesis that the Marinoan glacials are 580 Ma. Nevertheless, the implications of such an age assignment are important enough that this hypothesis should be tested. A datable ash within the Ediacaran type section in South Australia would be welcome.

3.3   The Ediacaran Period (635–542 Ma)

3.3.1   The Post-Marinoan Cap Carbonate Sequence

The beginning of the Ediacaran Period is now formally chronostratigraphically defined at the base of the Nuccaleena cap dolostone, above the Elatina glacials in Enorama Creek in the Flinders Ranges in South Australia (Knoll *et al.*, 2004, Knoll *et al.*, 2006). This definition is somewhat complicated by the ambiguity in the age of the Elatina glacials, as described above. For the purpose of this discussion, it will be assumed that the Nuccaleena cap dolostone is equivalent to the Maieberg and Doushantuo cap carbonates (Knoll *et al.*, 2006), and thus ca. 635 Ma (Condon *et al.*, 2005), although as just mentioned, radiometric confirmation is needed.

Two key U–Pb ages on the Doushantuo Formation (632.5 ± 0.5 and 635 ± 0.6 Ma; Condon et al., 2005) in south China have several important implications. First, they demostrate that the Keilberg (Maieberg Formation) and Doushantuo (basal member) cap dolostones are the same age (Hoffmann et al., 2004; Condon et al., 2005), and thus that Marinoan glaciation ended synchronously on two widely spaced cratons. Based on carbon-isotopic correlation between the Maieberg and Doushantuo formations (Condon et al., 2005), it appears that the Maieberg Formation was deposited in ~3–4 m.y (Fig. 6). These age constraints imply depositional rates of ~100 m my^{-1} in Namibia following the Marinoan glaciation and are probably far in excess of most cap carbonates sequences due to the preponderance of carbonate in this section. To the extent that this time scale is correct, it reinforces the hypothesis that the Marinoan glaciation lasted many millions of years (Hoffman et al., 1998a,b; Bodiselitch et al., 2005), for at reasonable subsidence rates on a passive margin and in the absence of major erosion of the platform (for which there is no evidence; Hoffman and Halverson, in press), far in excess of three million years is necessary to accommodate the 400 m-thick Maieberg Formation, which was deposited on a mature passive margin (Hoffman et al., 1998a,b). The rapid sedimentation rates for the Maieberg Formation, coupled with abundant carbonate content make it an ideal unit for high-resolution reconstruction of the evolution of post-Marinoan seawater (Hurtgen et al., 2006).

3.3.2 The Early Ediacaran

The early Ediacaran period is a crucial time period for understanding the evolution and diversification of acritarchs and animals (Knoll et al., 2004). Unfortunately, the geological record from the Marinoan cap carbonate sequence to the onset the 580 Ma Gaskiers glaciation is not well constrained. By far the most important succession, with respect to paleobiology, of Ediacaran age is the unusually fossil-rich Doushantuo Formation, which preserves fossilized embryos (Xiao et al., 1998), small bilaterians (Chen et al., 2004), and spiny acritarchs (Zhang et al., 1998), among other fossils. Unfortunately, the partitioning of time within this highly condensed section unit (~160 m) that spans over 80 m.y. (Condon et al., 2005) is poorly understood, despite the wealth of radiometric ages (Kaufman et al., 2005). Many authors assume that the oldest Doushantuo fossils predate the Gaskiers glaciation, based on a variety of arguments, including the stratigraphic position of unconformities within the Doushantuo (Xiao et al., 2004; Kaufman, 2005), Pb–Pb ages on the Doushantuo phosporites (Barfod et al., 2002; Chen et al., 2004), and molecular clock data, which suggest that Eumatozoa evolved between 634 and 604 Ma (Peterseon and Butterfield,

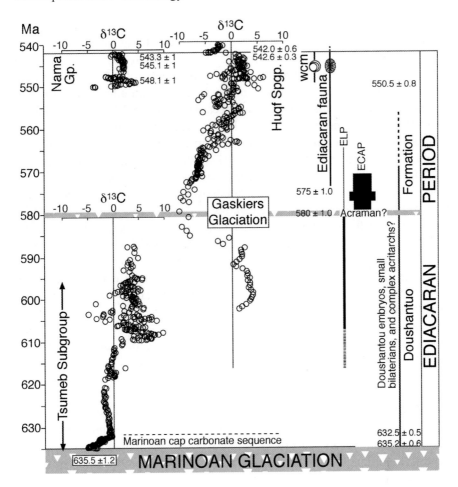

Figure 6. Summary of key $\delta^{13}C$ data from Namibia (Nama Group: Grotzinger *et al*., 1995; Saylor *et al*., 1998; Tsumeb Subgroup: Halverson *et al*., 2005) and Oman (Huqf Supergroup: Burns and Matter, 1992; Amthor *et al*., 2003; Cozzi *et al*., 2004), radiometric ages (in Ma), and biostratigraphic data for the Ediacaran Period. The age of the nadir of the Shuram/Wonaka anomaly is not well constrained, and has been estimated at 595 Ma by Le Guerroué *et al*. (2006) and 555 Ma by Condon *et al*. (2005). Here it is assumed that the actual age falls somewhere between these two estimates. wem = weakly calcified metazoa (i.e. *Namacalathus* and *Cloudina*). Relative species abundances of ELP (leiospheric) and ECAP (acanthomorphic) actritarchs are taken from Grey *et al*. (2003), who identified a major turnover spanning the Acraman Impact eject layer, roughly estimated at 580 Ma. Note, however, that some authors argue that the abundant ECAP actritarchs found in the Doushantuo Formation (e.g. Zhang *et al*., 1998; Xiao, 2004a, b), as well as other key fossils, predate the Gaskiers glaciation. See Table 1 for a key to the radiometric ages.

2005). On the other hand, Condon *et al.* (2005) argued instead that these Doushantuo assemblages post-date the Gaskiers glaciation based on their intepretation that the large negative $\delta^{13}C$ anomaly in the uppermost Douhsantuo, which they presume to be equivalent to the Wonoka-Shuram anomaly, only slightly predates an ash bed they dated (551.1 ± 0.7 Ma) above the anomaly; the same ash has been dated as 555.2 ± 6.1 Ma using U–Pb SHRIMP (Zhang *et al.*, 2005). The significance of these ages has been questioned on the basis of an unconformity separating the ash and the negative $\delta^{13}C$ anomaly (Kaufman, 2005). While these dates may support a relatively young age for the uppermost Doushantuo anomaly, the 555 Ma age occurs near the top of this anomaly and sedimentation rates were ostensibly very low in the Doushantuo Formation (Condon *et al.*, 2005), leaving open the possibility that the nadir of the anomaly could be many millions of years older. Furthermore, as pointed out by Zhang *et al.* (2005), this anomaly need not necessarily correlate with the Shuram-Wonoka anomaly, although based on the Oman record (Fig. 6) it is hard to envision what else it could correlate with. For now, it cannot be unequivocally verified that the Doushantuo acritarchs, fossilized embryos, and small bilaterians are post-Gaskiers. Resolving the controversy over the age of the Doushantuo biota relative to the Gaskiers glaciation and the Shuram-Wonoka anomaly is essential to unravelling the history of early animal evolution and its connection to the major climatic and biogeochemical perturbations during the Ediacaran Period.

Whereas the Doushantuo Formation is arguably the most important Ediacaran unit with regards to paleobiology and geochronology, the utility of its carbon isotope record is limited by its high organic content and extreme stratigraphic condensation (the entire formation, spanning from 635 to 550 Ma is about 160 m thick). In Namibia, where the early Ediacaran is spanned by the undated but carbonate-rich Tsumeb Subgroup (Hoffmann, 1989), the carbon isotope record shows generally low and smoothly varying $\delta^{13}C$ composition until the base of the Hüttenberg Formation (upper Tsumeb Subgroup), where $\delta^{13}C$ spikes to highly positive values (Fig. 6). Throughout the remainder of the Hüttenberg Formation, $\delta^{13}C$ fluctuates wildly, and it is not known if this is representative of the global ocean or local basin or diagenetic effects (Halverson *et al.*, 2005). Whereas the youngest sampled platform cabonates from the Hüttenberg Formation are isotopically positive, the youngest sediments in the foreslope equivalent (Kuiseb Formation; Hoffmann *et al.*, 2004) preserve a sharp downturn in $\delta^{13}C$, which Halverson *et al.* (2005) speculated might record the onset of a large $\delta^{13}C$ anomaly related to the Gaskiers glaciation. Highly negative $\delta^{13}C$ values are found in scant carbonates beneath the Gaskiers-equivalent Mortensnes Formation in northern Norway (Halverson *et al.*, 2005), which suggests that the Gaskiers glaciation, like the Marinoan glaciation, is preceded by a negative $\delta^{13}C$ anomaly. Since no more than one large $\delta^{13}C$ excursion between the Marinoan cap carbonate anomaly and the Precambrian–Cambrian boundary anomaly (Fig. 6) has ever been documented, Halverson *et al.* (2005) correlated the Gaskiers glaciation with an extreme and protracted $\delta^{13}C$ anomaly from late Proterozoic successions in South Australia (Wonoka Formation, Pell *et al.*, 1993; Calver *et al.*, 2000), Oman (Shuram-Kufai formations, Burns and Matter, 1993; Le Guerroué *et al.*, 2006), Death Valley (Johnnie Formation, Corsetti and Kaufman, 2003), and southern Namibia (Workman *et al.*, 2002).

As previously discussed, the age of this anomaly is the subject of much discussion. Halverson *et al.* (2005) suggested that the nadir predated the Gaskiers glaciation based on the occurrence of large, possibly glacially related paleovalleys that incised the Johnnie Formation and truncated the anomaly (Corsetti and Kaufman, 2003). However, Clapham and Corsetti (2005) have since argued that these paleovalleys are related to local tectonics rather than glacioeustasy, rendering this relative age constraint invalid. A pre-Gaskiers age (ca. 595 Ma) is independently argued by Le Guerroué *et al.* (2006) using a thermal subsidence model to invert stratigraphic height in the Nafun Group to time, based on the assumption that these sediments were deposited on a passive margin. In contrast, as discussed above, Condon *et al.* (2005) propose a much younger age of ca. 555 Ma based on correlation with the anomaly in the upper Doushantuo Formation.

While it appears that the unusually low $\delta^{13}C$ values must have spanned the Gaskiers glaciation (Fig. 6), Le Guerroué *et al.* (2006) argue against any causative connection between the Wonoka-Shuram anomaly and the Gaskiers glaciation based on the apparent immunity of the $\delta^{13}C$ record in the Nafun Group to the Gaskiers glaciation (according to their time scale). Nor has a persuasive explanation for this anomaly emerged. Assuming that the isotopic composition of mantle carbon has always remained ~ –6‰, the long duration and extremely low values (<–10‰) characteristic of the anomaly cannot be easily explained by our current understanding of what generates negative $\delta^{13}C$ anomalies (Melezhik *et al.*, 2005). Ostensibly, an enormous source of ^{13}C-depleted carbon, presumably organic carbon, was necessary to generate the anomaly. Rothman *et al.*, (2003) have proposed that large shifts in marine $\delta^{13}C$ can be driven by the remineralization of large, marine reservoirs of reactive organic matter, which transfers isotopically depleted carbon to the dissolved inorganic reservoir. This hypothesis was not formulated to explain such an extreme anomaly, nor can it explain how such a large pool of reactive carbon could have been stored in the oceans for tens of millions of years, but no other feasible model has yet been proposed to explain the Shuram-Wonoka anomaly.

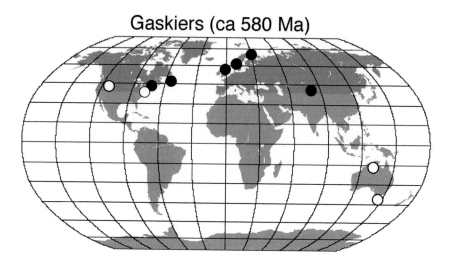

Figure 7. Distribution of glacial deposits of Gaskiers age: filled black circles represent glacial deposits of virtually certain Gaskiers age and open circles of possible Gaskiers age.

3.3.3 The Gaskiers Glaciation

The Gaskiers glaciation is precisely dated at 580 Ma and constrained to have lasted less than 1 m.y., at least in Newfoundland (Bowring *et al.*, 2003). Equivalent glacials deposits are also found further south in Avalonia, in the Boston basin, where the Squantum 'tillite' member is constrained to between 590 and 575 Ma (Thompson and Bowring, 2000). A similar age constraint exists for the possibly glacial Loch na Cille Boulder Bed in Scotland (Elles, 1934) and equivalent beds in Ireland (Condon and Prave, 2000), constrained to be younger than the 601 ± 4 Ma Tayvallich Volcanics (Dempster *et al.*, 2002). A U–Pb (LA–ICP–MS) age of 620 ± 14 from a detrital zircon in the Rendalen Formation (Hedmark Group) in southern Norway (Bingen *et al.*, 2005) confirms the suggestion by Knoll (2000), based on acritrarch biostratigraphy, that the Moelv Formation post-dates the Marinoan glaciation. However, the glacial origin of the Moelv Formation has not been rigorously substantiated.

Other likely post-Marinoan glacial deposits include the Mortensnes Formation in northern Norway (Halverson *et al.*, 2005) and the Hankalchough Formation in northwest China (Xiao *et al.*, 2004). If the seafloor precipitates and negative $\delta^{13}C$ anomaly preserved in the upper Scout Mountain Member of the Pocatello Formation in Idaho (Smith *et al.*, 1994, Lorentz *et al.*, 2004) are equivalent to the Marinoan cap carbonate, then incised valleys and associated diamictites in the overlying Caddy Canyon

Formation of northern Utah and southern Idaho (Levy et al., 1994; Christie Blick, 1997) may record the Gaskiers glaciation in the southern Cordillera. Possible Gaskiers-aged glacial deposits may also occur in eastern Laurentia, within the Fauquier Formation of northern Virginia (Kaufman and Hebert, 2003). Grey and Corkeron (1998) have suggested that that glacigenic Egan Formation in the southern Kimberly basin, northwestern Australia, also represents a post-Marinoan glaciation; however, correlation between the Kimberly region and central and South Australia remains unresolved (Knoll et al., 2006). Even a liberal tally of Gaskiers-aged glacial deposits leads to the conclusion that they are far less widespread than their predecessors (Fig. 7).

A thin and patchily preserved cap carbonate with negative $\delta^{13}C$ composition occurs above the Gaskiers glacials in Newfoundland (Myrow and Kaufman, 1999). Thin, isotopically negative carbonates also occur above the Hankalchough Formation in northwest China, but the $\delta^{13}C$ composition is highly variable (Xiao et al., 2004). The Moelv (Bingen et al., 2005) and Mortensnes (Halverson et al., 2005) formations in Norway lack a cap carbonate altogether. Even if the Cotton Breccia on King Island (Tasmania), which is overlain by a cap carbonate closely resembling those above Marinoan glacials (Calver and Walter, 2000), is proven to be Gaskiers age (Calver et al., 2004), it is clear that the Gaskiers glaciation differs markedly from the Marinoan glaciation in that it is not ubiquitously overlain by a transgressive cap dolostone. This conclusion is not necessarily surprising given evidence that the glaciation (in Newfoundland) lasted less than 1 m.y. This timing constraint on the Gaskiers glaciation is important, for by definition, snowball events last millions of years (Hoffman et al., 1998). Therefore, if it could be shown that the Gaskiers glaciation extended to low latitudes, then it would be apparent that snowball glaciation need not be invoked to account for all low-latitude glaciation. Reliable paleomagnetic constraints on the paleolatitude of Gaskiers glacial deposits are rare (Evans, 2000), but available data indicate middle to moderately high paleolatitudes (Bingen et al., 2005). Therefore, the snowball Earth hypothesis cannot be ruled out based on data from the Gaskiers glaciation. Since the high obliquity hypothesis for low-latitude glaciation (Williams, 1975) predicts no high-latitude glacial deposits, it seems that the Gaskiers glaciation more closely resembles the familiar Phanerozoic glaciations (Bingen et al., 2005). Indeed, considering the limited age constraints and variable $\delta^{13}C$ compositions spanning the presumed Gaskiers-aged glacials, it is entirely possible that the glaciation at ca. 580 Ma was diachronous or periodic.

The Acraman impact in South Australia, to the west of the Adelaide Rift Complex, is estimated to be 580 Ma (Grey et al., 2003), and was likely sufficiently catastrophic to have strongly perturbed the global environment

(Williams and Wallace, 2003). However, the age of the impact is poorly constrained and its precise temporal relationship to the Gaskiers glaciation is not established. The associated ejecta layer is found within the dominantly siliciclastic Bunyeroo Formation in the Adelaide Rift Complex (Gostin et al., 1986) and equivalent strata in the Officer Basin (Wallace et al., 1989), meaning that the event cannot be tied directly to a detailed carbonate $\delta^{13}C$ profile. However, $\delta^{13}C$ data from organic matter and thin carbonate beds from boreholes suggest that the impact occurred prior to a decline of ~4‰ and at the same level as a dramatic shift in acritarch assemblages in Australia, from simple spheroid (leiosphere), low diversity acritarchs (ELP) beneath the impact layer to large, complex (acanthomorphic), rapidly radiating actritarchs (ECAP) above (Calver and Lindsay, 1998, Grey et al., 2003) (Fig. 6).

The temporal link between the impact and the fossil turnover in Australia is unequivocal, but the occurrence of a diverse acanthomorphic acritarch assemblage in the Doushantuo Formation, perhaps pre-dating the Gaskiers glaciation and the Acraman impact (Xiao, 2004a; Xiao et al., 2004) suggests that either the timing of the impact and associated turnover in Australian acritarchs is inaccurate or that the absense of the acanthomorphic acritarchs below the impact ejecta horizon is a taphonomic or environmental artifact (Xiao, 2004a). Resolving this inconsistency is important to establishing whether the impressive Ediacaran diversification of the acanthomorphic acritarchs was the response to a mass extinction event (Grey et al., 2003), one of the Neoproterozoic glaciations (Marinoan or Gaskiers; Xiao et al., 2004), or the evolution of predatory Eumetazoa (Peterson and Butterfield, 2005).

3.3.4 The Terminal Proterozoic

The Ediacaran fauna are the protagonists in the final chapter of Neoproterozoic Earth history. The stratigraphic context of the Ediacaran biota and terminal Proterozoic biology in general are reviewed in more detail by Xiao (2004a) and Knoll et al. (2006). Irrespective of the precise timeframe for this anomaly, it is clear that late Neoproterozoic ocean was highly ^{13}C-depleted just prior to or leading into the first appearance of the Ediacara biota (Condon et al., 2005). Assuming that these very negative values were generated by ongoing oxidation of a large reservoir of ^{13}C-depleted reduced carbon, the recovery should have coincided with stabilizaiton or growth in atmospheric O_2 concentrations. Whether the $\delta^{13}C$ anomaly is directly linked to the appearance and diversification of complex bilaterians (Condon et al., 2005) is an intriguing question that warrants continued investigation of the timing of early animal evolution and its connection to late Proterozoic environmental change.

The final 13 million years of the Neoproterozoic are much better constrained geochronologically and biostratigraphically than any other period in the Precambrian. Detailed, multidisciplinary studies in the Nama Group in southern Namibia (Grotzinger et al., 1995, Saylor et al., 1998), Oman (Amthor et al., 2003), and Siberia (Bowring et al. 1993, Knoll et al. 1995) have permitted geochronological calibration of integrated chemostratigraphic and biostratigraphic records spanning the Precambrian–Cambrian boundary. For example, recent work in Oman has firmly pinned the Precambrian–Cambrian boundary, the nadir of the associated negative $\delta^{13}C$ anomaly, and the extinction of the weakly calcified metazoans (*Cloudina* and *Namacalathus*) to between 542.6 ± 0.3 and 542.0 ± 0.6 Ma (Amthor et al., 2003) (Fig. 6). Such a refined geological record should surely be the goal for the entire Ediacaran period.

4. CONCLUSIONS

The pieces of the Neoproterozoic puzzle are slowly but surely beginning to fall into place. A working outline of the chronology of the 458 million years of this era is now in place and tied to a composite carbon isotope record (Fig. 1). This record is far from complete, but is sufficient in its present state to highlight the most interesting intervals and features of the geological record. Terminal Proterozoic paleobiology and glacial intervals have deservingly garnered the lion's share of interest in the Neoproterozoic, but it is now apparent that the geological record in the first half of the era preserves equally fascinating stories about the evolution of the surface of the Earth. Whether or not the inertial interchange true polar wander (TPW) hypothesis to explain the pair of sharp isotopic shifts straddling the so-called Bitter Springs stage stands the test of new data, it is at least clear that these shifts were related to global-scale processes, but most likely not to glaciation. The possibility that a large negative $\delta^{13}C$ anomaly presaged the Sturtian glaciation, in addition to the Marinoan glaciation, the evidence for another somewhat older anomaly (ca. 750 Ma?), and the possible disconnection between the Wonoka/Shuram anomaly and the Gaskiers glaciation (Condon et al., 2005; Le Guerroué et al., 2006) further attenuates the implicit marriage between glaciations and major drops in the $\delta^{13}C$ composition of the oceans. It is tempting to envision global cooling as but one side effect of more fundamental perturbations to the earth's environment, with snowball glaciation resulting only in collaboration with other climate-influencing factors, such as paleogeography (Kirschvink, 1992; Schrag et al.,

2002), the weatherability of the continents (Donnadieu et al., 2004), or passage of the solar system through dense space clouds (Pavlov et al., 2005).

Much work remains to be done to fill in the Neoproterozoic chronology, as manifested by the inability to present a definitive composite $\delta^{13}C$ record. The more prominent gaps in our understanding of the chronology and data coverage include (but are not limited to) 1) the timing of the initial rise in marine $\delta^{13}C$ to the average of ~5‰ that characterizes the Tonian and Cryogenian periods, 2) the timing of the Sturtian glaciation—or perhaps, more accurately, the Pre-Marinoan glaciations—and documentation of the $\delta^{13}C$ record leading up to the first Cyrogenian glaciation; 3) whether or not a glaciation occurred between the Sturtian and Marinoan glaciations; and 4) a carbon isotope record spanning unequivocal evidence of the Gaskiers glaciation. Of course, precise radiometric ages are always in demand, and the older part of the record is especially impoverished of useful geochronological data. These holes in the record are sure to be patched as research continues into making sense of the extreme environment fluctuations and non-actualistic conditions that shaped the surface of the Neoproterozoic Earth.

ACKNOWLEDGEMENTS

I thank Shuhai Xiao and Alan J. Kaufman for inviting me to contribute this chapter. Adam Maloof and Matt Hurtgen provided $\delta^{13}C$ data from the Coates Lake Group, and Francis McDonald helped measure and sample the Little Dal Group. All carbon isotope data were analyzed in Dan Schrag's Laboratory for Geochemical Oceanography at Harvard University. New data presented in this paper were acquired via supported by the NASA Astrobiology Program and by NSF (Arctic Science and Earth History programs) grants to Paul Hoffman. Comments from Yves Goddéris, Anne Nédeléc, Shuhai Xiao and critical reviews by Malcolm Walter, Frank Corsetti, and Nate Lorentz significantly improved this manuscript.

REFERENCES

Aitken, J. D., 1981, Stratigraphy and sedimentology of the Upper Proterozoic Little Dal Group, Mackenzie Mountains, Northwest Territories, in: *Proterozoic Basins of Canada* (F. H. A. Campbell, ed.), *Geological Survey of Canada, Paper, 81–10*, pp: 47–71.

Aitken, J. D., 1991a, The Ice Brook Formation and post-Rapitan, Late Proterozoic glaciation, Mackenzie Mountains, Northwest Territories, *Geol. Surv. Can. Bull.* **404**: 1–43.

Aitken, J. D., 1991b, Two Late Proterozoic glaciations, Mackenzie Mountains, northwestern Canada, *Geology* **19**: 445–448.

Allen, P. A., Bowring, S. A., Leather, J., Brasier, M. D., Cozzi, A., Grotzinger, J. P., McCarron, G., and Amther, J. E., 2002, Chronology of Neoproterozoic glaciations: New nsigns from Oman, The16*th* Int. Sedimentol. Cong. Abstr. Vol., Johannesburg, pp. 7–8.

Allen, P. A.., and Leather, J., 2006, Post-marinoan marine siliciclastic sedimentation: The Masirah Bay Formation, Neoproterozoic Huqf Supergroup of Oman, *Precambrian Res.* **144**: 167–198.

Allen, P. A., Bowring, S., Leather, J. J., Brasier, M., Cozzi, A., Grotzinger, J. P., McCarron, G., and Amthor, J. E., 2002, Chronology of Neoproterozoic glacials: new insights from Oman, *The 16th Int. Sedimentol. Cong. Abstr. Vol.*, Johannesburg, pp 7–8.

Allen, P. A., Leather, J., and Brasier, M. D., 2005, The Neoproterozoic Fiq glaciation and its aftermath, Huqf Supergroup of Oman, *Bas. Res.* **160**: 507–534.

Allen, P. A., and Hoffman, P. F., 2005, Extreme winds and waves in the aftermath of a Neoproterozoic glaciation, *Nature* **433**: 123–127.

Amthor, J. E., Grotzinger, J. P., Schröder, S., Bowring, S. A., Ramezani, J., Martin, M. W., and Matter, A., 2003, Extinction of *Cloudina* and *Namacalathus* at the Precambrian–Cambrian boundary in Oman, *Geology* **31**: 431–434.

Asmerom, Y., Jacobsen, S., Knoll, A. H., Butterfield, N. J., and Swett, K., 1991, Strontium isotopic variations of Neoproterozoic seawater: Implications for crustal evolution, *Geochim. Cosmochim. Acta* **55**: 2883–2894.

Barfod, G. H., Albarede, F., Knoll, A. H., Xiao, S., Télouk, P., Frei, R., and Baker, J., 2002, New Lu–Hf and Pb–Pb age constraints on the earliest animal fossils, *Earth Planet. Sci. Lett.* **201**: 203–212.

Bartley, J. K., and Kah, L. C., 2004, Marine carbon reservoir, C_{org}–C_{carb} coupling, and the evolution of the Proterozoic carbon cycle, *Geology* **32**: 129–133.

Bartley, J. K., Semikhatov, M. A., Kaufman, A. J., Knoll, A. H., Pope, M. C., and Jacobsen, S. B., 2001, Global events across the Mesoproterozoic–Neoproterozoic boundary: C and Sr isotopic evidence from Siberia, *Precambrian Res.* **111**: 165–202.

Bingen, B., Griffin, W. L., Torsvik, T. H., and Saeed, A., 2005, Timing of late Neoproterozoic glaciation on Baltica constrained by detrital geochronology in the Hedmark Group, southeast Norway, *Terra Nova* **17**: 593–596.

Bodiselitsch, B., Koeberl, C., Master, S., and Reimold, W. U., 2005, Estimating duration and intensity of Neoproterozoic snowball glaciations from Ir anomalies, *Science* **308**: 239–242.

Bowring, S., Myrow, P., Landing, E., Ramezani, J., and Grotzinger, J., 2003, Geochronological constraints on terminal Proterozoic events and the rise of the Metazoans, *Geophys. Res. Abstr.* **50**: 13219.

Bowring, S. A., Grotzinger, J. P., Isachsen, C. E., Knoll, A. H., Pelechaty, S., and Kolosov, P., 1993, Calibrating rates of Early Cambrian evolution, *Science* **261**: 1293–1298.

Brasier, M., McCarron, G., Tucker, R., Leather, J., Allen, P., and Shields, G., 2000, New U–Pb zircon dates for the Neoproterozoic Ghubrah glaciation and for the top of the Huqf Supergroup, Oman, *Geology* **28**: 175–178.

Brasier, M. D., and Lindsay, J. F., 1995, A billion years of environmental stability and the emergence of eukaryotes: New data from northern Australia, *Geology* **26**: 555–558.

Buick, R., Des Marais, D. J., and Knoll, A. H., 1995, Stable isotopic composition of carbonates from the Mesoproterozoic Bangemall Group, northwestern Australia, *Chem. Geol.* **123**: 153–171.

Burns, S. J., and Matter, A., 1993, Carbon isotopic record of the latest Proterozoic from Oman, *Ecl. Geol. Helv.* **86**: 595–607.

Caldeira, K., and Kasting, J. F., 1992, Susceptibility of the early Earth to irreversible glaciation caused by carbon dioxide clouds, *Nature* **359**: 226–228.

Calver, C. R., 2000, Isotope stratigraphy of the Ediacaran (Neoproterozoic III) of the Adelaide rift complex, Australia, and the overprint of water column stratification, *Precambrian Res.* **100**: 121–150.

Calver, C. R., Black, L. P., Everard, J. L., and Seymour, D. B., 2004, U–Pb zircon age constraints on late Neoproterozoic glaciation in Tasmania, *Geology* **32**: 893–896.

Calver, C. R., and Lindsay, J. F., 1998, Ediacaran sequence and isotope stratigraphy of the Officer Basin, South Australia, *Australia J. Earth Sci.* **45**: 513–532.

Calver, C. R., and Walter, M. R., 2000, The late Neoproterozoic Grassy Group of King Island, Tasmania: correlation and palaeogeographic significance, *Precambrian Res.* **100**: 299–312.

Chen, D., Dong, W., Zhu, B., and Chen, X. P., 2004, Pb-Pb ages of Neoproterozoic Doushantuo phosphorites in South China: Constraints on early metazoan evolution and glaciation events, *Precambrian Res.* **132**: 123-132.

Chen, J. Y., Bottjer, D. J., Oliveri, P., Dornbos, S. Q., Gao, F., Ruffins, S., Chi, H., Li, C. W., and Davidson, E. H., 2004, Small bilaterian fossils from 40 to 55 million years before the Cambrian, *Science* **305**: 218–222.

Christie-Blick, N., 1997, Neoproterozoic sedimentation and tectonics in west-central Utah, *Proterozoic to Recent Stratigraphy, Tectonics and Volcanology, Utah, Nevada, Southern Idaho and Central Mexico: Brigham Young University Geology Studies* **42**, *Part I*, pp. 1–30.

Christie-Blick, N., Sohl, L. E., and Kennedy, M. J., 1999, Considering a Neoproterozoic snowball Earth, *Science*, **284**: online.

Clapham, M. E., and Corsetti, F. A., 2005, Deep valley incision in the terminal Neoproterozoic (Ediacaran) Johnnie Formation, eastern California, USA: Tectonically or glacially driven? *Precambrian Res.* **141**: 154–164.

Condon, D., Zhu, M., Bowring, S., Jin, Y., Wang, W., and Yang, A., 2005, From the Marinoan glaciation to the oldest bilaterians: U–Pb ages from the Doushantou Formation, China, *Science* **308**: 95–98.

Condon, D. J., and Prave, A. R., 2000, Two from Donegal: Neoproterozoic glacial episodes on the northeast margin of Laurentia, *Geology* **28**: 951–954.

Corsetti, F. A., and Kaufman, A. J., 2003, Statigraphic investigations of carbon isotope anomalies and Neoproterozoic ice ages in Death Valley, California, *Geol. Soc. Amer. Bull.* **115**: 916–932.

Cozzi, A., Allen, P. A., and Grotzinger, J. P., 2004, Understanding carbonate ramp dynamics using $\delta^{13}C$ profiles: examples from the Neoproterozoic Buah Formation of Oman, *Terra Nova* **16**: 62–67.

Crittenden, M. D. Jr., Schaeffer, F. E., Trimble, D. E., and Woodward, L. A., 1971, Evidence for two pulses of glaciation during the Late Proterozoic in northern Utah and southern Idaho, *Geol. Soc. Amer. Bull.* **82**: 581–602.

Dempster, T. J., Rogers, G., Tanner, P. W. G., Bluck, B. J., Muir, R. J., Redwood, S. D., Ireland, T. R., and Patterson, B. A., 2002, Timing of deposition, orogenesis and glaciation within the Dalradian rocks of Scotland: constraints from U–Pb zircon ages, *J. Geol. Soc. London* **159**: 83–84.

Donnadieu, Y., Fluteau, F., Ramstein, G., Ritz, C., and Besse, J., 2003, Is there a conflict between Neoproterozoic glacial deposits and the snowball Earth interpretation: an improved understanding with numerical modeling, *Earth Planet. Sci. Lett.* **208**: 101–112.

Donnadieu, Y., Goddéris, Y., Ramstein, G., Nédelec, A., and Meert, J., 2004, A 'snowball earth' climate triggered by continental break-up through changes in runoff, *Nature* **428**: 303–306.

Dunn, P. R., Thomson, B. P., and Rankama, K., 1971, Late Pre-Cambrian glaciation in Australia as a stratigraphic boundary, *Nature* **231**: 498–502.

Elles, G. L., 1934, The Loch na Cille Boulder Bed and its place in the Highland Succession, *Quart. J. Geol. Soc.* **91**: 111–147.

Evans, D. A. D., 2000, Stratigraphic, geochronological, and paleomagnetic constraints upon the Neoproterozoic climatic paradoxes, *Amer. J. Sci.* **300**: 347–443.

Fairchild, I. J., Spiro, B., Herrington, P. M., and Song, T., 2000, Controls on Sr and C isotope compositions of Neoproterozoic Sr-rich limestones of East Greenland and North China, in: *Carbonate Sedimentation and Diagenesis in an Evolving Precambrian World* (J. P. Grotzinger and N. P. James, eds.), *SEPM Special Publications 67*, Tulsa, pp. 297–313.

Fairchild, I. J., and Hambrey, M. B., 1995, Vendian basin evolution in East Greenland and NE Svalbard, *Precambrian Res.* **73**: 217–333.

Fanning, C. M., and Link, P. K., 2004, U–Pb SHRIMP ages of Neoproterozoic (Sturtian) glaciogenic Pocatello Formation, southeastern Idaho, *Geology* **32**: 881–884.

Fanning, C. M., Ludwig, K. R., Forbes, B. G., and Preiss, W. V., 1986, Single and multiple grain U–Pb zircon analyses for the Early Adelaidean Rook Tuff, Willouran Ranges, South Australia, *Abst. Geol. Soc. Australia* **15**: 71–72.

Farmer, J., Vidal, G., Moczydlowskia, M., Strauss, H., Ahlberg, P., and Siedlecka, A., 1992, Ediacaran fossils from the Innerelv Member (late Proterozoic) of the Tanafjorden area, northeastern Finnmark, *Geol. Mag.* **129**: 181–195.

Frimmel, H. W., Klötzi, U. S., and Siegfried, P. R., 1996, New Pb–Pb single zircon age constraints on the timing of Neoproterozoic glaciation and continental break-up in Namibia, *J. Geol.* **104**: 459–469.

Gehling, J. G., Narbonne, G. M., and Anderson, M. M., 2000, The first named Ediacaran body fossil, *Aspidellaterranovica, Palaeontology* **43**: 427–456.

Gorin, G. E., Racz, L. G., and Walter, M. R., 1982, Late Precambrian–Cambrian sediments of Huqf Group, Sultanate of Oman, *Amer. Assoc. Petrol. Geol. Bull.* **66**: 2609–2627.

Goscombe, B., Hand, M., Gray, D., and Mawby, J., 2003, The metamorphic architecture of a transpressive orogen: the Kaoko Belt, Namibia, *J. Pet.* **44**: 679–711.

Gostin, V. A., Haines, P. W., Jenkins, R. J. F., and Compston, W., 1986, Impact ejecta horizon within late Precambrian shales, Adelaide geosyncline, South Australia, *Science* **233**: 542–544.

Grey, K., and Corkeron, M., 1998, Late Neoproterozoic stromatolites in glacigenic successions of the Kimberly region, Western Australia: evidence for a younger Marinoan glaciation, *Precambrian Res.* **92**: 65–87.

Grey, K., Walter, M. R., and Calver, C. R., 2003, Neoproterozoic biotic diversification: Snowball Earth or aftermath of the Acraman impact, *Geology* **31**: 459–462.

Grotzinger, J. P., Bowring, S. A., Saylor, B. Z., and Kaufman, A. J., 1995, Biostratigraphic and geochronologic constraints on early animal evolution, *Science* **270**: 598–604.

Halverson, G. P., Dudas, F. Ö., Maloof, A. C., and Bowring, S. A., in press. Evolution of the $^{87}Sr/^{86}Sr$ composition of Neoproterozoic seawater. *Palaeogeogr. Palaeoclimatol. Palaeoecol.*

Halverson, G. P., Hoffman, P. F., and Schrag, D. P., 2002, A major perturbation of the carbon cycle before the Ghaub glaciation (Neoproterozoic) in Namibia: prelude to snowball Earth? *Geochem., Geophys., Geosys.* **3**: 10.1029/2001GC000244.

Halverson, G. P., Hoffman, P. F., Schrag, D. P., Maloof, A. C., and Rice, A. H., 2005, Towards a Neoproterozoic composite carbon isotope record, *Geol. Soc. Amer. Bull.* **117**: 1181–1207.

Halverson, G. P., Maloof, A. C., and Hoffman, P. F., 2004, The Marinoan glaciation (Neoproterozoic) in northeast Svalbard, *Bas. Res.* **16**: 297–324.

Halverson, G. P., Maloof, A. C., Schrag, D. P., Dudas, F. Ö., and Hurtgen, M., in press, Stratigraphy and geochemistry of a ca 800 Ma negative carbon isotope stage in northeastern Svalbard, *Chem. Geol.*

Harlan, S. S., Heaman, L., LeCheminant, A. N., and Premo, W. R., 2003, Gunbarrel mafic magmatic event: a key 780 Ma time marker for Rodinia plate reconstructions, *Geology* **31**: 1053–1056.

Hill, A. C., Arouri, K., Gorjan, P., and Walter, M. R., 2000, Geochemistry of marine and nonmarine environments of a Neoproterozoic cratonic carbonate/evaporite: the Bitter Springs Formation, central Australia, in: *Carbonate Sedimentation and Diagenesis in an Evolving Precambrian World* (J. P. Grotzinger and N. P. James, eds.), SEPM Special Publications 67, Tulsa, pp. 327–344.

Hoffman, P. F., 1991, Did the breakout of Laurentia turn Gondwana inside out? *Science* **252**: 1409–1412.

Hoffman, P. F., 2000, Comment: Vreeland Diamictites—Neoproterozoic glaciogenic slope deposits, Rocky Mountains, northeast British Columbia, *Bull. Can. Pet. Geol.* **48**: 360–363.

Hoffman, P. F., and Halverson, G. P., in press, Otavi Group of the northern platform and the northern margin zone, in: *Handbook on the Geology of Namibia* (R. McG. Miller, ed.), Geological Survey of Namibia, Windhoek.

Hoffman, P. F., Hawkins, D. P., Isachsen, C. E., and Bowring, S. A., 1996, Precise U–Pb zircon ages for early Damaran magmatism in the Summas Mountains and Welwitschia Inlier, northern Damara belt, Namibia, *Com. Geol. Surv. Namibia* **11**: 47–52.

Hoffman, P. F., Kaufman, A. J., and Halverson, G. P., 1998a, Comings and goings of global glaciations on a Neoproterozoic tropical platform in Namibia, *GSA Today* **8**: 1–9.

Hoffman, P. F., Kaufman, A. J., Halverson, G. P., and Schrag, D. P., 1998b, A Neoproterozoic snowball Earth, *Science* **281**: 1342–1346.

Hoffman, P. F., and Schrag, D. P., 2002, The snowball Earth hypothesis: testing the limits of global change, *Terra Nova* **14**: 129–155.

Hoffmann, K., 1989, New aspects of lithostratigraphic subdivision and correlation of late Proterozoic to early Cambrian rock of the southern Damara Belt and their correlation with the central and northern Damara Belt and Gariep Belt, *Com. Geol. Surv. Namibia* **5**: 59–67.

Hoffmann, K. H., Condon, D. J., Bowring, S. A., and Crowley, J. L., 2004, A U–Pb zircon date from the Neoproterozoic Ghaub Formation, Namibia: Constraints on Marinoan glaciation, *Geology* **32**: 817–820.

Hoffmann, K. H., and Prave, A. R., 1996, A preliminary note on a revised subdivision and regional correlation of the Otavi Group based on glaciogenic diamictites and associated cap dolostones, *Com. Geol. Surv. Namibia* **11**: 81–86.

Hofmann, H. J., Narbonne, G. M., and Aitken, J. D., 1990, Ediacaran remains from intertillite beds in northwestern Canada, *Geology* **29**: 1091–1094.

Hurtgen, M. T., Halverson, G. P., Arthur, M. A., and Hoffman, P. F., 2006, Sulfur cycing in the aftermath of a 635-ma snowball glaciation: Evidence for a syn-glacial sulfidic deep ocean, *Earth Planet. Sci. Lett.* **245**: 551–570.

Ikeda, T., and Tajika, E., 1999, A sudy of the energy balance climate model with CO-dependent outgoing radiation: implications for the glaciation during the Cenozoic, *Geophys. Res. Lett.* **26**: 349–352.

James, N. P., Narbonne, G. M., Dalrymple, R. W., and Kyser, T. K., 2005, Glendonites in Neoproterozoic low-latitude, interglacial, sedimentary rocks, northwest Canada: Insights ino the Cryogenian ocean and Precambrian cold-water carbonates, *Geology* **33**: 9–12.

Jefferson, C. W., and Parrish, R. R., 1989, Late Proterozoic stratigraphy, U/Pb zircon ages and rift tectonics, Mackenzie Mountains, northwestern Canada, *Can. J. Earth Sci.* **26**: 1784–1801.

Jefferson, C. W., and Ruelle, J. C. L., 1986, The Late Proterozoic Redstone Copper Belt, Mackenzie Mountains, Northwest Territories, in: *Mineral Deposits of Northern Cordillera* (J. A. Morin, ed.), *Special Volume 37,* The Canadian Institute of Mining and Metallurgy, pp. 154–168.

Jiang, G., Sohl, L. E., and Christie-Blick, N., 2003, Neoproteroozic stratigraphic comparison of the Lesser Himalaya (India) and Yangtze block (south China): Paleogeographic implications, *Geology* **31**: 917–920.

Kaufman, A. J., 2005, The calibration of Ediacaran time, *Science* **308**: 59–60.

Kaufman, A. J., and Hebert, C. L., 2003, Stratigraphic and radiometric constraints on rift-related volcanism, terminal Neoproterozoic glaciation, and animal evolution, *Geol. Soc. Amer. Abstr. Progr.* **356**: 516.

Kaufman, A. J., and Knoll, A. H., 1995, Neoproterozoic variations in the C-isotopic composition of seawater, *Precambrian Res.* **73**: 27–49.

Kaufman, A. J., Knoll, A. H., and Narbonne, G. M., 1997, Isotopes, ice ages, and terminal Proterozoic Earth history, *Proc. Nat. Acad. Sci. USA* **95**: 6600–6605.

Kendall, B. S., and Creaser, R. A., 2004, Re–Os depositional age of Neoproterozoic Aralka Formation (Amadeus basin, Australia) revisited, *Geol. Soc. Amer. Abstr. Progr.* **36**: 459.

Kendall, B. S., Creaser, R. A. Ross, G., and Selby, D., 2004. Constraints on the timing of Marinoan "snowball Earth" glaciation by ^{187}Re–^{187}Os dating of a Neoproterozoic, post-glacial black shale in western Canada, *Earth Plan. Sci. Lett.* **222**: 729–740.

Kendall, B. S., Creaser, R. A., and Selby, D., 2005, Re–Os depositional age of Neoproterozoic post-glacial black shales in Australia: evidence for diachronous Neoproterozoic glaciations, *Geol. Soc. Amer. Abstr. Progr.* **37**: 42.

Kennedy, M. J., Runnegar, B., Prave, A. R., Hoffman, K. H., and Arthur, M., 1998, Two or four Neoproterozoic glaciations? *Geology* **26**: 1059–1063.

Key, R. M., Liyungu, A. K., Njamu, F. M., Somwe, V., Banda, J., Mosley, P. N., and Armstrong, R. A., 2001, The western arm of the Lufilian Arc in NW Zambia and its potential for copper mineralization, *J. Afr. Earth Sci.* **33**: 503–528.

Kilner, B., MacNiocaill, C., and Brasier, M., 2005, Low-latitude glaciation in the Neoproterozoic of Oman, *Geology* **33**: 413–416.

Kirschvink, J. L., 1992, Late Proterozoic low-latitude glaciation: The snowball Earth, in: *The Proterozoic Biosphere* (J. W. Schopf and C. Klein, eds.), Cambridge University Press, Cambridge, pp 51–52.

Knoll, A. H., 1994, Proterozoic and Early Cambrian protists: Evidence for accelerating evolutionary tempo, *Proc. Nat. Acad. Sci. USA* **91**: 6743-6750.

Knoll, A. H., 2000, Learning to tell Neoproterozoic time, *Precambrian Res.*, **100**: 3–20.

Knoll, A. H., Grotzinger, J. P., Kaufman, A. J., and Kolosov, P., 1995, Integrated approaches to terminal Proterozoic stratigraphy: An example froom the Olenek Uplift, northeastern Siberia, *Precambrian Res.* **73**: 251–270.

Knoll, A. H., Hayes, J. M., Kaufman, A. J., Swett, K., and Lambert, I. B., 1986, Secular variation in carbon isotope ratios from Upper Proterozoic successions of Svalbard and east Greenland, *Nature* **321**: 832–837.

Knoll, A. H., Swett, K., 1990, Carbonate deposition during the late Proterozoic era: an example from Spitsbergen. *Am. J. Sci.* ***290A***: *104–132*.

Knoll, A. H., and Walter, M. R., 1992, Latest Proterozoic stratigraphy and Earth history, *Nature* **356**: 673–677.

Knoll, A. H., Walter, M. R., and Christie-Blick, N., 2004, A new period for the geological time scale, *Nature* **305**: 621–622.

Knoll, A. H., Walter, M. R., Narbonne, G. M., and Christie-Blick, N., 2006, The Ediacaran Period: a new addition to the geologic time scale, *Lethaia* **39**: 13-30.

Kuznetsov, A. B., Gorkhov, I. M., Semikhatov, M. A., Melnikov, N. N., and Kozlov, V. I., 1997, Strontium isotopic composition in the limestone of the Inzer Formation, Upper Riphean type sections, Southern Urals, *Transact. Russ. Acad. Sci. (Earth Sci. Sect.)* **353**: 319–324.

Leather, J., Allen, P. A., Brasier, M. D., and Cozzi, A., 2002, Neoproterozoic snowball Earth under scrutiny: Evidence from the Fiq glaciation of Oman, *Geology* **30**: 891–894.

Le Guerroué, E., Allen, P. A., Cozzi, A., Etienne, J. L., and Fanning, C. M., 2006, 50 Myr recovery from the largest negative $\delta^{13}C$ excursion in the Ediacaran ocean, *Terra Nova* **18**: 147–153.

Le Guerroué, E., Allen, P. A., Cozzi, A., 2005, Two distinct glacial successions in the Neoproterozoic of Oman. *GeoArabia* **10**: 17–34.

Levy, M., Christie-Blick, N., and Link, P. K., 1994, Neoproterozoic incised valleys of eastsern Great Basin, Utah and Idaho: fluvial response to changes in depositional base level, in: *Incised-valley Systems: Origin and Sedimentary Sequences* (R. W. Dalrymple, R. Boyd and B. A. Zaitlin, eds.), *SEPM Special Publication No. 51*, Tulsa pp. 369–382,

Li, Z. X., and Powell, C. McA., 2001, An outline of the palaeogeographic evolution of the Australasian region since the beginning of the Neoproterozoic, *Earth Sci. Rev.* **53**: 237–277.

Lorentz, N. J., Coresetti, F. A., and Link, P. K., 2004, Seafloor precipitates and C-isotope stratigraphy from the Neoproterozoic Scout Mountain Member of the Pocatello Formation, southeast Idaho: implications for Neoproterozoic Earth system behavior, *Precambrian Res.* **130**: 57–70.

Lund, K. L., Aleinikoff, J. N., Evans, K. V., and Fanning, C. M., 2003, SHRIMP U–Pb geochronology of Neoproterozoic Windermere Supergroup, central Idaho: Implications for rifting of western Laurentia and synchroneity of Sturtian glacial deposits, *Geol. Soc. Amer. Bull.* **115**: 349–372.

Maloof, A. C., Halverson, G. P., Kirschvink, J. L., Weiss, B., Schrag, D. P., and Hoffman, P. F., 2006, Combined paleomagnetic, isotopic and stratigraphic evidence for true polar wander from the Neoproterozoic Akademikerbreen Group, Svalbard, *Geol. Soc. Amer. Bull.* **118**:1099–1124.

Martin, M. W., Grazhdankin, D. V., Bowring, S. A., Evans, D. A. D., Fendonkin, M. A., and Kirschvink, J. L., 2000, Age of Neoproterozoic bilaterian boday and trace fossils, White Sea, Russia: implications for metazoan evolution, *Science* **288**: 841–845.

Master, S., Rainaud, C., Armstrong, R. A., Phillips, D., and Robb, L. J., 2005, Provenance ages of the Neoproterozoic Katanga Supergroup (Central African Copperbelt), with implications for basin evolution, *J. Afr. Earth Sci.* **42**: 41–60.

McKirdy, D. M., Burgess, J. M., Lemon, N. M, Yu, X., Cooper, A. M., Gostin, V. A., Jenkins, R. J. F., and Both, R. A., 2001, A chemostratigraphic overview of the late Cryogenian interglacial sequence in the Adelaide Fold-Thrust Belt, South Australia, *Precambrian Res.* **106**: 149–186.

McMechan, M. E., 2000, Neoproterozoic glaciogenic slope deposits, Rocky Mountains, northeast British Columbia, *Bull. Can. Pet. Geol.* **48**: 246–261.

Melezhik, V. A., Fallick, A. E., and Pokrovsky, B. G., 2005, Enigmatic nature of thick sedimentary carbonates depleted in ^{13}C beyond the canonical mantle value: The challenges to our understanding of the terrestrial carbon cycle, *Precambrian Res.* **137**: 131–165.

Mound, J. E., Mitrovica, J. X., Evans, D. A. D., and Kirschvink, J. L., 1999, A sea-level test for inertial interchange true polar wander events, *Geophy. J. Int.* **136**: F5–F10.

Myrow, P. M., and Kaufman, A. J., 1999, A newly discovered cap carbonate above Varanger-age glacial deposits in Newfoundland, Canada, *J. Sed. Res.* **69**: 784–793.

Narbonnne, G. M., and Gehling, J. G., 2003, Life after snowball: The oldest complex Ediacaran fossils, *Geology* **31**: 27–30.

Pavlov, A. A., Hurtgen, M. T., Kasting, J. F., and Arthur, M. A., 2003, Methane-rich Proterozoic atmosphere? *Geology* **31**: 87–90.

Pavlov, A. A., Toon, O. B., Pavlov, A. K., Bally, J., and Pollard, D., 2005, Passing through a giant molecular cloud: Snowball glaciations produced by interstellar dust, *Geophys. Res. Lett.* **32**: L03705, 10.1029/2004GL021890.

Pell, S. D., McKirdy, D. M., Jansyn, J., and Jenkins, R. J. F., 1993, Ediacaran carbon isotope stratigraphy of South Australia—An initial study, *Trans. Royal Soc. S. Austral.* **117**: 153–161.

Peterson, K. J, and Butterfield, N. J., 2005, Origin of the Eumetazoa: Testing ecological predictions of molecular clocks against the Proterozoic fossil record, *Proc. Nat. Acad. Sci. USA* **102**: 9547–9552.

Preiss, W. V., 2000, The Adelaide Geosyncline of South Australia and its significance in Neoproterozoic continental reconstruction, *Precambrian Res.* **100**: 21–63.

Rainbird, R. H., Jefferson, C. W., and Young, G. M., 1996, The early Neoproterozoic sedimentary Succession B of northwestern Laurentia: correlations and paleogeographic significance, *Geol. Soc. Amer. Bull.* **108**: 454–470.

Rothman, D. H., Hayes, J. M., and Summons, R. E., 2003, Dynamics of the Neoproterozoic carbon cycle, *Proc. Nat. Acad. Sci. USA* **100**: 124–129.

Saylor, B. Z., Kaufman, A. J., Grotzinger, J. P., and Urban, F., 1998, A composite reference section for terminal Proterozoic strata of southern Namibia, *J. Sed. Res.* **68**: 1223–1235.

Schaefer, B. F., and Burgess, J. M., 2003, Re–Os isotopic age constraints on deposition in the Neoproterozoic Amadeus Basin: implications for the 'Snowball Earth', *J. Geol. Soc. London* **160**: 825–828.

Schrag, D. P., Berner, R. A., Hoffman, P. F., and Halverson, G. P., 2002, On the initiation of a snowball Earth, *Geochem. Geophys. Geosys.* **31**: 10.1029/2001GC000219.

Semikhatov, M. A., Kuznetsov, A. B., Gorokhov, I. M., Konstantinova, G. V., Melnikov, N. N., Podkovyrov, V. N., and Kutyavin, E. P., 2002, Low $^{87}Sr/^{86}Sr$ ratios in seawater of the Grenville and post-Grenville time: determining factors, *Strat. Geol. Correl.* **10**: 1–41.

Shields, G., and Veizer, J., 2002, Precambrian marine carbonate isotope database: Version 1.1, *Geochem. Geophys. Geosys.* **3**: 10.1029/2001GC000266.

Smith, L. H., Kaufman, A. J., Knoll, A. H., and Link, P. K., 1994, Chemostratigraphy of predominantly siliciclastic Neoproterozoic successions: a case study of the Pocatello Formation and lower Brigham Group, Idaho, USA, *Geol. Mag.* **131**: 301–314.

Sohl, L. E., Christie-Blick, N., and Kent, D. V., 1998, Paleomagnetic polarity reversals in Marinoan (ca. 600 Ma) glacial deposits of Australia: implications for the duration of low-latitude glaciations in Neoproterozoic time, *Geol. Soc. Amer. Bull.* **111**: 1120–1139.

Thompson, M. D., and Bowring, S. A., 2000, Age of the Squantum 'tillite', Boston basin, Massachusetts: U–Pb zircon constraints on terminal Neoproterozoic glaciation, *Amer. J. Sci.* **300**: 630–655.

Trindade, R. I. F., Font, E., D'Agrella-Filho, M. S., Nogueira, A. C. R., and Riccomini, C., 2003, Low-latitude and multiple geomagnetic reversals in the Neoproterozoic Puga cap carbonate, Amazon craton, *Terra Nova* **15**: 441–446.

Vidal, G., and Moczydlowska-Vidal, M., 1997, Biodiversity, speciation, and extinction trends of Proterozoic and Cambrian phytoplankton, *Paleobiology* **23**: 230-246.

Wallace, M. W., Gostin, V. A., and Keays, R. R., 1989, Discovery of the Acraman impact ejecta blanket in the Officer Basin and its stratigraphic significance, *Australia J. Earth Sci.* **36**: 585–587.

Walter, M. R., Veevers, J. J., Calver, C. R., Gorjan, P., and Hill, A. C., 2000, Dating the 840–544 Ma Neoproterozoic interval by isotopes of strontium, carbon, and sulfur in seawater and some interpretive models, *Precambrian Res.* **100**: 371–433.

Williams, G. E., 1975, Late Precambrian glacial climate and the Earth's obliquity, *Geol. Mag.* **112**: 441–465.

Williams, G. E., and Wallace, M. W., 2003, The Acraman asteroid impact, South Australia: magnitude and implications for late Vendian environment, *J. Geol. Soc. London* **160**: 545–554.

Wingate, M. T. D., Campbell, I. H., Compston, W., and Gibson, G. M., 1998, Ion microprobe U–Pb ages for Neoproterozoic basaltic magmatism in south-central Australia and implications for the breakup of Rodinia, *Precambrian Res.* **87**: 135–159.

Wingate, M. T. D., and Giddings, J. W., 2000, Age and palaeomagnetism of the Mundine Well dyke swarm, Western Australia: implications for an Australia-Laurentia connection at 755 Ma, *Precambrian Res.* **100**: 335–357.

Workman, R. K., Grotzinger, J. P., and Hart, S. R., 2002, Constraints on Neoproterozoic ocean chemistry from C and B analyses of carbonates from the Witvlei and Nama groups, Namibia, in *Goldschmidt Conference Procedings* (Davos, Switzerland) pp. A847.

Xiao, S., 2004a, Neoproterozoic glaciations and the fossil record, in: *The Extreme Proterozoic: Geology, Geochemistry, and Climate* (G. Jenkins, M. A. S. McMenamin, C. McKay, and L. Sohl, eds.), *Geophysical Monograph Series 146*, American Geophysical Union, pp. 199–214.

Xiao, S., 2004b, New multicellular algal fossils and acritarchs in Doushantuo chert nodules (Neoproterozoic; Yangze Gorges, South China), *J. Paleontol.* **78**: 393–401.

Xiao, S., Bao, H., Wang, H., Kaufman, A. J., Zhou, C., Li, G., Yuan, X., and Ling, H., 2004, The Neoproterozoic Quruqtagh Group in eastern Chinese Tianshan: evidence for a post-Marinoan glaciation, *Precambrian Res.* **130**: 1–26.

Xiao, S., Zhang, Y., and Knoll, A. H., 1998, Three-dimensional preservation of algae and animal embryos in a Neoproterozoic phosphorite, *Nature* **391**: 553–558.

Yoshioka, H., Asahara, Y., Tojo, B., and Kawakami, S., 2003, Systematic variations in C, O, and Sr isotopes and elemental concentrations in Neoproterozoic carbonates in Namibia: implications for glacial to interglacial transition, *Precambrian Res.* **124**: 69–85.

Zhang, S., Jiang, G., Zhang, J., Song, B., Kennedy, M. J., and Christie-Blick, N., 2005, U–Pb sensitive high-resolution ion microprobe ages from the Doushantuo Formation in south China: Constraints on late Neoproterozoic glaciations, *Geology* **33**: 473–476.

Zhang, Y., Yin, L., Xiao, S., and Knoll, A. H., 1998, Permineralized fossils from the terminal Proterozoic Doushantuo Fm., South China, *Paleontol. Soc. Mem.* **50**: 1–52.

Zhou, C., Tucker, R., Xiao, S., Peng, Z., Yuan, X., and Chen, Z., 2004, New constraints on the ages of Neoproterozoic glaciations in south China, *Geology* **32**: 437–440.

Chapter 9

On Neoproterozoic Cap Carbonates as Chronostratigraphic Markers

FRANK A. CORSETTI and NATHANIEL J. LORENTZ

Department of Earth Science, University of Southern California, Los Angeles, CA 90089-0740, USA.

1. Introduction	273
1.1. "Two Kinds" of Cap Carbonates	276
2. Key Neoproterozoic Successions	277
2.1. Southeastern Idaho	277
2.2. Oman	282
2.3. South China	283
2.4. Namibia	284
2.5. Tasmania	284
2.6. Conterminous Australia	285
2.7. Newfoundland	285
2.8. Northwestern Canada	286
3. Discussion	286
3.1. Global Correlations, Cap Carbonates, and New Radiometric Constraints	286
3.2. Intra-continental Marinoan-style Cap Carbonates ~100 m.y. Apart	288
3.3. Is it Time to Abandon the Terms Sturtian and Marinoan?	290
4. Conclusion	290
References	291

1. INTRODUCTION

A robust global stratigraphy is required in order to understand the evolution of the biosphere on our planet. In Phanerozoic strata, a long-established stratigraphic code is used to define type sections to which all other sections can (theoretically) be correlated; ultimately, we take advantage of evolution and use key fossil occurrences to guide the correlations, with help from chemostratigraphic methods, radiometric calibration, magnetostratigraphy, and such. However, correlating pre-Phanerozoic stratigraphic successions is difficult, given the near absence of useful biostratigraphic information, radiometric calibration, and unaltered chemo- and magneto-stratigraphic information. For example, the evidence for Neoproterozoic low latitude glaciation suggests a climate deterioration of possibly unprecedented magnitude a few tens of millions of years before the Cambrian radiation of metazoa (e.g., Kirschvink, 1992; Kaufman et al., 1997; Hoffman et al., 1998; Kennedy et al., 1998; Knoll, 2000; Walter et al., 2000; Hoffman and Schrag, 2002; Halverson et al., 2004), but workers cannot agree upon the number and timing of the glacial events. Unfortunately, the available Neoproterozoic biostratigraphy is too coarse to resolve separate glacial events (e.g., Knoll, 1994; Vidal and Moczydlowska-Vidal, 1997; Grey et al., 2003), most of the commonly implemented chemostratigraphic techniques generally produce equivocal correlations (compare Knoll, 2000; Walter et al., 2000), and, until very recently, there were few directly dated Neoproterozoic glacial units. Here, we will investigate the correlation of Neoproterozoic glacial deposits and examine how some long-standing correlations might change in light of the latest chronometric data. We do not mean to downplay the importance of chemostratigraphic techniques here. While widely implemented in Neoproterozoic sections, chemostratigraphic profiles—and in particular $\delta^{13}C$ profiles—are ultimately ambiguous. Thus, we will focus on radiometrically-constrained sections in order to minimize undue interpolation, interpretation, and ambiguity.

Neoproterozoic successions commonly contain at least one, but very rarely more than two, glacial deposits (cf., Kennedy et al., 1998; but see Xiao et al., 2004). However, it is currently widely believed that there were at least three great "ice ages" in Neoproterozoic time, an older interval ca. 750–700 (Brasier et al., 2000; Allen et al., 2002; Fanning and Link, 2004), a middle interval ca. 635 Ma (Hoffmann et al., 2004; Zhou et al., 2004; Zhang et al., 2005), and a younger interval ca. 580 Ma (Bowring et al., 2002; Calver et al., 2004). Names have been applied to the glacial intervals: the older is commonly termed the Sturtian, the middle the Marinoan, and the youngest the Gaskiers, each based on a key locality (the two former from Australia, the latter from Newfoundland). Other names have been used, but at this point only serve to obfuscate useful dialogue. For example, the Varanger, or Varangian, named for deposits in Scandinavia, was once thought to be analogous to Marinoan, although the correlations are not clear.

Figure 1. Typical cap carbonates from around the world. (A) Maieberg Formation cap carbonate overlies the Ghaub Formation diamicite, Otavi Group, Namibia. (B) Cap carbonate overlying the El Chiquerio Formation, San Juan, Peru. Note alternating bands of organic rich-organic poor laminae. (C) Conglomerate (diamictite) and Dolomite Member (cap carbonate), Ibex Formation, Death Valley, United States.

The glacial deposits are capped by enigmatic carbonates that record negative $\delta^{13}C$ values (e.g., Kennedy, 1996; Kaufman et al., 1997; Hoffman et al., 1998; Prave, 1999; James et al., 2001; Hoffman and Schrag, 2002; Corsetti and Kaufman, 2003; Rodrigues-Nogueira et al., 2003; Halverson et al., 2004; Lorentz et al., 2004; Porter et al., 2004; Xiao et al., 2004; Fig. 1). The interpretation of the driving force behind cap carbonate deposition forms the cornerstone in various Neoproterozoic glacial and post-glacial hypotheses (e.g., Kaufman et al., 1997; Hoffman et al., 1998; Kennedy et al., 2001). The lithologic and isotopic characteristics of the cap carbonates have been the focus of much study, and their striking similarity from continent to continent promotes the impression that they might prove useful in correlation when other means are absent, which is usually the case in Neoproterozoic strata. New radiometric age constraints, however, reveal a more complex pattern in cap carbonate temporal distribution, implying that correlation by cap carbonate characteristics deserves careful scrutiny.

1.1 "Two Kinds" of Cap Carbonates

An unofficial Neoproterozoic correlation scheme has emerged based primarily on the lithologic and carbon isotopic characteristics of the cap carbonates that overlie glacial deposits around the world. The lithologic character of the cap carbonates falls into two groups (as defined in the influential paper by Kennedy et al., 1998). One group is associated with the Sturtian interval and the other with the Marinoan; some workers suggest the Gaskiers glaciation was not as severe as the Sturtian and Marinoan, and thus give it subsidiary importance in the overall glacial-cap carbonate scheme; we will further investigate this concept in the discussion section. The Sturtian group of cap carbonates is characterized by (among other things) dark, organic-rich, finely laminated carbonates with rhythmic laminae, and some contain roll-up structures (Fig. 2). In particular, negative basal $\delta^{13}C$ values climb rapidly to mildly positive values within a few meters to tens of meters of stratigraphic section. The Marinoan group of cap carbonates is generally characterized by a lighter coloration and the presence of unusual features, including seafloor fans (pseudomorphs of aragonite and/or barite), tubestones, sheetcrack cements, and tepee-like structures (Fig. 3). The $\delta^{13}C$ values are negative at the base of the cap carbonate and continue to record negative values up-section. Hereafter, we will use the terms Sturtian-style and Marinoan-style to describe the cap carbonates in any given section.

The aforementioned characters were assembled from 12 cap carbonate successions around the world and examined using parsimony analysis in order to test the informal pattern noted above (Kennedy et al., 1998). The resulting "cladogram" confirmed the pattern (see fig. 4 of Kennedy et al.,

1998, p. 1062). As a result, the lithologic and carbon isotopic characteristic of cap carbonates have been widely implemented to assign age control where chronometric data are absent (e.g., in most Neoproterozoic successions). Although it was not likely the intent of these workers, others have embraced the bipartite glacial-cap carbonate scheme, and on as little evidence as the color of the cap carbonate or shape of the $\delta^{13}C$ profile, have assigned ages to unconstrained glacial-cap carbonate couplets. New chronometric data from Idaho, Oman, China, Namibia Tasmania/Australia, Newfoundland, and northwestern Canada will allow the lithologic/isotopic pattern of cap carbonate occurrence to be tested more rigorously (Fig. 4). The basic lithostratigraphy will be outlined for each region with specific focus on the stratigraphic context of the radiometric dates and the style of cap carbonate in each section. All referenced dates are U–Pb zircon ages unless otherwise noted (e.g., northwestern Canada).

Figure 2. Sturtian-style cap carbonate, Rasthof Formation, Otavi Group, Namibia. Note finely laminated organic rich carbonate with intricate rollup structures.

2. KEY NEOPROTEROZOIC SUCCESSIONS

2.1 Southeastern Idaho

Neoproterozoic strata of southeast Idaho include the partly glaciogenic Pocatello Formation, the Blackrock Canyon Limestone, and part of the Brigham Group (e.g., Link *et al.*, 1993) (Fig. 5A). Glacial diamictites of the

Scout Mountain Member of the Pocatello Formation represent the oldest glacial units in the region. The Pocatello Formation is divided into three members: the Bannock Volcanic Member, the Scout Mountain Member, and the (informal) upper member (Link, 1983).

The Scout Mountain Member contains two glaciogenic diamictite units separated by sandstones, siltstones, and a massive cobble conglomerate (Ludlum, 1942; Crittenden et al., 1971; Trimble, 1976; Crittenden et al., 1983; Link, 1983; Link et al., 1994). The diamictites have been considered stade deposits within a single glaciation (Crittenden et al., 1983), but the actual duration of the glaciation is not known. Iron-rich turbidites occur in the interval immediately below the uppermost diamictite south of the Portneuf Narrows, near Pocatello, Idaho (Link, 1983). A rhyolite clast within the upper Scout Mountain Member diamictite has been dated at 717 ± 4 Ma (Fanning and Link, 2004), constraining the diamictites to be younger than ca. 717 Ma. A thin, finely laminated pink dolostone with consistently negative $\delta^{13}C$ values lies in depositional contact with the uppermost diamictite of the Scout Mountain Member of the Pocatello Formation (Link, 1983; Smith et al., 1994) (Fig. 6A–B). The cap dolostone is truncated by a minor but regional incision surface with several meters of erosive relief, and is overlain by a ~100-meter thick transgressive, cyclic, but upward-fining section of sandstone, siltstone, and very minor carbonates. Siliciclastics through this interval display dewatering structures and occasional climbing ripples, indicating relatively rapid sedimentation. The most prominent carbonate unit in the succession, termed the "carbonate and marble unit" by Link (1983), tops the Scout Mountain Member and is light gray to pink limestone, records negative $\delta^{13}C$ values that decline up-section, and contains seafloor fans (pseudomorphs after aragonite; Fig. 6C–E) (Lorentz et al., 2004). The thin cap dolostone and the carbonate and marble unit thus fit the description of "Marinoan" style carbonates. An ash near the base of the fan-bearing carbonate unit has been dated at 667 ± 5 Ma and likely approximates the depositional age of the carbonate (Fanning and Link, 2004). Extensive investigation of the strata beneath the carbonate and marble unit suggests that the section is continuous and devoid of obvious hiatal surfaces. The

Figure 3. (on Page 279) Marinoan-style cap carbonate facies. (A) Seafloor fans (pseudomorphs of aragonite), Otavi Group, Namibia. (B) Tubestones from the Noonday Dolomite cap carbonate, Death Valley, California. Bedding dips to the right and the tubes define the vertical direction. (C) Bedding plane view of Noonday tubestones. (D) Sheetcrack cement (cf. stromatactis) from the Noonday Dolomite. (E) Polished slab of Noonday Dolomite tubestone. The darker areas comprise the sediment filled tube-structures, and the light colored material is the "host-rock" for the tubes.

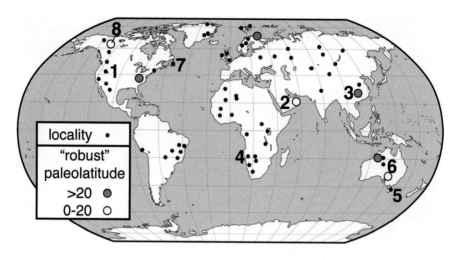

Figure 4. Global distribution of Neoproterozoic glacial deposits (after Evans, 2000; Hoffman and Schrag, 2002). Robust paleolatitudes follow the convention of Evans (2000). The localities with radiometric age control discussed in the text are: 1—Idaho, 2—Oman, 3—South China, 4—Namibia, 5—Tasmania, 6—Conterminous Australia, 7—Newfoundland, and 8—Northwest Canada.

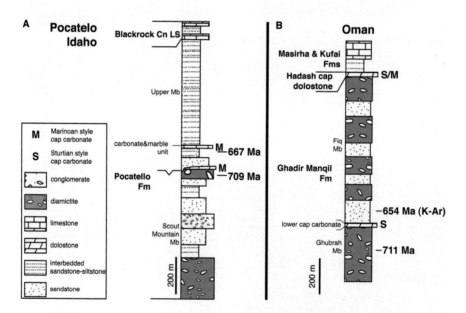

Figure 5. Generalized stratigraphic successions from (A) southeast Idaho and (B) Oman. Idaho column adapted from Link (1983) and Link *et al.* (1993); Oman column adapted from Braiser *et al.* (2000), Leather *et al.* (2002), and Allen *et al.* (2004).

On Neoproterozoic Cap Carbonates as Chronostratigraphic Markers 281

Figure 6. Cap carbonate and cap carbonate-like facies from the Pocatello Formation, Idaho. (A–B) Thin, well-laminated dolostone cap carbonate in contact with the Pocatello Formation diamictites. The pink coloration and consistent isotopic profile fit the Marinoan-style cap carbonate, but the underlying diamictites are dated at 709 Ma and constrained via an overlying ash to be older than 667 Ma (thus, not "Marinoan" in age). (C) Carbonate and Marble unit, Scout Mountain Member, Pocatello Formation, lies above an ash dated at 667 Ma with no obvious intervening hiatus. Thus, we assume that the age of the ash closely approximates the age of the carbonate unit. (D–E) Small seafloor fans from the Carbonate and Marble unit (D from outcrop, E is polished). The seafloor fans, coupled with the declining negative $\delta^{13}C$ profile, find affinities with Marinoan-style cap carbonates. However, the age of ca. 667 is inconsistent with a "Marinoan" age.

(informal) upper member of the Pocatello Formation is composed of greater than 600 meters of laminated argillite/shale, with minor siltstone and quartzite (Crittenden *et al.*, 1971; Trimble, 1976; Crittenden *et al.*, 1983; Link, 1983).

The Bannock Volcanic Member exists as a lenticular body intercalated with the Scout Mountain Member and is composed of metabasalts and volcanic breccias. The chemistry of the Bannock Volcanic Member is consistent with intra-plate, rift-related volcanism (Harper and Link, 1986). Fanning and Link (2004) dated an epiclastic crystal tuff bed of the Bannock Volcanic Member at 709 ± 5 Ma, constraining the age of the sub- and superjacent glacial units to be ca. 709 Ma. Thus, the radiometric dates from the Idaho succession provide important constraints on the timing of this phase of Neoproterozoic glaciation: The thin cap carbonate was deposited between 709 Ma and 667 Ma, and the carbonate and marble unit was deposited ca. 667 Ma.

Evidence for a younger glaciation is inferred from the incised valleys of the Caddy Canyon Quartzite ~2000 meters above the glacial deposits in the Pocatello Formation (Christie-Blick and Levy, 1989). The Browns Hole Formation, 500 to 1000 m above the Caddy Canyon Quartzite, contains an extrusive unit dated at 580 ±7 Ma ($^{40}Ar-^{39}Ar$ date recalculated by Christie-Blick and Levy, 1989). The putatively glacial incised valleys are therefore constrained between 667 Ma and 580 Ma. No demonstrably glaciogenic strata or cap carbonate are associated with the Caddy Canyon Quartzite.

2.2 Oman

At least two glaciations are recognized from the Neoproterozoic Huqf Supergroup in Oman (e.g., Braiser et al., 2000) (Fig. 5B). The Ghubrah Member of the Huqf Supergroup consists of glaciogenic diamictite and is constrained to be 723 +16/–10 Ma by Braiser et al. (2000) and ca. 711 Ma by Allen et al. (2002). The Ghubrah Member is overlain by an interval of organic-rich Sturtian-style cap carbonate(s) that records a negative to positive $\delta^{13}C$ profile (Braiser et al., 2000). The superjacent Fiq Member of the Huqf Supergroup records periodic glaciation overlain by the Hadash cap dolostone, part of the Massirah Bay cap carbonate sequence (Leather et al., 2002). Gorin et al. (1982) provide a K/Ar constraint of 654 ± 12 Ma from within the Fiq Member. Based on descriptions by Allen et al. (2004), the Hadash Dolostone appears to contain qualities of both Marinoan-style and Sturtian-style cap carbonates. For example, they report microbial roll-up structures, grey carbonates, and carbonate "stringers" consistent with Sturtian-style cap carbonates combined with C-isotope profile that declines up section and thus is most consistent with Marinoan-style cap carbonates. Thus, the "Sturtian" style cap carbonate above the Ghubrah diamictites was deposited after ca. 711 Ma and before 654 Ma (although this K/Ar date may be less robust), and the Sturtian/Marinoan-style cap carbonate superjacent to the Fiq member was deposited after ca. 654 Ma.

2.3 South China

At least two episodes of glaciation are noted in south China (Fig. 7A). The glacial Changan and Tiesiao Formations represent at least one glaciation; it is unclear whether these units represent discrete glaciations, as the Changan Formation does not have a known cap carbonate. The Tiesiao Formation is overlain by the Datangpo Formation, the base of which consists of a finely laminated, organic-rich rhodochrosite cap carbonate (Zhou et al., 2004). An ash dated at 663 ± 4 Ma lies just above the Mn-rich cap carbonate (Zhou et al., 2004). The Datangpo Formation is overlain by the glaciogenic Nantuo Formation and subsequent Doushantuo Formation. The basal ~5 meters of the Doushantuo Formation contains classic Marinoan-style cap carbonate features, including pseudo-tepee structures, sheetcrack cements, and bladed barite cements (e.g., Jiang et al., 2003). The predicted negative carbon isotope anomaly is present, and will be discussed in detail in a subsequent section. Several dates are available for the basal Doushantuo Formation. Condon et al. (2005) report an age of ca. 635 Ma for the basal Doushantuo Formation and an age of ca. 550 Ma for the top. Zhang et al. (2005) provide corroborating dates. Therefore, the Mn-rich cap carbonate in the basal Datangpo Formation above the Tiesiao Formation glacial strata was deposited just prior to ca. 663 Ma and the basal Doushantuo cap carbonate above the glacial Nantuo Formation was deposited ca. 635 Ma.

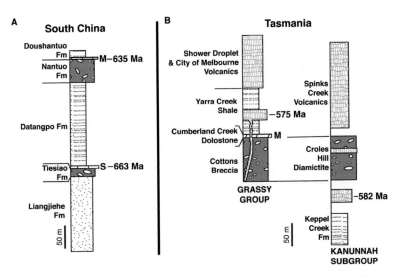

Figure 7. Generalized stratigraphic successions from (A) South China and (B) Tasmania. South China column adapted from Zhou et al. (2004); Tasmania column adapted from Calver et al. (2004).

2.4 Namibia

Two Neoproterozoic glacial intervals are recognized from the Otavi Group, Namibia; although they are not directly dated, correlative units are. The Sturtian-style Rasthof Formation cap carbonate overlies the older Chuos Formation glacial deposit (Hoffman *et al.*, 1998). The Maieberg Formation, the cap carbonate atop the younger Ghaub Formation, is one of the best studied in the world and is considered by some a model example of a Marinoan-style cap carbonate, with spectacular seafloor fans, tubestones, sheetcrack cements and a declining trend in $\delta^{13}C$ values throughout the cap carbonate (Hoffman *et al.*, 1998). The well-characterized units in the Otavi Group are not themselves radiometrically constrained, but it is thought that the Swakop Group, a metamorphosed slope to basinal facies to the south of the Otavi platformal deposits, can be correlated to the Otavi platformal units. The Swakop Group contains a metamorphosed dropstone-bearing unit assigned to the Ghaub Formation dated at 635 ± 1.2 Ma (Hoffmann *et al.*, 2004). A 0.5–2-m-thick, buff to tan meta-dolostone overlies the dropstone-bearing strata, and, in places, directly on brecciated mafic flow tops dated at 635 Ma. Where best developed, the lower 10–30 cm of the dolostone is laminated and locally contains sheetcrack cements. No $\delta^{13}C$ data were presented for the thin cap carbonate, but the presence of sheetcrack cements and the potential correlation to the Maiberg Formation to the north is most consistent with a Marinoan-style cap carbonate deposited ca. 635 Ma. It is interesting to note that, in general, deeper water facies record thinner cap carbonates versus the platformal facies.

2.5 Tasmania

New radiometric control is available from King Island, Tasmania, where the glaciogenic Cottons Breccia and Cumberland Creek Dolostone cap carbonate are intruded by the Grimes Intrusive Suite dated at 575 ± 3 Ma, considered close to the depositional age based upon the nature of the contact between the sediments and the intrusive units (Calver *et al.*, 2004) (Fig. 7B). The Cumberland Creek Dolostone is also a classic Marinoan-style cap carbonate (Calver and Walter, 2000), characterized by pale pinkish-gray laminated dolostone with declining $\delta^{13}C$ values throughout the cap carbonate. The Croles Hill Diamictite in northwestern Tasmania, correlative to the Cottons Breccia, is younger than 582 ± 4 Ma, thus supporting a ca. 580 Ma depositional age for the Cumberland Creek cap carbonate (Calver *et al.*, 2004). The application of these dates to the Cumberland Creek cap carbonate depends on the correlation between the Croles Hill Diamictite and the Cottons Breccia, which is subject of current debate. Initial studies of detrital zircons from the Cottons Breccia/Cumberland Creek Dolostone transition show a 635 Ma affinity (e.g., Raub *et al.*, 2005). It follows that the Cottons Breccia/Cumberland Creek Dolostone may be ca. 635 Ma and thus not correlative with the Croles Hill Diamictite. The question then lies in how fast a new zircon can become detritus and find its way into the Cottons Breccia/Cumberland Creek Dolostone transition. Because of this complication, the data from the Tasmanian sections should be considered carefully until further observations clarify the situation.

2.6 Conterminous Australia

Two glaciations are usually inferred from conterminous Australian and related sections. Using the Adelaide Geosyncline for reference, the Sturt/Appila glaciogenic deposit with cap carbonate represents the older glacial pulse (e.g., Walter *et al.*, 2000). The Elatina glaciogenic deposit represents the younger and is overlain by the Marinoan-style Nuccaleena cap carbonate (e.g., Kennedy, 1996; Kennedy *et al.*, 1998) with a declining $\delta^{13}C$ trend up-section, sheetcrack cements, pseudo-tepee structures and barite fans. Recently, the base of the Nuccaleena Formation cap carbonate was chosen as the GSSP for the newly erected Ediacaran Period, thus placing even more global significance on the Australian sections. No U–Pb dates are available for the conterminous Australian sections. However, using refined Re–Os techniques, Kendall and Creaser (2005) provides a minimum Re–Os age for the Areyonga Formation, thought to correlate to the Sturtian diamictites, of 657.2 ± 5.4 Ma from the overlying Aralka Formation. Because of depositional hiatus between the Aralka and overlying Olympic/Pioneer glacials, this date does not bear strongly on the "Marinoan" deposits from Australia, except to say that they are younger than 657 Ma by some duration that includes an unknown unconformity (Kendall and Creaser, 2005).

2.7 Newfoundland

The Gaskiers Formation is composed of deeper water glaciomarine deposits (Eyles, 1990). The succession contains a ~0.5 m thick, $\delta^{13}C$-depleted cap carbonate of unknown style (Myrow and Kaufman, 1999). Like the thin cap carbonate in the basinal facies of the Ghaub Formation in Namibia, the Gaskiers cap carbonate is thin when compared to other cap carbonates precipitated in shallower paleoenvironments. This carbonate is the only carbonate known from the section (Myrow and Kaufman, 1999; Bowring *et al.*, 2002). This cap carbonate and the underlying diamictite are constrained to be ca. 580 Ma, were deposited in less than 1 m.y., and precede occurrences of Ediacaran fossils dated at ca. 575 Ma (Bowring et al., 2002).

2.8 Northwestern Canada

James et al. (2001) described a succession from the Mackenzie Mountains (northwestern Canada) nearly lithologically identical to the Pocatello succession: the glaciogenic Icebrook Formation is capped by the pink Ravensthroat Formation cap dolostone, which is incised by an erosional surface and followed by the limestone-dominated Hayhook Formation containing seafloor-precipitated fans (the Ravensthroat/Hayhook designations are informal formations that collectively comprise the Tepee Dolomite in the region; James et al., 2001). These carbonates also record negative $\delta^{13}C$ values becoming increasingly more negative up-section. Finally, black shales of the Sheepbed Formation follow the Hayhook Formation. The Icebrook, Ravensthroat, and Hayhook Formations are not currently constrained by radiometric dates. However, they have been considered "Marinoan" in age by various workers (e.g., Hoffman and Schrag, 2002) based on their lithologic character (seafloor fans, pseudo tepee structures, declining $\delta^{13}C$ trend). The Old Fort Point Formation in the Canadian Cordillera, which is currently correlated with the Tepee Dolomite cap carbonate above the Icebrook glacials, is dated via Re–Os as ca. 607 Ma (Kendall et al. 2004). If the correlation from the Old Fort Point Formation to the Tepee Dolomite is correct, then the Icebrook-Ravensthroat glacial-cap carbonate couplet could be as young as 607 Ma.

3. DISCUSSION

3.1 Global Correlations, Cap Carbonates, and New Radiometric Constraints

Glacial deposits overlain by cap carbonates from southeastern Idaho (Fanning and Link, 2004) and Oman (Allen et al., 2002) are now known to have been deposited ca. 710 Ma. The age of glacial termination is unknown, but in Idaho the cap carbonate above the glacial units is constrained to be older than 667 Ma (how much older is not known). An episode of cap carbonate formation preceded by glaciation occurred ca. 670 Ma as demonstrated by the Mn-cap carbonate in the basal Datangpo Formations (663 ± 4 Ma) above the Tiesiao diamictites, but as above, the duration of the preceding glacial period is not known. Interestingly, the Pocatello Formation's "carbonate and marble" unit is dated at 667 ± 5 Ma, identical to the basal Datangpo Formation within analytical error. Although Lorentz *et al.* (2004) preferred to interpret the "carbonate and marble" as a cap-like carbonate independent of glacial processes, one might interpret the preceding transgressive succession as glacial outwash and thus linking the carbonate and marble unit to glaciation. For example, Lund *et al.* (2003) dated putative glacial metadiamictites in central Idaho at ca. 685 Ma (not discussed above because they lack cap carbonates). The dates from the central Idaho diamictites suggest that glaciation may have 1) continued from ca. 710 Ma through 685 Ma until cap carbonate deposition ca. 667, or 2) there were two episodes of glacial-cap carbonate formation, once ca. 710 Ma and another ca. 685–667 Ma. Ultimately, the data do not permit a more precise interpretation.

At least two additional late Neoproterozoic glacial episodes are apparent: one recorded in Namibia and China dated at 635 Ma (e.g., Hoffmann *et al.*, 2004) and one in Tasmania and Newfoundland dated at ca. 580 Ma (e.g., Bowring *et al.*, 2002; Calver *et al.*, 2004). Calver *et al.* (2004) consider a correlation that would make the Cumberland Creek cap carbonate in Tasmania older, but reject it as unduly complex. As above, the duration of the glacial interval ca. 635 Ma is not known. In China, it is clear that the glacial interval could have been no longer than 28 million years (the difference between the ashes associated with the Mn-rich basal Datangpo Formation and the Doushantuo Formation), but the actual duration is not apparent. The duration of the final glaciation is well constrained: Bowring *et al.* (2002) demonstrate it can be no longer than 1 million years, based on dates from ash beds that bracket the glacial deposits.

Interesting patterns emerge when the radiometric dates are combined with the style of cap carbonate at each locality (Fig. 8). The Neoproterozoic cap carbonate(s) from Idaho have Marinoan-style characteristics and were deposited between 709 Ma and 667 Ma. The Maieberg Formation cap carbonate in Namibia and the basal Doushantuo Formation also have Marinoan-style features, but they were deposited ca. 635 Ma. The Tepee Dolomite in the Canadian Cordillera could represent Marinoan-style deposition ca. 607 Ma. The Cumberland Creek cap carbonate in Tasmania has Marinoan characteristics but was deposited ca. 580 Ma (according to Calver *et al.*, 2004). Thus, while some cap carbonates with Marinoan style features were clearly synchronous (China, Namibia, perhaps others), some were not.

Cap carbonates with Sturtian and Marinoan characteristics were precipitated after synchronous glaciations (Fig. 8). For example, the Sturtian-style cap carbonate above the Ghubrah Member of the Huqf

Supergroup that was deposited after a glaciation dated at ca. 711 Ma. The Marinoan style cap carbonates atop the Scout Mountain diamictites in Idaho were deposited in the aftermath of glaciation dated at ca. 709 Ma. The glacial deposits from Idaho and Oman convincingly represent the same glacial interval, but they are associated with cap carbonates of different character. Similarly, the dark-colored, organic-rich, finely-laminated Datangpo Formation cap carbonate, deposited ca. 663 Ma, would best fit the mold of a Sturtian cap carbonate. However, the carbonate and marble unit in Idaho, with pink coloration, declining $\delta^{13}C$ trend, and seafloor fans, also precipitated ca. 667 (within analytical error of the Datangpo cap carbonate), fits the description of a Marinoan cap carbonate. Both were arguably contemporaneous. The fact that synchronous glacial deposits are overlain by cap carbonates with dissimilar characteristics supports the concept that correlation via cap carbonate style is unwise.

In the interest of completeness, if we accept an alternate correlation not favored by Calver *et al.* (2004), it is possible that the Namibian and Tasmanian cap carbonates could be of similar age, and given that the 607 Ma date on the Canadian succession is a minimum age, it is permissible to consider its deposition ca. 635 Ma, as well. However, these cap carbonates would still be 30 to 75 m.y. younger than the Marinoan-style carbonates in Idaho. Collectively, these Marinoan-style cap carbonates of greatly different ages suggest that intercontinental correlation via cap carbonate characteristics alone is unwise and potentially misleading, as shown in Fig. 8.

3.2 Intra-continental Marinoan-style Cap Carbonates ~100 m.y. Apart

Both the Idaho and Mackenzie Mountains successions record: 1) pink, $\delta^{13}C$-depleted dolostone in depositional contact with underlying glaciogenic rocks; 2) an erosional surface; 3) deposition of a $\delta^{13}C$-depleted, fan-bearing limestone; and subsequent shale/argillite deposition. The lithologic and isotopic characteristics of both carbonate units in the Idaho succession match known Marinoan-style cap carbonates. However, they were deposited between 709 and 667 Ma (Fanning and Link, 2004), a time associated with the Sturtian interval rather than Marinoan interval (Fanning and Link, 2004; Zhou *et al.*, 2004). Alternatively, the Icebrook Formation/Tepee Dolomite correlation to the Old Fort Point Formation suggests Marinoan-style deposition ca. 608 Ma. That two Neoproterozoic successions from the same continental margin can be nearly lithologically identical and yet reasonably interpreted as deposited up to ~100 m.y. apart should serve as dissuasion

On Neoproterozoic Cap Carbonates as Chronostratigraphic Markers   289

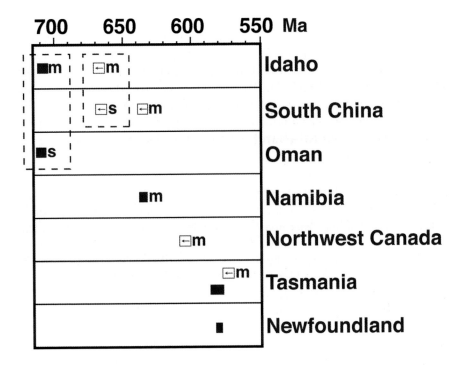

Figure 8. Age distribution and cap carbonate style for well-dated glacial units using criteria outlined in the text. Note that Sturtian and Marinoan style cap carbonates co-occur ca. 709 Ma and ca. 667 Ma (dashed boxes). Marinoan-style cap carbonates occur at least five separate times between 709 Ma and 580 Ma. Even if the Tasmanian section is considered older than ca. 580 Ma (not favored by Calver *et al.*, 2004) and the "carbonate and marble" unit in Idaho is considered younger than 667 Ma (not favored by the sedimentological evidence), Marinoan-style cap carbonates still occurred at least three different times.

toward using cap carbonates as chronostraigraphic markers. Given the incompleteness of the stratigraphic record, we acknowledge that it is permissible to hypothesize an unrecognized, cryptic hiatus below the Old Fort Point Formation, such that 608 Ma represents a minimum age and the

Icebrook-Ravensthroat Formations could be older, as discussed above. However, the relationship of the Icebrook Formation, with simple Ediacaran fossils below it (Hofmann, 1990) and more complex Ediacaran fossils above it (Narbonne and Aiken, 1995) is more consistent with an age that significantly post-dates 635 Ma. To our knowledge, no Ediacaran fossils are known from other, well-constrained units that predate 635 Ma.

3.3 Is it Time to Abandon the Terms Sturtian and Marinoan?

When faced with two glacial units in a given, undated succession, it has been commonplace to assign the older strata to the "Sturtian" glacial interval and the younger to the "Marinoan" glacial interval. However, it is now clear that there was at least one additional glaciation (if not more) in Neoproterozoic time: the Gaskiers event, ca. 580 Ma. Thus, in the absence of radiometric dates, it will be unclear which of the three glacial intervals are represented in any given succession.

Some would suggest that the Gaskiers is a minor glaciation compared to the others, and was not global in extent. The reasoning is model driven: the duration of the Gaskiers event was too short to qualify as a snowball event, which require ~5–10 million years of ice cover to ultimately drive cap carbonate deposition (see Hoffman et al., 1998). However, the Gaskiers deposit does have a cap carbonate, albeit thin, that records negative $\delta^{13}C$ values. Recall that the basinal facies of the basal Doushantuo Formation, an accepted cap carbonate atop an accepted major glaciation, is also thin (Jiang et al., 2003), as is the basinal Swakop Group cap carbonate in Namibia (Hoffmann et al., 2004). Thus, we question the concept that the Gaskiers glaciation was somehow subsidiary to the previous glacial events based on the thinness of its cap carbonate and/or its incompatibility with the theoretical requirements of any given paradigm. Sturtian and Marinoan were terms originally used locally for certain deposits in Australia. As the "original" Sturtian and Marinoan in Australia are not dated via U–Pb, and cap carbonate style is misleading, we suggest that the broad use of the terms Sturtian and Marinoan outside of Australia should be abandoned.

4. CONCLUSION

The generally held notion regarding interregional correlation of cap carbonates seems robust in most known examples where two glacial units are present: the older cap carbonate in the succession is Sturtian-style and the younger cap carbonate is Marinoan-style, but detailed investigation where radiometric constraints are available paints a more complex picture. The straightforward scenario of two glaciations distinguishable by their cap carbonates has understandably developed strong support over the last few years, but the application of such information has likely gone far beyond the original intent of the preliminary observations. As correlation of Neoproterozoic strata is difficult given the absence of a useful biostratigraphy, any scheme that appears to work is attractive. However, the new radiometric dates suggest that Marinoan-style cap carbonates are not unique to one post-glacial period, but rather appeared at least three times, if not more, between ca. 710 Ma and 580 Ma. It is conceivable that the repetitive nature of Neoproterozoic glacial episodes fostered similar glacial/post-glacial conditions, and therefore similar cap carbonates, at different times during the late Neoproterozoic. The occurrence of multiple glaciations overlain by similar cap carbonates makes the Neoproterozoic interval all the more interesting, but the comfort engendered by the simplistic two glacial model must be replaced with a more realistic view that the correlation of cap carbonates, in the absence of other features, should be avoided. Perhaps the Neoproterozoic glacial record could be somewhat analogous to the Pleistocene glacial record: four (or fewer) glacial advances are commonly recorded at any given terrestrial section (Nebraskan, Kansan, Illinoisan, and Wisconsin), but the more complete deep sea $\delta^{18}O$ record reveals greater than 20 advances/retreats (cf., Balco et al., 2005).

REFERENCES

Allen, P. A., Bowring, S., Leather, J., Brasier, M., Cozzi, A., Grotzinger, J. P., McCarron, G., and Amthor, J. J., 2002, Chronology of Neoproterozoic glaciations: New insights from Oman, *16th International Sedimentological Congress Abstracts*, Johannesburg, South Africa.

Allen, P.A., Leather, J.W., and Brasier, M.D., 2004, The Neoproterozoic Fiq glaciation and its aftermath, Huqf supergroup of Oman, *Bas. Res.* **16**: 507–534.

Balco, G., Rovey, C. W., and Stone, J. O., 2005, The first glacial maximum in North America, *Science* **307**: 222.

Bowring, S. A., Myrow, P. M., Landing, E., and Ramezani, J., 2002, Geochronological constraints on Neoproterozoic events and the rise of metazoans, *Astrobiology* **2**: 457–458.

Brasier, M., McCarron, G., Tucker, R., Leather, J., Allen, P. A., and Shields, G., 2000, New U–Pb zircon dates for the Neoproterozoic Ghubrah glaciation and for the top of the Huqf Supergroup, Oman, *Geology* **28**: 175–178.

Calver, C. R., Black, L. P., Everard, J. L., and Seymour, D. B., 2004, U–Pb zircon age constraints on late Neoproterozoic glaciation in Tasmania, *Geology* **10**: 893–896.

Calver, C. R., and Walter, M. R., 2000, The late Neoproterozoic Grassy Group of King Island, Tasmania; correlation and palaeogeographic significance, *Precambrian Res.* **100**: 299–312.

Christie-Blick, N., and Levy, M., 1989, Stratigraphic and tectonic framework of upper Proterozoic and Cambrian rocks in the Western United States, in: *Late Proterozoic and Cambrian Tectonic, Sedimentation, and Record of Metazoan Radiation in the Western United States* (N. Christie-Blick, M. Levy, J. F. Mount, P. W. Signor, and P. K. Link, eds.), American Geophysical Union, Washington, DC, p. 113.

Condon, D., Zhu, M., Bowring, S., Jin, Y., Wang, W., and Yang, A., 2005, From the Marinoan glaciation to the oldest bilaterians: U–Pb ages from the Doushantou Formation, China, *Science* **308**: 95–98.

Corsetti, F. A., and Kaufman, A. J., 2003, Stratigraphic investigations of carbon isotope anomalies and Neoproterozoic ice ages in Death Valley, California, *Geol. Soc. Am. Bull.* **115**: 916–932.

Crittenden, M. D., Jr., Christie-Blick, N., and Link, P. K., 1983, Evidence for two pulses of glaciation during the late Proterozoic in northern Utah and southeastern Idaho, *Geol. Soc. Am. Bull.* **94**: 437–450.

Crittenden, M. D., Jr., Schaeffer, F. E., Trimble, D. E., and Woodward, L. A., 1971, Nomenclature and correlation of some upper Precambrian and basal Cambrian sequences in western Utah and southeastern Idaho, *Geol. Soc. Am. Bull.* **82**: 581–602.

Evans, D. A. D., 2000, Stratigraphic, geochronological, and paleomagnetic constraints upon the Neoproterozoic climatic paradoxes, *Amer. J. Sci.* **300**: 347–443.

Eyles, N., 1990. Marine debris flows; late Precambrian "tillites" of the Avalonian-Cadomian orogenic belt. *Palaeogeogr. Palaeoclimatol. Palaeoecol.* **79**: 73–98.

Fanning, C. M., and Link, P. K., 2004, U–Pb SHRIMP age of Neoproterozoic (Sturtian) glaciogenic Pocatello Formation, southeastern Idaho, *Geology* **10**: 881–884.

Gorin, G. E., Racz, L. G., and Walter, M. R., 1982, Late Precambrian–Cambrian sediments of Huqf Group, Sultanate of Oman, *AAPG Bull.* **66**: 2609–2627.

Grey, K., Walter, M. E., and Calver, C. R., 2003, Neoproterozoic biotic diversification: Snowball Earth or aftermath of the Acraman impact? *Geology* **31**: 459–462.

Halverson, G. P., Maloof, A. C., and Hoffman, P. F., 2004, The Marinoan glaciation (Neoproterozoic) in northeast Svalbard, *Bas. Res.* **16**: 297–324.

Harper, G. D., and Link, P. K., 1986, Geochemistry of Upper Proterozoic rift-related volcanics, northern Utah and southeastern Idaho, *Geology* **14**: 864–867.

Hoffman, P. F., Kaufman, A. J., Halverson, G. P., and Schrag, D. P., 1998, A Neoproterozoic snowball Earth, *Science* **281**: 1342–1346.

Hoffman, P. F., and Schrag, D. P., 2002, The snowball Earth hypothesis: testing the limits of global change, *Terra Nova* **14**: 129–155.

Hoffmann, K. H., Condon, D. J., Bowring, S. A., and Crowley, J. L., 2004, U–Pb zircon date from the Neoproterozoic Ghaub Formation, Namibia: Constraints on Marinoan glaciation, *Geology* **32**: 817–820.

Hofmann, H.J., Narbonne, G.M., and Aitken, J.D., 1990. Ediacaran remains from intertillite beds in northwestern Canada. *Geology* **18**: 1199–1202.

James, N. P., Narbonne, G. M., and Kyser, T. K., 2001, Late Neoproterozoic cap carbonates: Mackenzie Mountains, northwestern Canada: precipitation and global glacial meltdown, Can. J. Earth Sci. **38**: 1229–1262.

Jiang, G., Kennedy, M. J., and Christie-Blick, N., 2003, Stable isotopic evidence for methane seeps in Neoproterozoic postglacial cap carbonates, *Nature* **426**: 822–826.

Kaufman, A. J., Knoll, A. H., and Narbonne, G., M., 1997, Isotopes, ice ages, and terminal Proterozoic Earth history, *Proc. Nat. Acad. Sci. USA* **94**: 6600–6605.

Kendall, B.S., Creaser, R.A.R.G.M., and Selby, D., 2004. Constraints on the timing of Marinoan snowball Earth glaciation by (super 187) Re/ (super 187) Os dating of a Neoproterozoic, post-glacial black shale in Western Canada. *Earth Planet. Sci. Lett.* **222**: 729–740.

Kendall, B.S., and Creaser, R.A., 2005. Re–Os depositional ages for Neoproterozoic post-glacial black shales in Australia; evidence for diachronous Neoproterozoic glaciations, *Geol. Soc. Amer. Annu. Meeting Abstr. with Prog.* **37**(7): 42.

Kennedy, M. J., 1996, Stratigraphy, sedimentology, and isotopic geochemistry of Australian Neoproterozoic postglacial cap dolostones; deglaciation, $\delta 13C$ excursions, and carbonate precipitation, *J. Sed. Res.* **66**: 1050–1064.

Kennedy, M. J., Christie-Blick, N., and Sohl, L. E., 2001, Are Proterozoic cap carbonates and isotopic excursions a record of gas hydrate destabilization following Earth's coldest intervals? *Geology* **29**: 443–446.

Kennedy, M. J., Runnegar, B., Prave, A. R., Hoffmann, K. H., and Arthur, M. A., 1998, Two or four Neoproterozoic glaciations? *Geology* **26**: 1059–1063.

Kirschvink, J. L., 1992, Late Proterozoic low-latitude global glaciation; the snowball Earth, in: *The Proterozoic Biosphere: A Multidisciplinary Study* (J. W. Schopf and C. Klein, eds.), Cambridge University Press, Cambridge, pp. 51–52.

Knoll, A. H., 1994, Proterozoic and Early Cambrian protists; evidence for accelerating evolutionary tempo, *Proc. Nat. Acad. Sci. USA* **91**: 6743–6750.

Knoll, A. H., 2000, Learning to tell Neoproterozoic time, *Precambrian Res.* **100**: 3–20.

Leather, J., Allen, P. A. B. M. D., and Cozzi, A., 2002, Neoproterozoic snowball Earth under scrutiny; evidence from the Fiq Glaciation of Oman, *Geology* **30**: 891–894.

Link, P. K., 1983, Glacial and tectonically influenced sedimentation in the Upper Proterozoic Pocatello Formation, southeastern Idaho, in: *Tectonic and Stratigraphic Studies in the Eastern Great Basin* (D. M. Miller, V. R. Todd, and K. A. Howard, eds.), Geological Society of America Memoir 157, pp. 165–181.

Link, P. K., Christie-Blick, N., Devlin, W. J., Elston, D. P., Horodyski, R. J., Levy, M., Miller, J. M. J., Pearson, R. C., Prave, A., Stewart, J. H., Winston, D., Wright, L. A., and Wrucke, C. T., 1993, Middle and Late Proterozoic stratified rocks of the western U.S. Cordillera, Colorado Plateau, and Basin and Range Province, in: *The Geology of North America (DNAG)* (P. K. Link, ed.), Geological Society of America, pp. 463–595.

Link, P. K., Miller, J. M. G., and Christie-Blick, N., 1994, Glacial-marine facies in a continental rift environment: Neoproterozoic rocks of the western United States Cordillera, in: *International Geological Correlation Project 260: Earth's Glacial Record* (M. Deynoux, J. M. G. Miller, E. W. Domack, N. Eyles, I. J. Fairchild, and G. M. Young, eds.), Cambridge University Press, Cambridge, U.K., pp. 29–59.

Lorentz, N. J., Corsetti, F. A., and Link, P. K., 2004, Seafloor precipitates and C-isotope stratigraphy from the Neoproterozoic Scout Mountain Member of the Pocatello Formation, southeast Idaho: implications for Neoproterozoic earth system behavior, *Precambrian Res.* **130**: 57–70.

Ludlum, J. C., 1942, Pre-Cambrian formations at Pocatello, Idaho, *J. Geology* **50**: 85–95.

Lund, K., Aleinikoff, J.N., Evans, K.V., and Fanning, C.M., 2003. SHRIMP U–Pb geochronology of Neoproterozoic Windermere Supergroup, central Idaho: Implications for rifting of western Laurentia and synchroneity of Sturtian glacial deposits. Geol. Soc. Am. Bull. **115**: 349–372.

Myrow, P. M., and Kaufman, A. J., 1999, A newly discovered cap carbonate above Varanger-age glacial deposits in Newfoundland, Canada, J. Sed. Res. **69**: 784–793.

Narbonne, G.M., and Aitken, J.D., 1995. Neoproterozoic of the Mackenzie Mountains, northwestern Canada. *Precambrian Res.* **73**: 101–121.

Porter, S. M., Knoll, A. H., and Affaton, P., 2004, Chemostratigraphy of Neoproterozoic cap carbonates from the Volta Basin, West Africa, P *Precambrian Res.* **130**: 99–112.

Prave, A. R., 1999, Two diamictites, two cap carbonates, two δ 13C excursions, two rifts; the Neoproterozoic Kingston Peak Formation, Death Valley, California, *Geology* **27**: 339–342.

Rodrigues-Nogueira, A., Riccomini, C., Nóbrega-Sial, A., Veloso-Moura, C., and Fairchild, T., 2003, Soft-sediment deformation at the base of the Neoproterozoic Puga cap carbonate (southwestern Amazon craton, Brazil): Confirmation of rapid icehouse to greenhouse transition in snowball Earth, *Geology* **31**: 613–616.

Raub, T.D., Evans, D.A.D., Wingate, M.T.D., Calver, C.R. & Izard, C.F., 2005. Detrital zircon geochronology of the Australian Marinoan glacial interval. Supercontinents and Earth Evolution Symposium, Perth, Australia, September 26–30. *Geol. Soc. Australia Abstr.* **81**: 146.

Smith, L. H., Kaufman, A. J., Knoll, A. H., and Link, P. K., 1994, Chemostratigraphy of predominantly siliciclastic Neoproterozoic successions; a case study of the Pocatello Formation and lower Brigham Group, Idaho, USA, Geol. Mag. **131**: 301–314.

Trimble, D. E., 1976, *Geology of the Michaud and Pocatello Quadrangles, Bannock and Power Counties, Idaho*, United States Geological Survey.

Vidal, G., and Moczydlowska-Vidal, M., 1997, Biodiversity, speciation, and extinction trends of Proterozoic and Cambrian phytoplankton, *Paleobiology* **23**: 230–246.

Walter, M. R., Veevers, J. J., Calver, C. R., Gorjan, P., and Hill, A. C., 2000, Dating the 840–544 Ma Neoproterozoic interval by isotopes of strontium, carbon, sulfur in seawater, and some interpretative models, *Precambrian Res.* **100**: 371–433.

Xiao, S., Bao, H. W. H., Kaufman, A. J. Z. C., Li Guoxiang, Y. X., and Ling, H., 2004, The Neoproterozoic Quruqtagh Group in eastern Chinese Tianshan; evidence for a post-Marinoan glaciation, *Precambrian Res.* **130**: 1–26.

Zhang, S., Jiang, G., Zhang, J., Song, B., Kennedy, M. J., and Christie-Blick, N., 2005, U–Pb sensitive high-resolution ion microprobe ages from the Doushantuo Formation in south China: Constraints on late Neoproterozoic glaciations, *Geology* **33**: 473–476.

Zhou, C., Tucker, R. D., Xiao, S., Peng, Z., Yuan, X., and Chen, Z., 2004, New constraints on the ages of Neoproterozoic glaciations in south China, *Geology* **32**: 437–440.

Index

Acraman Impact, 43, 255
Acritarchs, 2, 8, 14, 24, 25, 28, 29, 31, 32, 33, 36, 38, 39, 40, 41, 42, 43, 44, 45, 58, 59, 61, 75, 76, 78, 82, 254, 256, 260
Acropora millepora, 168
Algae, 2, 5, 8, 12, 13, 14, 24, 42, 57, 58, 61, 63, 64, 66, 67, 69, 77, 79, 81, 82, 118, 140, 205, 206, 209, 210, 211, 217, 219
Amoebozoans, 2, 3, 205, 206, 207, 208, 210
Angiosperm, 213
Anhuiphyton, 65, 69
Animals, 2, 4, 10, 14, 25, 38, 45, 58, 61, 65, 69, 76, 81, 82, 92, 94, 95, 96, 99, 108, 116, 138, 146, 160, 161, 166, 168, 176, 177, 180, 185, 186, 200, 205, 206, 208, 209, 210, 213, 214, 215, 216, 217, 218, 219, 221, 232, 254
Annelids, 165, 179, 180, 181
Anomalophyton, 64, 69, 78
Arcella, 7
Arcella conica, 7
Archaeonassa, 115, 123, 135, 136, 144, 145
Artacellularia kellerii, 38
Arthropods, 40, 139, 162, 165, 173, 174, 175, 179, 180, 181, 182, 210
Ascomycota, 214
Aspidella, 4, 11, 103, 105, 117, 121, 140
Asterichnus, 144
Aulichnites, 123, 136
Ausia, 105
Australia, 5, 8, 41, 43, 93, 96, 98, 103, 104, 115, 122, 126, 135, 231, 234, 238, 240, 243, 244, 247, 249, 252, 253, 259, 260, 273, 274, 277, 280, 285, 290
Avalon Assemblage, 91, 101, 103, 107
Avalonia, 98, 258

Baculiphyca, 64, 66, 69, 74, 80
Baculiphyca taeniata, 64, 69, 74, 80
Bangiomorpha pubescens, 67, 80
Basidiomycota, 214
Bavlinella, 35, 43
Bavlinella faveolata, 35
Beltanelliformis, 60, 67, 70, 115, 120, 136, 144
Bergaueria, 123, 136, 144
Bilaterian, 96, 99, 100, 102, 103, 105, 109, 116, 160, 161, 162, 163, 165, 168, 169, 170, 171, 173, 174, 179, 181, 182, 183, 185, 186, 187, 188, 215
Bilinichnus, 115, 123, 137, 144
Biodiversity, 24, 25
Biomarkers, 8, 14, 95, 97
Biostratigraphy, 7, 92, 233, 258, 274, 291
Bioturbation, 82, 92, 110, 117, 147
Body size, 28, 31, 32, 33, 74, 221
Bodyplans, 160, 161, 163, 185
Bootstrap, 29, 208
Botrydium, 63
Botryocladia, 63
Bradgatia, 101, 102, 107
Briareus borealis, 38
Brooksella, 123, 144
Buchholzbrunnichnus, 124, 144

Caenorahabditis elegans, 173
Cambrian explosion, 23, 44, 200, 215, 216, 218, 221
Canada, 5, 7, 58, 63, 67, 96, 123, 125, 126, 128, 129, 130, 133, 134, 236, 238, 239, 242, 247, 248, 249, 250, 273, 277, 280, 286
Cap carbonates, 236, 239, 240, 247, 252, 253, 254, 257, 258, 259, 273, 275, 276, 277, 278, 281, 282, 283, 284, 285, 286, 287, 288, 289, 290
Carpediemonas, 206
Catenasphaerophyton, 66, 140
Caulerpa, 82
Chambalia, 64
Charnia, 4, 11, 101, 102, 103, 104, 107, 108
Charniodiscus, 4, 11, 101, 102, 103, 104, 105, 107, 109
Chelicerates, 182
China, 4, 5, 8, 43, 58, 62, 63, 64, 65, 66, 67, 69, 94, 96, 99, 109, 119, 125, 126, 129, 131, 132, 137, 140, 144, 145, 236, 242, 247, 252, 254, 258, 259, 273, 277, 280, 283, 287
Chondrites, 115, 119, 124, 137
Chordates, 165, 166, 175
Chromalveolates, 2, 3, 7, 206
Chuaria, 58, 60, 61, 62, 63, 67, 69, 74
Cloudina, 81, 105, 119, 121, 127, 131, 132, 142, 143, 145, 146, 255, 261
Cochlichnus, 115, 124, 135, 137, 147
Competition, 80
Convergence, 9, 25, 30, 45, 58, 59, 67, 170
Corophioides, 144
Crustaceans, 182
Cucullus, 65

Curvolithus, 120
Cyanobacteria, 5, 13, 58, 220
Cymatiosphaera wanlongensis, 38
Daltaenia, 64
Dasysphaeridium trichotum, 38
Derbesia, 61, 67
Deuterostomes, 161, 162, 164, 165, 166, 171, 172, 175, 178, 179, 185, 210, 215
Diamictite, 236, 238, 246, 247, 252, 253, 275, 278, 282, 285
Dickinsonia, 93, 98, 102, 103, 104, 105, 107, 117, 121, 135, 142, 147
Dickinsonid trace fossils, 116, 142
Dictyosphaera delicata, 38
Dictyostelium, 205
Didymaulichnus, 115, 125, 137
Dissimilarity, 23, 29, 32, 33, 39
Doushantuo Formation, 4, 64, 65, 66, 67, 69, 71, 80, 81, 96, 98, 108, 236, 254, 256, 257, 260, 283, 287, 290
Doushantuo-Pertatataka acritarchs, 43, 44, 45
Doushantuophyton, 64, 69, 74, 78
Doushantuophyton lineare, 64, 69, 74
Drosophila, 161, 162, 170, 172, 173, 174, 176, 178, 179, 180, 181, 182, 183, 184, 187

Ecdysozoa, 163, 164, 165, 171, 215
Echinoderms, 109, 165, 181
Ediacara, 23, 24, 33, 35, 38, 43, 44, 45, 60, 75, 91, 92, 93, 94, 95, 96, 98, 99, 101, 102, 103, 105, 106, 108, 118, 121, 123, 124, 125, 126, 127, 128, 129, 130, 131, 132, 134, 135, 142, 143, 145, 146, 147, 215, 260
Ediacaran, 4, 7, 8, 9, 27, 33, 35, 38, 43, 44, 45, 59, 60, 63, 64, 66, 67, 68, 69, 71, 74, 75, 76, 77, 78, 79,

Index

80, 81, 82, 91, 92, 93, 94, 95, 96, 97, 98, 99, 101, 104, 105, 108, 109, 115, 116, 117, 118, 119, 120, 121, 122, 123, 135, 136, 137, 138, 139, 140, 141, 142, 143, 144, 145, 146, 160, 185, 187, 231, 232, 234, 240, 242, 250, 253, 254, 256, 260, 261, 285, 286, 290
Edysozoans, 215
Ellipsophysa, 62, 63, 67, 69
Enteromorphites, 64, 69, 74
Enteromorphites siniansis, 69
Eoporpita, 107, 142
Eosaccharomyces ramosus, 5, 7
Ernietta, 105
Eukaryotes, 1, 2, 3, 7, 8, 10, 13, 14, 24, 42, 43, 61, 77, 97, 108, 179, 199, 203, 205, 206, 207, 208, 209, 210, 212, 218, 219
Excavates, 3, 10, 206, 207
Eye, 159, 176, 185

Fabiformis baffinensis, 38
Fischerella, 64
Flabellophyton, 65, 69
Flatworms, 165, 167, 171, 215
Fungi, 2, 4, 5, 10, 14, 24, 58, 61, 169, 200, 205, 206, 208, 209, 210, 212, 213, 214, 218, 219, 220, 221

Gaojiashania, 119, 123, 130, 137
Gaskiers glaciation, 43, 69, 231, 234, 242, 252, 253, 254, 256, 257, 258, 259, 260, 261, 262, 276, 290
Giardia, 3, 205, 206, 208
Global glaciations, 24, 41, 42, 201, 203, 219, 220
Gnatichnus, 141
Gordia, 125, 131, 138
Greenland, 245

Grypania, 2, 63, 67, 200
Gymnodinium catenatum, 13
Gymnosperm, 213
Gyrolithes, 115, 138, 144

Harlaniella, 115, 125, 126, 138
Helminthoida, 126, 138, 139
Helminthoidichnites, 115, 124, 126, 134, 135, 138, 145, 146, 147
Helminthopsis, 126, 130, 138, 144, 145
Helminthorhaphe, 138
Hemichordates, 165, 175, 181
Herbivory, 44, 77, 79, 81, 82, 83
Hiemalora, 123, 142
Hornworts, 212
Hox cluster, 161, 162, 173, 188
Huangshanophyton, 65, 69
Hydra, 161, 168

Idaho, 234, 236, 246, 247, 258, 273, 277, 278, 280, 281, 282, 286, 287, 288, 289
Intrites, 9, 11, 123, 126, 136, 140, 144
Ivesheadia, 102

Jacutianema solubila, 67

Kimberella, 93, 94, 98, 103, 104, 107, 141, 145, 147, 185
Konglingiphyton, 64, 69
Konservat-Lagerstätten, 58

Lagerstätten, 58, 91, 95, 96, 108, 200
Laminaria, 82
Leiosphaeridia sp., 35
Liverworts, 212
Lockeia, 115, 126, 127, 139, 144
Longfengshania, 61, 62, 63, 69, 74, 80
Longifuniculum, 64, 69

Lophophorates, 165
Lophotrochozoa, 164, 165, 171, 215
Lophotrochozoans, 165, 215

Macroalgae, 57, 58, 59, 60, 61, 62, 66, 67, 69, 70, 74, 75, 76, 77, 78, 79, 80, 81, 82
Majasphaeridium, 38
Marinoan Glaciation, 43, 69, 231, 232, 238, 239, 247, 249, 250, 251, 252, 254, 257, 258, 259, 261
Mawsonites, 142
Medvezichnus, 127, 144
Melanocyrillium hexodiadema, 6, 7, 11
Melicerion poikilon, 7, 10
Meniscate trace fossils, 116, 142
Metazoa, 161, 165
Metazoan, 44, 79, 95, 102, 136, 160, 161, 162, 163, 166, 168, 169, 170, 171, 172, 186, 187, 189, 215
Miaohephyton, 64, 67, 69
Microbial mats, 70, 72, 92, 93, 96, 108, 140
Molecular clock, 2, 76, 95, 97, 99, 187, 200, 201, 202, 203, 212, 213, 214, 216, 254
Molluscs, 109, 141, 165
Monomorphichnus, 115, 127, 139, 144
Morphological disparity, 23, 25, 29, 33, 35, 38, 39, 40, 41, 42, 43, 44, 45, 59, 71, 74, 79, 82, 108
Morphospace, 31, 33, 36, 40, 45, 59, 70, 73, 74, 75
Mosses, 212
Muensteria, 127, 142
Multicellularity, 165
Multifronsphaeridium pelorium, 38
Myriapods, 182

Nama Assemblage, 91, 105, 106, 107
Namacalathus, 105, 106, 255, 261
Namibia, 96, 97, 105, 106, 118, 122, 123, 124, 125, 127, 128, 129, 130, 131, 132, 133, 134, 135, 137, 143, 145, 234, 236, 238, 239, 240, 245, 247, 248, 249, 250, 252, 254, 255, 256, 261, 273, 275, 277, 278, 280, 284, 285, 287, 290
Nantuo glaciation, 43
Nematodes, 165, 173, 215
Nematostella vectensis, 168
Nemiana, 105, 107, 120, 136
Nenoxites, 127, 144, 146
Neonereites, 9, 11, 127, 128, 139, 140, 144
Nereites, 128, 139, 144
Newfoundland, 4, 93, 94, 125, 128, 129, 133, 138, 234, 250, 253, 258, 259, 273, 274, 277, 280, 285, 287
Non-metric multidimensional scaling (MDS), 30, 35, 70
Nostoc, 61
Nucellosphaeridium magnum, 38

Octoedryxium truncatum, 38
Oldhamia, 128, 132, 135, 143
Oman, 234, 236, 240, 246, 251, 255, 256, 257, 261, 273, 277, 280, 282, 286, 288
Opisthokonts, 2, 3, 205, 206, 207, 209
Orbisiana, 66, 140
Oxygen, 12, 59, 80, 83, 95, 108, 199, 200, 218, 219, 221

Palaeoarcella athanata, 6, 7, 11
Palaeopascichnus, 7, 9, 11, 115, 120, 129, 138, 139, 144
Palaeophragmodictya, 104

Palaeophycus, 115, 129, 140
Palaeovaucheria clavata, 67
Paleoecology, 58, 66, 92, 95, 97, 104, 108, 109
Paleovaucheria, 8
Parachuaria simplicis, 61
Paralongfengshania, 62, 69, 74
Pararenicola, 65, 80
Parasitism, 81, 82
Parmia, 65
Parvancorina, 103, 104, 107
Paulinella, 9
Photosynthesis, 1, 2, 3, 78, 203, 219
Phyllodicites, 139
Phytophthora, 208
Planolites, 115, 119, 120, 130, 131, 132, 140, 141
Plants, 2, 3, 25, 169, 200, 205, 206, 209, 210, 212, 213, 217, 218, 220, 221
Platyhelminthes, 171
Platyneris, 177, 179
Podocoryne, 168, 169, 170
Podocoryne carnea, 168
Polychaete, 177, 179
Porphyra, 69, 78
Predation, 81
Priapulids, 143, 165
Principal Components Analysis (PCA), 30
Prokaryotes, 43, 97, 200, 203, 204, 205, 219, 220
Proterocladus major, 67
Protists, 11, 13, 24, 25, 40, 118, 140, 169, 205, 208, 209, 210, 212, 218
Protoarenicola, 64, 65, 80
Protoarenicola baiguashanensis, 64, 80
Protostome, 161, 165, 179, 185
Pteridinium, 99, 105, 106, 109
Pterospermella solida, 38

Pterospermopsimorpha pileiformis, 38

Radhakrishnania, 62
Radulichnus, 93, 94, 107, 115, 127, 131, 139, 140, 141
Randomization, 31, 35, 39, 71, 72, 74
Rangea, 103, 105, 107
Rangeomorpha, 101
Retisphaeridium brayense, 38
Rhizarians, 2, 3, 9, 206
Ruyang Group, 5

Sabellidites cambriensis, 66
Saccoglossus kowalevskii, 175
SAS/IML, 23, 29, 30, 31
Satka squamifera, 38
Scalarituba, 139
Schizofusa sinica, 38
Seirisphaera, 63, 66
Sellaulichnus, 136, 144
Shuiyousphaeridium macroreticulatum, 38
Simia simica, 38
Sinianella uniplicata, 38
Sinocyclocyclicus, 100
Sinosabellidites, 65, 69
Sinospongia, 65
Skolithos, 116, 132, 141
Snowball Earth, 41, 42, 92, 95, 201, 220, 251, 252, 259
South Australia, 43, 94, 102, 103, 115, 118, 123, 124, 125, 126, 127, 128, 129, 130, 131, 132, 134, 135, 139, 141, 142, 143, 231, 234, 240, 248, 249, 252, 253, 257, 259
Spirorhaphe, 132, 138
Sponges, 96, 97, 99, 108, 109, 163, 165, 166, 167
Spriggina, 103

Star-shaped trace fossils, 116, 142
Stelloglyphus, 132, 142, 144
Stromatolites, 92, 93, 140, 249
Strongelocentrotus purpuratus, 179
Sturtian glaciation, 231, 245
Suberites douncula, 166
Suketea, 62
Surface/volume ratio, 59, 74, 77, 78, 79, 80, 82
Suzmites, 132, 144
Svalbard, 236, 238, 239, 240, 243, 244, 245, 246, 247
Swartpuntia, 105, 106, 107, 109
Symbiosis, 81
Syringomorpha, 132, 144

Taenidium, 132, 142
Tappania, 5, 38
Tardigrades, 165
Tasmania, 234, 252, 259, 273, 277, 280, 283, 284, 287
Tasmanites volkovae, 38
Tawuia, 61, 62, 63, 67, 69, 74, 80
Thallophyca ramosa, 64
Thecatovalvia annulata, 38
Thectardis, 102, 107
Tiering, 43, 98, 99, 102, 104, 105, 108, 109
Torrowangea, 116, 133, 141
Trace fossils, 9, 93, 99, 102, 103, 105, 115, 116, 117, 118, 119, 120, 121, 122, 123, 135, 136, 137, 138, 139, 140, 141, 142, 143, 144, 145, 146, 186, 200
Trachelomonas, 10
Tracheophytes, 212
Treptichnids, 116, 143

Treptichnus, 128, 133, 134, 135, 139
Tribrachidium, 103, 104, 107, 121, 135
Tripedalia cytosphora, 170
Trypanosoma, 208

Ulva, 69, 78, 79
Urbilateria, 159, 161, 162, 163, 165, 180, 188

Valonia, 61, 63
Valvimorpha annulata, 38
Vendichnus, 133, 144
Ventogyrus, 105
Vernanimalcula, 98, 100
Vertebrates, 161, 162, 170, 172, 173, 174, 175, 176, 177, 178, 179, 180, 181, 182, 183, 184, 216
Vimenites, 133, 144

White Sea Assemblage, 91, 102, 103, 107
Wynniatt Formation, 5, 7

Xenopus, 162, 179

Yelovichnus, 9, 11, 120, 133, 140, 144
Yorgia, 93, 98, 103, 107, 142
$\delta^{13}C$, 12, 41, 231, 232, 233, 234, 236, 237, 238, 239, 240, 242, 243, 244, 245, 247, 248, 249, 250, 253, 255, 256, 257, 258, 259, 260, 261, 262, 274, 276, 278, 281, 282, 284, 285, 286, 288, 290